Disaster Risk Reduction in South Asia

Disaster Risk Reduction in South Asia

Editors

Pardeep Sahni
Professor, Public Administration
Indira Gandhi National Open University
New Delhi

Madhavi Malalgoda Ariyabandu
Programme Manager
Disaster Mitigation
Intermediate Technology Development Group (ITDG), South Asia
Sri Lanka

PHI Learning Private Limited

Delhi-110092
2013

Rs. 450.00

DISASTER RISK REDUCTION IN SOUTH ASIA
Edited by Pardeep Sahni and Madhavi Malalgoda Ariyabandu

ISBN-978-81-203-2200-4

Eighth Printing ... ... July, 2013

Published by Asoke K. Ghosh, PHI Learning Private Limited, Rimjhim House, 111, Patparganj Industrial Estate, Delhi-110092 and Printed by Rajkamal Electric Press, Plot No. 2, Phase IV, HSIDC, Kundli-131028, Sonepat, Haryana.

Contents

Acknowledgements

Disasters, whether natural or man-made, are like unwelcome guests and South Asia has played unfortunate host to them innumerable times. A tough mesh of poverty, unplanned urbanization, chronic malnutrition and nightmarish population densities trap its people. The major part of the region is vulnerable to earthquakes, floods, droughts, landslides, snow avalanche and forest fires. Since time immemorial, natural disasters regularly take toll of the human, social and physical capital of the South Asian region. Frequent occurrence of natural disasters is one of the major factors that impedes the general progress of the region.

Consequent upon each disaster, the governments, NGOs, community bodies, etc., undertake extensive relief and rehabilitation operations. Yet the outcome of such operations remains far from satisfactory and the benefits seldom reach direct victims. Community participation and a proactive approach are, of late, being practised as the basic elements necessary for any disaster mitigation or risk reduction activity to succeed. Although some of the South Asian countries have begun to implement strategies on these lines, there still is the need for an all-round paradigm shift to proactive measures in the approach to disaster mitigation. Only then can the economic development and progress of the South Asian region be ensured. It is in this spirit that the contributions comprising this book have been presented.

Twenty-seven contributors, both academicians and practitioners, investigate the challenges that the region faces and how changes can be effected at the community, society, government and non-government levels to foster a culture of preparedness. The overall focus is on risk reduction through prevention, mitigation, response, recovery, rehabilitation and reconstruction and on fostering a culture of preparedness to combat the challenges that lie ahead.

It has been very thoughtful and kind of our experienced and dedicated contributors that our effort has materialised. Each contribution reflects a serious and meaningful point of concern to be put to practice for reducing disaster risk. We are much beholden to all our worthy contributors.

We are indebted to the management and the Disaster Mitigation Programme team of ITDG, South Asia, for providing us with all help and assistance in bringing out this volume. Our special thanks are also due to Dr. K.B. Rokaya, Country Director, ITDG, Nepal, Alok Rajouria,

Manager, Programme Development, ITDG, Nepal, Dr. E. Vayunandan, Dr. Alka Dhameja, and Dr. Uma Medury.

We are thankful to Ms Upasana, Manav, and Stuti for all their cooperation towards completion of this work.

We are much obliged to the Conflict and Humanitarian Affairs Department (CHAD) of DFID UK for the financial support extended in producing this volume as a part of the Project on 'Livelihood Options for Disaster Risk Reduction in South Asia'.

We owe a deep sense of gratitude to Mr. Pankaj Khare, Mr. Kamal Kant Verma, and Mrs. Asha Arora for their substantive help in word-processing, copy editing, and format editing of all the contributions. We also sincerely thank PHI Learning, New Delhi, for their excellent editorial and production efforts in bringing out this volume in record time.

Pardeep Sahni
Madhavi Malalgoda Ariyabandu

About the Editors

Dr. Pardeep Sahni is Professor of Public Administration at the Indira Gandhi National Open University (IGNOU). He also holds the additional charge of the Head, Material Production and Distribution Division of the university. Prior to joining IGNOU, he served with the Punjabi University, Patiala and the Haryana Institute of Public Administration, Chandigarh. He has over 32 years of experience in teaching and research. In addition to being the author of seven books, he has published a number of research papers and articles in reputed Indian and foreign journals. He received his M.Phil. and Ph.D. from Punjabi University, Patiala.

Prof. Sahni, along with his colleagues, has shown pioneering initiative in developing courses and undertaking extension activities in the field of disaster management. He has also been a consultant in disaster management for the Disasters Emergency Committee, UNDP and the World Bank. He is involved in the development of Postgraduate Diploma in Disaster Management and the Masters Programme in Public Administration, both courses being conducted by IGNOU.

Ms. Madhavi Malalgoda Ariyabandu, holds an M.Sc. in 'Agricultural Economics' and an M.Sc. in 'Agronomy'. With her long and varied experience in the field of disaster management, she has accumulated extensive knowledge of disaster and conflict issues. She is the Programme Manager, Disaster Mitigation, at the Intermediate Technology Development Group (ITDG), South Asia. She is also in charge of the South Asia Regional Initiative on Livelihood Options for Disaster Risk Reduction, a programme of work with a holistic approach to address the issues of disasters and vulnerability, implemented in Bangladesh, India, Nepal, Pakistan and Sri Lanka. The programme has identified the linkages between disaster risk and livelihoods, and demonstrated livelihood options which have the potential to reduce disaster risk. Ms. Ariyabandu is the author of many books, articles and papers in the area of disaster mitigation.

Contributors

Dr. Alka Dhameja
Reader
Public Administration Faculty
Indira Gandhi National Open University
New Delhi, India

Mr. Amjad Bhatti
Executive Coordinator
Journalist Resource Centre
Islamabad, Pakistan

Dr. Anselm Smolka
Manager Head
Geophysical and Geological Research Group
Munich Reinsurance Company
80791, Munich, Germany

Mr. Balkrishna Lohani
Chairman
Bhandara VDC, Chitwan
Nepal

Ms. Dhanushi Senanayake
ITDG, South Asia
5, Lionel Edirisinghe Mawatha
Kirulapone
Colombo 5, Sri Lanka

Dr. Dolly Mathew
Reader
Public Administration Faculty
Indira Gandhi National Open University
New Delhi, India

Dr. E. Vayunandan
Reader
Public Administration Faculty
Indira Gandhi National Open University
New Delhi, India

Dr. Earl James Goodyear
International Disaster Management Consultant
3273 Aldoro Avenue, Spring Hill
Florida 34069, USA

Colonel George N. Ritchie
Former Director
Cranfield Disaster Preparedness Centre
Shrivenhan
United Kingdom

Prof. Hafiza Khatun
Department of Geography and Environment
University of Dhaka
Dhaka, Bangladesh

Mr. Harish Chandra Acharya
Programme Coordinator
Centre for Policy Studies
Kathmandu, Nepal

Ms. Hina Shah
Director
International Centre for Entrepreneurship and Career Development
Bhopal, Ahmedabad, India

Mr. Hrusikesh Harichandan
Disaster Management Cell
Orissa State Red Cross Society
Bhubaneswar, Orissa, India

Prof. Janki Andharia
Head
Department of Urban and Rural Community Development
Tata Institute of Social Sciences
Mumbai, India

Dr. Kulwant Singh
United Nations Development Programme
Jaipur, Rajasthan, India

Mr. Kul Chandra Shreshtha
Ministry of Home Affairs
Government of Nepal
Kathmandu, Nepal

Mr. Kumar Prasad Poudyal
Director General
Drug Control and Disaster Management Departmant
Government of Nepal
Kathmandu, Nepal

Ms. Madhavi Malalgoda Ariyabandu
Programme Manager
Disaster Mitigation
Intermediate Technology Development Group, South Asia
5, Lionel Edirisinghe Mawatha
Kirulapone, Colombo 5, Sri Lanka

Dr. Manoj K. Arora
Department of Electrical Engineering and Computer Sciences
Syracuse University
New York 13244, USA

Mr. Meen Bahadur Poudyal
Ministry of Home Affairs, Government of Nepal
Kathmandu, Nepal

Prof. N. Vinod Chandra Menon
Programme Officer, Emergency
India Country Office, UNICEF
New Delhi, India

Prof. Pardeep Sahni
Chairman
Public Administration Faculty
Indira Gandhi National Open University
New Delhi, India

Mr. Pir Abdul Haque Jan Sirhindi
District Coordination Officer
Tharparkar, Sindh, Pakistan

Colonel Rajive Kohli
President
Centre of Disaster and Development
Jung Pura, New Delhi, India

Mr. S.K. Swamy
Director
Natural Disaster Management Division
Ministry of Home Affairs, Government of India
New Delhi, India

Dr. Uma Medury
Reader
Public Administration Faculty
Indira Gandhi National Open University
New Delhi

Mr. Zubair Murshed
Training Manager
Asian Disaster Preparedness Centre
Kiong Luang, Pathumithani 1212o, Thailand

1 Introduction

— Pardeep Sahni
— Madhavi Malalgoda Ariyabandu

Disasters have been around since well before recorded history and they appear to be increasing in both frequency and intensity. Disasters are like unwelcome guests, which leave permanent impressions of their visits on the victims. Broad categories of disasters can be natural, either rapid onset such as earthquakes, cyclones, floods and landslides or slow onset such as droughts, desertification and environmental degradation; and man-made, such as conflicts, wars, industrial accidents, transport accidents, terrorism and crowd incidents. Complex emergencies, usually a combination of war or conflict, form another category. There are two key components of any disaster: (a) hazard(s) and, (b) the population that is vulnerable to the hazard.

South Asia is among the world's most vulnerable regions to both natural and man-made disasters. A tough mesh of poverty, rampant and unplanned urbanisation, chronic malnutrition and nightmarish population densities have trapped its people. It is said that in South Asia, you can set your clock by the diurnal rhythm of the floods. The four seasons arrive and depart in tandem with four major kinds of natural disasters: floods, earthquakes, cyclones and droughts. Other catastrophic events such as landslides, avalanches, blizzards and fires occur less frequently and threaten fewer people (Sapir).

Since times immemorial, natural disasters have destroyed the human, social and physical capital of the South Asian subcontinent. The region recorded 15 out of the 40 major disasters in the world from 1970 to 2000. Disasters have not only disrupted the normal life of affected communities and countries in the subcontinent but have also halted developmental efforts, as the funds earmarked for new initiatives were transferred to relief, response and rehabilitation work. By and large, the response of the major stakeholders has been reactive rather than proactive. Consequent upon each disaster, the South Asian governments,

NGOs, charities, communities, etc., have responded with relief and rehabilitation activities. The idea of having such structural and non-structural measures that could reduce the risk of disasters did not gain much currency in South Asia in the past. In consonance with the need of the hour, some efforts are now being made all over the world, including South Asia, for disaster risk reduction by adhering to detailed, comprehensive and participatory strategies. Some of the countries in the continent have taken certain steps towards disaster risk reduction but there is still a long way to go as the result have not been encouraging. There is need for concerted and well-planned efforts for risk reduction through risk identification, risk sharing and risk transfer.

Disaster risk, per se, incorporates hazard (a source of danger) multiplied by value (anything of human value including life, property and/or livelihood) multiplied by vulnerability (extent of exposure to hazard impact) (Van Oppen, 2001). A disaster may be defined as the realisation of a hazard that severely impacts things of human value (Smith, 1996). A hazard could have direct and/or indirect impact. For example, hazardous events such as earthquakes and cyclones are not disasters on their own. They become disasters when they impact or effect things of human value. Thus, disaster risk can be reduced by any such technique that reduces the impact of the hazardous event.

FROM RELIEF TO MITIGATION—TOWARDS A CULTURE OF DISASTER PREPAREDNESS

Disasters and *development* have almost always been dealt with separately. No possible linkages between the two were understood or foreseen till recent times. Disasters were taken to happen in a vacuum, with no relationship with the social, economic or institutional context. As a result, the course of action followed was of emergency management and relief with welfare orientation. Interventions are made once a disaster occurs, with the objective of returning the situation to as it was before the event. Each year, the budgets for emergency management are shrinking and people who are forced to live in the same conditions fall deeper into the trap of poverty and become more vulnerable.

Responding to emergencies is no doubt an important aspect of disaster management planning. However, the absence of other important components such as *disaster preparedness* based on the root causes of the event and a sustainable approach towards relief and rehabilitation, is cause for concern. Today, there is a strong need to move from an *emergency management culture* to a culture of disaster preparedness.

In recent times, there has been extensive discussion and debate on the relationship between socio-economic forces and disasters. Although, the discussions have reached the highest levels, such as the International Decade of Natural Disaster Reduction (IDNDR), application remains elementary. The body of research, and documented evidence based on real experiences is still small. The changeover from a relief culture to laying a foundation of preparedness is a mammoth task requiring commitment, awareness at all levels, knowledge, capacity and financial resources.

Holistic development planning based on adequate analysis will not leave room for hazards and disasters. *Disaster risk management* is an integral part of a comprehensive development plan for any geographical area or sector. Such an approach will lead to a reduction in relief and rehabilitation costs. The money thus saved can be invested in development ventures. Further, the currently applied approaches towards relief and rehabilitation can be made long-term and sustainable. Analysis of the root causes of disasters, and risk and vulnerability analysis are the keys to reaching the goal.

In South Asia there have been a handful of initiatives to test and document community-based disaster preparedness moves. The Conflict and Humanitarian Affairs Department (CHAD) of the Department For International Development (DFID) is supporting a regional programme wherein livelihood support and capacity building of communities for risk management are taken as the basis for measuring disaster preparedness. The programme has resulted in over seven community level demonstrations on disaster risk reduction and has also brought about many lessons from a regional perspective. A wider advocacy programme will take the successful and good practices to more communities and bring increased investment from local and national governments.

ECHO Disaster preparedness wing (Dip ECHO) has launched a community-based flood preparedness programme across South Asia by supporting nine projects in the region. This programme is expected to result in communities better prepared to manage floods, community-based early warning systems, awareness of the existing government institutions and propogation of useful lessons for wider replication. National disaster management centres supported by the United Nations Development Programme (UNDP), function in India and Sri Lanka. Training and awareness-building programmes are also carried out to introduce government and NGO staff to the principles and practices of community centred disaster preparedness. In addition, there are national government programmes, which are often limited and ad-hoc.

The foundation of the Intermediate Technology Development Group's (ITDG) South Asian programme of work is disaster risk management. Using the *Duryog Nivaran* network as its vehicle, ITDG demonstrates community-based risk management at the grassroots levels, co-opts local and national government structures into taking the good practices forward and advocates shifts in investment at the regional and international levels. ITDG is actively working with the media to achieve sensitive disaster reporting over sensational reporting. Its programme of work includes producing tools for disaster preparedness planning for practitioners and policy makers aiming to bring about long-term policy changes. Its current work includes the Livelihood Options for Disaster Risk reduction programme supported by CHAD and a part of the Dip ECHO flood preparedness programme.

CHALLENGES FOR SOUTH ASIA

South Asia faces enormous challenges in the task of creating conditions for risk management instead of emergency management and in achieving a culture of preparedness. This amounts to a major shift in conceptualisation, planning and action. Changes are required at all levels: community, planning and decision-making levels.

Gradual integration of disaster risk analysis into development plans is a must. There is also a strong need to change perceptions. For instance, communities are not mere victims, they are resources since they possess enormous knowledge and capabilities for coping with and managing risk. A change of perception is also required within the communities which have been burdened with a dependency mentality for long.

Emergency management, relief and rehabilitation plans can be made more fruitful by focusing on sustainable goals instead of merely returning the communities to the same status as before. Analysis and planning based on disaster-development linkages is fundamental to achieving this. One such example is turning enormous amounts of drought relief distributed annually by the national governments into drought preparedness investment by introducing measures such as well-planned water harvesting structures, merging water shed management with forestry programmes and by reviving traditional methods of water preservation.

The process can be bottom-up. There is a need to create and set up as many examples as possible at the community level to demonstrate that organised communities with their capacities enhanced to meet the challenges of nature can manage disaster risk effectively. Investments on infrastructure, such as, locally appropriate early warning systems, shelters and flood management structures can realise their maximum potential when the awareness levels and the capacities of the communities to utilise them are high. The main players in this venture can be the local governments, the community-based organisations (CBOs) and the non-governmental organisations (NGOs) whose mandate is to ensure poverty eradication and sustainable livelihood.

Advocacy, taking lessons from successful experiences and incorporating them in larger plans, making investments for larger impacts, building capacity and knowledge are some other steps in the process.

TOWARDS DISASTER RISK REDUCTION

There are now a number of leading agencies and forums around the globe, which are actively making efforts towards disaster risk reduction.

Some of the initiatives undertaken and plans made by such agencies and forums towards disaster risk reduction are as follows:

International Strategy for Disaster Risk Reduction

It is a fact that hazards are inevitable and the elimination and/or reduction of all risks are impossible. Of course, there are a number of technical measures, traditional practices and public experience that can reduce the extent or severity of natural disasters. The International Strategy for Disaster Risk Reduction (ISDR) programme aims at enabling all communities to become resilient to the effects of natural, technological and environmental hazards and reducing the compound risks they pose to social and economic vulnerabilities within modern societies.

The ISDR document lists the following major goals and objectives of this programme:

Goals

- Increase public awareness of the risks that natural, technological and environmental hazards pose to modern societies.
- Obtain commitment by public authorities to reduce risks to people, their livelihoods, social and economic infrastructure and environmental resources.
- Engage public participation at all levels of implementation to create disaster-resistant communities through increased partnership and expanded risk reduction networks at all levels.
- Reduce the economic and social losses of disasters as measured, for example, by Gross Domestic Product.

Objectives

- Stimulate research and application, provide knowledge, convey experience, build capabilities and allocate necessary resources for reducing or preventing severe and recurrent impacts of hazards, for those people who are most vulnerable.
- Increase opportunities for organisations and multi-disciplinary relationships to foster more scientific and technical contributions to the public decision-making process in matters of hazard, risk and disaster prevention.
- Develop a more proactive interface between management of natural resources and risk reduction practices.
- Form a global community dedicated to making risk and disaster prevention a public value.
- Link risk prevention and economic competitiveness issues to enhance opportunities for greater economic partnerships.
- Complete comprehensive risk assessments and integrate them with development plans.

- Develop and apply risk reduction strategies and mitigation measures with supporting arrangements and resources for disaster prevention at all levels of activity.
- Identify and engage designated authorities, professionals drawn from the widest possible range of expertise and community leaders to develop increased partnership activities.
- Establish risk monitoring capabilities, and early warning systems as integrated processes, with particular attention to emerging hazards with global implications such as those related to climate variation and change, at all levels of responsibility.
- Develop sustained programmes of public information and institutionalised educational components pertaining to hazards and their effects, risk management practices and disaster prevention activities, for all ages.
- Establish internationally and professionally agreed standards/methodologies for the analysis and expression of the socio-economic impacts of disasters on societies.
- Seek innovative funding mechanisms dedicated to sustained risk and disaster prevention activities.

The World Bank

The World Bank, through its Disaster Management Facility (DMF), has been actively reducing human suffering and economic loss caused by natural and man-made disasters. It has been working to ensure that disaster prevention and mitigation are integral aspects of development. The DMF acts by providing technical support to World Bank operations, promoting capacity-building and establishing partnerships with the international and scientific community working on disaster issues. The following are the specific objectives of the DMF:

(i) To improve the management of disaster risk in member countries and reduce vulnerability in the World Bank portfolio.

(ii) To promote sustainable projects and initiatives that incorporate effective prevention and mitigation measures.

(iii) To promote the inclusion of risk analysis in World Bank operations, analysis and country assistance strategies.

(iv) To promote training in the areas of disaster prevention, mitigation and response.

(v) To identify policy, institutional and physical interventions aimed at reducing catastrophic losses from natural disasters through structural and non-structural measures, community involvement and partnerships with the private sector.

DMF undertakes the above activities through education, training, support and partnerships. To this effect, the key functions of the DMF include the following:

(i) Technical support and guidance to member countries and to the World Bank staff in operations on lending and on the preparation of country assistance strategies and sectoral work programmes to reduce risks from natural and technological disasters.

(ii) Partnerships with the international and scientific communities and NGOs are being established to promote dialogue on disaster management issues, collaborate in activities and receive inputs for World Bank programmes. A key partnership initiative is the *ProVention Consortium*, launched in February, 2000, to reduce disaster risk in developing countries and make disaster prevention and mitigation an integral part of development efforts. The ProVention Consortium functions as a network to share knowledge and connect and leverage resources aimed at reducing disaster risk.

(iii) Examining the World Bank's disaster assistance portfolio to extract lessons for future operations.

(iv) Identification and dissemination of the good practices of disaster management of the World Bank and other agencies.

(v) Training in the areas of disaster prevention, mitigation and response.

It would be appropriate to say that disaster risks are a culmination of hazardousness and vulnerability that vary over seasonal to decadal time-scales as well as geographically. Risks can be managed through risk identification, risk reduction and risk transfer. A World Bank paper entitled *What is Risk Management*? explains the components in the following manner:

Risk Identification

Any effective strategy to manage disaster risk must begin with an identification of the hazards and who is vulnerable to them. This involves information on the nature and extent of risk that characterises a particular location, including information on the nature of particular physical hazards obtained through hazard assessments, as well as information and data on the degree of exposure of a population and its environment to such hazards. In this way informed decisions can be made on where to invest and how to design sustainable projects that will withstand the impacts of potential disasters. Hazard mapping and the utilisation of Geographic Information Systems (GIS) are among the ways in which risk information may be organised for the benefit of potential users. A more complete understanding of the full economic, financial, and social impacts of disasters on a country also helps to demonstrate the importance of including risk reduction measures in development plans.

Risk Reduction

Disasters result when an extreme natural or technological event coincides with a vulnerable human settlement. Reducing disaster risk requires that all stakeholders change their perceptions and behaviour to place a high priority on safety in planning and development. Effective risk reduction involves mitigation measures in hazard prone developing countries. Such measures include land use planning, structural design and construction practices, and disaster warning systems. In addition to employing scientific and technical knowledge, risk reduction may also involve overcoming the socio-economic, institutional and political barriers to the adoption of effective risk reduction strategies and measures in developing countries. This may be accomplished through projects analysing the possible roles of government, non-government and private sector organisations in risk reduction, local and regional workshops and conferences aimed at heightening the awareness of stakeholders to the threat of natural disasters and what can be done about it, and educational and training activities that increase the understanding of policy makers, decision makers and practitioners about disaster management.

Risk Sharing and Transfer

The private insurance sector contributes important funding for natural disaster reconstruction in developing countries, but it has made fewer inroads in the markets of developing countries. In emerging economies, the state and the individual bear much of the cost of disasters. As a result, ad-hoc funds transfers to respond to disaster emergencies disrupt planned development activities. Such diversion of development funding postpones progress towards long-term economic and social improvement. Tools have to be developed to assist the very poor to more effectively manage disaster risk. This includes micro-finance mechanisms that can deal with covariate risks such as disasters and that build social capital and encourages risk mitigation for the very poor. In addition to that, measures may include safety nets and calamity funds, and informal mechanisms.

Intermediate Technology Development Group (ITDG)

The ITDG has been engaged in working on disaster risk reduction in Latin America, Asia and elsewhere. ITDG is an international development organisation, whose mandate is to introduce appropriate technologies. With over 30 years of experience in working with disadvantaged communities in seven developing countries in the world, ITDG has been able to clearly analyse and establish the linkages between disasters and development for the purpose of disaster risk reduction. The group points out that no developmental intervention is valid or can be successful unless risk and vulnerability to disasters is addressed

(Livelihood Options for Disaster Risk Reduction, 2002). On the bases of empirical research in Bangladesh, India, Nepal, Pakistan and Sri Lanka, ITDG in a paper titled *Disaster Risk Reduction Strategy* has highlighted the following key issues:

(i) There is a growing frequency of small- and medium-scale disasters related to patterns of human environmental intervention, indicating that disaster risk is accumulating. These disasters often have a larger cumulative impact than the spectacular but occasional large catastrophes; but are generally ignored by both national and international disaster management or development policies and represent a challenge for disaster risk reduction rather than emergency preparedness and response.

(ii) The primary responsibility for managing this kind of disaster risk at local level lies with municipal governments, community organisations and NGOs, rather than national and international organisations. In highly vulnerable regions, such organisations often have very low institutional capabilities for disaster risk reduction.

(iii) Most conventional disaster management programmes and projects focus on emergency management rather than disaster risk reduction, on large catastrophes rather than small- and medium-scale disasters and on national and international rather than local organisations.

(iv) As such they fail to address risk accumulation processes in highly vulnerable regions. Without the participation of vulnerable communities and other stakeholders in their design and implementation, disaster management plans, projects and policies often prove to be unsustainable. By contrast, where disaster management is focused on risks rather than on emergencies, and when it builds on the existing coping strategies of vulnerable social groups, it can become a vehicle for facilitating sustainable development.

(v) Despite the recent International Decade of Natural Disaster Reduction, disaster risk reduction has not been given sufficient political priority either at the national or international level and there is little evidence to show that disaster risk considerations are being incorporated in development policies, programmes and projects. On the contrary, ill-planned development continues to generate risk, leading to an increasing demand for emergency relief and humanitarian assistance.

ITDG, with the financial support of the Conflict and Humanitarian Affairs Department (CHAD) of DFID, UK, is at present implementing a project called *Livelihood Options for Disaster Risk Reduction in South Asia*. The project, implemented in five disaster prone countries of South

Asia has taken a holistic approach to address the issues of disasters and vulnerability. The five major components of the project are:

(i) Research to identify implications of disaster risk on livelihoods.
(ii) Formulate strategies to strengthen livelihoods that reduce disaster risk.
(iii) Build capacity of stakeholders through a community-based approach to disaster management.
(iv) Pilot demonstration of identified risk-reducing strategies.
(v) Advocacy and policy to influence a paradigm shift based on the *alternate approach* to deal with disaster.

The project has studied selected communities living with different disasters in various countries in South Asia. Possible livelihood strengthening and risk reduction approaches identified by research have been/are being formulated and implemented with communities in seven locations in India, Pakistan and Sri Lanka.

ITDG is also implementing the *Alternative Strategies for Community-based Flood Preparedness in South Asia*, project as part of the Dip ECHO Flood Preparedness programme. The project aims to have pilot demonstrations in three locations in Bangladesh, Nepal and in Pakistan.

ProVention Consortium

The *ProVention Consortium*, comprising 43 governments, international organisations, academic institutions, academic institutions, the private sector, and civil society organisations was launched in February, 2000. The member organisations of this consortium are: African Development Bank; Asian Development Bank; Inter-American Development Bank; The World Bank; Ministry of Construction, Japan; Royal Ministry of Foreign Affairs, Norway; Organisation of American States; Pan-American Health Organisation; World Food Programme; United Nations Development Programme; World Meteorological Organisation; United Nations Office for the Coordination of Humanitarian Affairs; United Nations Environmental Programme; The World Conservation Union; Federal Emergency Management Agency; Ministry of Land Infrastructure and Transport, Japan; World Institute for Disaster Risk Management; International Institute for Applied Systems Analysis; Earthquake Disaster Mitigation Research Center, Japan; National Research Institute for Earth Science and Disaster Prevention; Southwest Research Institute; International Federation of Red Cross and Red Crescent Societies; Swiss Agency for Development and Cooperation; Asian Disaster Preparedness Center; Asian Disaster Reduction Center, Japan; Earthquakes and Megacities Initiative (EMI); International Strategy for Disaster Reduction; Global Fire Monitor Center, University Freiburg; Middle Eastern Technical University, Turkey; Columbia Earth Institute, University of Columbia;

University of Kyoto, Japan; Wharton School, University of Pennsylvania; Natural Hazards Center at the University of Colorado; CESIR, University of Stanford; Cenapred Mexico; The National Research Council; Lloyd's; Cemex; Voice; Munich Re Group; Renaissance Re Insurances; Swiss Re, Global Reinsurer; Grameen Bank, Bangladesh. The major aim of the consortium is to plan and undertake steps for reducing the impact of disasters in developing countries. The consortium works through sharing knowledge and resources for reducing disaster risks. Besides stressing on synergy, it emphasises coordination for concerted efforts. The major objectives of the consortium include the following:

- To promote a culture of safety through education and training among leaders and citizens of developing countries.
- To support public policy that can reduce the risk of natural and technological disasters within developing countries.
- To support pilot projects and to disseminate information about the proven *best practices* to mitigate the scope and frequency of disasters.
- To develop the abilities of governments to minimise disasters and to respond effectively when they occur.
- To forge links between public and private sectors, between the scientific community and policy makers, between donors and victims so that all stakeholders work together to strengthen the economy, reduce pain and suffering and promote the common good.

The Disaster Risk Reduction Hemispheric Conference

The Disaster Risk Reduction Hemispheric Conference, held at San Jose, Costa Rica, on December 4–6, 2001 has recognised the need to develop, implement and sustain shared comprehensive disaster management strategies and programmes to reduce the vulnerability of populations and national economies to natural and man-made disasters, besides maintaining or quickly restoring minimum levels of consumption, income and production at the household and community levels in the aftermath of disasters, including irregular population settlements. It has acknowledged the need to expand the community of stakeholders at the regional, national and local levels engaged in the formulation of early warning systems. While emphasising sustainable development as an important facet of managing disasters, the conference highlighted the following strategies for effective risk management and response operations in the event of disasters:

(i) Develop the capacity to forecast, prepare for and mitigate the potential impacts of natural and man-made hazards; promote vulnerability reduction; adopt and enforce better building codes and standards; ensure appropriate land-use practices; make an inventory of and evaluate the vulnerability of critical facilities and infrastructure; estimate climate change variability and sea-level rise and assess their possible impacts; and in pursuit of the above, create the requisite legal framework and establish the cooperative mechanisms to access and share advances in science and technology and their application in the early warning, preparedness for and mitigation of these hazards.

(ii) Promote the exchange of information on the vulnerability of infrastructure exposed to disasters as well as the early warning capacity, particularly in the border areas of the countries of the Americas, in order to design specific prevention measures in the fields of engineering and legislation with the aim of reducing the socio-economic impact of natural disasters.

(iii) Establish or strengthen, where appropriate, partnerships with all relevant actors, including the private sector, technical and professional associations, regional institutions, civil society, educational and research institutions and other multilateral coordinating agencies such as the Office for the Coordination of Humanitarian Affairs (OCHA), in the development and implementation of disaster management policies and programmes at the national and community levels, and promote greater awareness and effective integration of these policies and programmes among national policy makers, local authorities, communities and media, and promote the insurance and reinsurance of the social and economic infrastructure as well as the decentralisation of information and decision-making.

(iv) Promote the exchange of knowledge and experiences regarding the combat against inappropriate practices in the exploitation of natural resources and unsustainable patterns of consumption, including the problems of waste management, which increase the vulnerability of people to natural disasters.

(v) Promote the development of telecommunications for humanitarian assistance; actively encourage greater use and interoperability of telecommunications and other technologies and information systems that allow the observation and monitoring of different natural phenomena; use early warning systems such as remote sensing imagery, Geographic Information Systems (GIS) based data necessary to address and prevent emergencies; promote the compatibility of these systems in the planning and response to emergency operations among governments, specialised agencies, relevant international organisations, and NGOs.

(vi) Consider the creation of a hemispheric system for prevention and mitigation of disasters that would include, among others, a specialised database containing the best information available on the characteristics, experiences, strengths and weaknesses of national and regional agencies responsible for disaster prevention and mitigation and provide a new framework for technical cooperation and research aimed at creating a hemispheric culture of prevention and solidarity.

(vii) Adopt and support, as appropriate, initiatives aimed at promoting capacity building at all levels, such as the transfer and development of technology for prevention—risk reduction, awareness, preparedness, mitigation—and response to natural and other disasters, as well as for the rehabilitation of affected areas.

(viii) Promote mechanisms that incorporate risk management and risk reduction methods in public and private development investments.

Earlier the focus was more on disaster relief and response. Not much attention was paid to disaster prevention and preparedness. With the passage of time and realisation of the fact that disasters can be managed better if the emphasis is more on preparedness and coordinated efforts by the concerned stakeholders, there has been a change in the perspective of disaster management. Initiatives are now being taken for disaster risk reduction through various structural and non-structural measures. In consonance with the need of the hour, some efforts are now being made in South Asia for disaster risk reduction by adhering to detailed, comprehensive and participatory strategies. The present volume is also a step in this direction wherein an effort has been made to record some of the initiatives and thought processes on disaster risk reduction. The various chapters included in this volume focus on disaster risk reduction and have been contributed by experts from South Asian and other countries.

Following the above general introduction to the subject, here is a summary of the various chapters.

In Chapter 2, *Bringing together Disaster and Development—Concepts and Practice, Some Experience from South Asia,* disaster risk is looked upon as a part of the dynamic forces at play in the process of development. Disasters need to be seen in the context of where they take place. No development plan is complete unless contributory factors to disaster risk are appropriately addressed. The overall objective of development and disaster management is to reduce socio-economic vulnerability. Some of the current concepts and definitions of disasters and development are discussed. An attempt is made to demonstrate the application of disaster management at the institutional and community levels. The analysis is based on the research carried out by the ITDG South Asia under the *Livelihood Options for Disaster Risk Reduction in South Asia* project and the discussion held at certain other forums. The institutional arrangements pertaining to development plans and disaster management, and disaster development linkages are also covered. Special reference has been made to floods in the context of development. The chapter also contains comparative studies of managing droughts in Sri Lanka and Pakistan and refers to the handling of drought situations in some Indian states. It concludes with the thought that disaster management needs a transition from *emergency management* to *risk management*, while *risk reduction* needs to be a part and parcel of development thought and action.

Chapter 3, *The Principle of Risk Partnership and the Role of Insurance in Risk Mitigation,* emphasises the potential role of insurance in risk mitigation. There is an urgent need for action to cope with the substantive amount of losses emanating from disasters. The principle of *risk partnership* has been identified and distinctions drawn between insured persons, primary insurers, reinsures, capital markets and government/public authorities as role players in the context of natural disaster relief. Each of these parties has its own cost and responsibility in managing the risks arising from natural disasters. The chapter also looks upon earthquake insurance as a tool of risk mitigation. The use of earthquake insurance as a motivating tool within the framework of loss mitigation programmes has been increasingly discussed in recent years. Yet, the actual implementation is much below potential. The chapter is illustrated with examples from Turkey, New Zealand and other countries.

Risk Reduction through Managing Disasters and Crises, Chapter 4, looks at the tragedy of natural and man-made disasters as embedded in human memory. It is important to recognise that the risk of loss of lives, livelihood, and agriculture infrastructure and development investments is much too significant to be consigned to a disaster management plan that embraces only emergency relief measures. The chapter highlights the agenda for disaster preparedness of the International Federation of Red Cross and Red Crescent Society. Creating a culture for prevention is essential to address everyday hazards and the consequences of disasters. Reducing risk to critical hazards faced by a general population is no longer the sole responsibility of the national government. Humanitarian organisations share a similar responsibility is disaster preparedness, but in a role complementary to that of the government. The chapter refers to the Ouagadougou Declaration of September 2000, and the Fribourg Forum held in June 2000. The purpose of the forums was to obtain policy guidance, political support and commitment necessary to improve coordination and cooperation in international emergency humanitarian assistance. The following objectives and measures should be part of a crisis management pact: to reaffirm and ensure respect for international humanitarian law and the principles guiding emergency humanitarian assistance in crisis; to support and improve coordination at all levels of the state through coherent policies and complimentary structures; to improve coordination and cooperation between neighbouring and partner states in crisis; to improve coordination and cooperation in crisis prevention and response by civil society humanitarian agencies; and to provide a secure environment for the conduct and delivery of humanitarian assistance. For millions of people, humanitarian assistance offers a vital lifeline. It is an expression of solidarity and humanity. The old Russian saying that people only cross themselves when they hear the sound of thunder is quoted to emphasise that it is time to produce the sound of thunder for all disaster stakeholders to hear and consider measures that can be taken to create a safer world in the 21st century.

The definitional issues and disaster policy in South Asia are referred to in Chapter 5, *Disaster Risk Reduction through Livelihood Concerns and Disaster Policy in South Asia.* An attempt is made to address these issues in relation to livelihoods and identity the contours of disaster policy in South Asian countries besides suggesting advocacy options, mainly drawn from the ITDG Research project. Inferences are drawn from ten case studies on droughts, landslides, floods, cyclones, land erosion and arsenic contamination conducted in five South Asian countries under the ITDG Project, to state that disaster prone communities at risk in South Asian countries are living virtually at subsistence levels with very limited livelihood options and opportunities. The chapter also highlights the policy initiatives to be taken at the regional and community level. There is need to adopt an integrated and holistic approach for effectively managing disasters and thus reducing risks emanating out of these for the ultimate benefit of the society, polity and community.

Information Technology has penetrated all disciplines and made the management of information and knowledge more efficient, cost-effective and virtually real time. The phenomenal expansion of telecommunication facilities in developing countries in the recent past makes the integration of information technology with telecommunication interfaces easier. Chapter 6, *Applications of Information Technology in Disaster Risk Reduction* looks at a major breakthrough of information technology applications in disaster management—the design and development of Geographical Information Systems permitting the development of base maps at micro levels. The chapter refers to Relational Database Management Systems, Management Information Systems, Decision Support Systems, Knowledge Bases, Expert Systems, Simulation Modelling and Scenario Analysis, etc. The issues of awareness and dissemination of information for disaster risk reduction are discussed. Mitigation, prevention and preparedness campaigns can be carried out through various means. There is need for documenting the best practices and coping strategies of disaster prone communities so that these can be disseminated to create greater awareness among affected communities elsewhere through the Internet, intranets and extranets. Appropriate use of various applications could substantively and qualitatively reduce disaster risk.

A categorisation of disasters and attitudinal insensitivites is given in Chapter 7, *Trigger Mechanism—The Concept for Emergency Response Plan for Disaster Risk Reduction.* The *Trigger Mechanism* envisages that on receiving signals of disasters happening or likely to happen, all activities required for the mitigation process are energised and activated simultaneously without any loss of time and the management of the event is made feasible. The primary objective of the mechanism is to undertake immediate rescue and relief operations and stabilise the mitigation process as quickly as possible. The mechanism requires planners to identify disasters and their probability, evolve signal/warning mechanisms, identify the activities and sub-activities, define the level of response, specify authorities, determine the response kind, work out individual activity plans, have quicker response teams, undergo preparedness drills, provide appropriate delegations and have alternative plans. There are two important components of the mechanism—the authority; and the coordination, command and control activities. The chapter also presents the Trigger Mechanism network, including the response to be initiated at the national level and right down to the community level.

Chapter 8 titled *Role of Remote Sensing in Disaster Risk Reduction*, is structured around the thought that natural hazards or disasters can strike any part of the world, any time, often without any warning and trigger colossal loss of life and property. In developing countries, every year thousands of people lose their lives and property worth billions of rupees is damaged due to the occurrence of a number of natural disasters. These include cyclones, floods, droughts, volcanic eruptions, landslides and earthquakes, etc. A disaster management programme may be divided into the following phases: pre-disaster planning phase, disaster preparedness or early warning phase, monitoring phase, emergency response or damage assessment phase, and recovery and relief phase. Remote sensing and GIS, a powerful set of tools for collecting, storing, retrieving, transforming and displaying spatial data from the real world for a particular set of purposes, can play a crucial role in furnishing important information in each of these phases. The chapter also focuses on the utility of remote sensing and GIS in some commonly occurring hazards like floods, cyclones, earthquakes, landslides and forest fires. A case study of landslide hazard zonation (LHZ) of the Bhagirathi valley using remote sensing and GIS has also been referred to in the paper. The chapter concludes by saying that it is imperative that many countries should involve themselves in developing disaster risk management methodologies adaptable to local factors. The methodologies developed by different countries may be shared with the international community through the Internet for better understanding of natural disasters.

Chapter 9, *Disaster Risk Reduction by Education, Information and Public Awareness,* states that disasters happen at the interface of natural occurrences with people, their livelihoods, economies and infrastructure. As people play a crucial role in disaster risk reduction, it is imperative that they have appropriate awareness, training and orientation to meet their responsibilities. For ensuring effective disaster risk reduction plans, the community needs to be encouraged to increase its self-reliance in both preparedness and response. Through educational, awareness and training programmes, the community can be prepared to minimise the impacts of disasters. There has been a great reservoir of human, material and financial support in the developed nations in response to the needs of disaster prone developing nations. This has manifested itself in commitment to working with and amongst the affected populations, invariably the poor and the disadvantaged, when disaster strikes. There are four points of responsibility for action in relation to improved standards of disaster preparedness, better disaster management and in the development of self-help and self-reliance in communities: Governments at all levels; news and information media; administrators and professionals; and the people, NGOs, and community-based organisations and networks. To be effective in their purpose of creating appropriate standards of disaster preparedness, government policies must be translated into action. The chapter takes stock of the role of policies, existing public awareness programmes, community-based programmes, the mass media, etc. in disaster risk reduction. The chapter highlights the role of preparedness in risk reduction and concludes that one penny spent now in disaster reduction activity will be returned thousand-fold in terms of reduced relief needs.

Chapter 10, *Partnership in Health and Disaster Management for Risk Reduction in South Asia*, refers to the health and disaster situation in South Asian countries besides exploring the possible areas of technical cooperation for risk reduction. The underlying factors contributing to the high prevalence of communicable diseases include poverty, malnutrition, ignorance, an unsanitary environment and lack of drinking water. The public health infrastructure in the continent needs to be strengthened in terms of good public health policy, better system of healthcare centres, and training of the involved persons. Disaster management policies in South Asian countries are quite weak. There is no proactive approach with systematic, multifaceted and long-term policies built on regional and sectoral linkages. Disasters in South Asia have affected the community in many ways. Partnerships in health and emergency assistance in South Asia can provide a broad inter-governmental framework for building a cooperative approach so that the precondition for the success of disaster management policies can be met and the health of the people protected and improved upon. The broad objectives of such a partnership programme can be to: improve collaboration in the national public health and disaster management effort; improve coordination and sustainability in public health strategies towards disaster management; and strengthen the public health infrastructure and its capacity. The partnership can be formalised by the setting up of a Regional Health and Disaster Management Centre and a Regional Health Disaster Information Network. The chapter also suggests how these two should operate.

Chapter 11, *Implications of Macro Level Development Planning on*

Disaster Risk Reduction, looks at disaster risk reduction in Nepal. Helping the needy is the basic philosophy of the Nepalese. Though Nepal started its process of planned development in the 1950s, disaster related issues could not be properly addressed until much later. The chapter contains an analysis of the Natural Disaster Relief Act, 1982, subsequent upon which disaster management activities were geared up in Nepal. The major thrusts of the Plan of Action on Disaster Management formulated in 1996 are disaster preparedness, mitigation, response, and rehabilitation. The chapter also looks at the financial arrangements made in Nepal for reducing disaster risk. Areas that need to be worked on for effective disaster management have also been pointed out. Though Nepal has initiated several multi-sectoral programmes for disaster management, yet, poor coordination, low level of public awareness, and lack of resources are the major bottlenecks in managing disasters effectively. There is need for initiating measures at the micro and macro levels by integrating disasters with development for better results.

Chapter 12, *Disaster Risk Reduction: A Preparedness Approach*, brings to light the fact that the entire approach of reviewing and assessing the effects of disasters, not just in technical and scientific terms, but in a holistic manner, encompassing humanitarian, social and economic concerns, is assuming significance. It has led to growing awareness and concerted government efforts towards better disaster preparedness. The greater goal is to reduce the risks posed by disasters. Risk reduction is a task that has to be shared by a variety of actors. The initiatives of civil and community level organisations need to be taken cognisance of to strengthen risk management. Disaster risk reduction needs to be looked upon as a convergence of prevention, preparedness and mitigation measures. To support this point the concept and significance of disaster preparedness for effective risk reduction is analysed. The chapter also highlights the essentials of disaster preparedness and points out that preparedness and prevention strategies are the need of the hour and there is a shift from post-disaster assistance to pre-disaster preparedness, from readiness to mitigation, from dependency to self-reliance, from individual aid to restoration of community services and from relief to rehabilitation. The key strategies for disaster preparedness include: application of information technology; planning for building disaster resilient communities; promotion of sustainable livelihood strategies; strengthening of disaster preparedness capacities and partnership for disaster preparedness. The chapter concludes by saying that disaster risk reduction shall be meaningful and effective if it is multi-faceted, accompanied by strong political will and the implementation capability is supported by adequate policy and system reform.

Community Capacity Building for Risk Reduction in South Asia, Chapter 13, notes the paradigm shift from a traditional relief and disaster preparedness focus where communities are considered *victims* and *beneficiaries* of assistance from outside experts, to a more holistic and long-term approach incorporating vulnerability reduction and risk management concerns as parts of the development planning process. This paradigm shift has been accompanied by a growing realisation that disaster mitigation is most effective at the community level. The approach recognises that when disasters occur, the people of the community are the first to respond. It is always the immediate relatives, neighbours and other community members who come to help the victims and their families. It is important that capacities of the disaster prone communities are recognised and integrated with externally designed risk reduction programmes and projects. Using top-down interventions alone for disaster mitigation is insufficient because such interventions often pay little attention to addressing community dynamics, perceptions and needs, and ignore the potential of local resources and capacities and may, in some cases, even increase people's vulnerability. Moreover, local communities are often either unaware of these formal disaster mitigation interventions or they find them inappropriate due to their lack of recognition of the community's vulnerabilities and capacities. The management of risk reduction includes—building vulnerability capacities; empowering people; being contextually specific and putting a premium on organising communities. The strategies for community capacity building are: community risk assessment, preparedness, mitigation, emergency response and continuing responses. There are no straightforward solutions for risk reduction. Rather, there is need for concerted efforts, at different levels and across different sectors to improve the understanding of the linkages and to devise effective mechanisms for disaster risk reduction.

Chapter 14, *Disaster Risk Reduction through Capacity Building of the Community and Panchayati Raj Institutions*, is based on a study conducted in two districts of Orissa which were severely hit by cyclones in October 1999. People and agricultural lands in a number of districts, blocks, *gram panchayats* and villages were affected by the super cyclone. The chapter focuses on the need for capacity building in the community as unless people are prepared and made aware, risk reduction becomes difficult. There is a need to build the capacity of *Panchayati Raj* Institutions (PRIs) and the elected bodies at the village, block and district levels for disaster risk reduction. The chapter refers to community preparedness, coordination within the community, community awareness, and community participation as measures for capacity building. The need for having a holistic approach to community-based disaster preparedness is emphasised. Capacity building measures identified during the study include—preparation of community action plans, strengthening educational activities, formation of active community disaster task forces and creation of alternative livelihood options. PRIs can be strengthened by enhancing their knowledge base. Given the large number of PRI members, it will be cost-effective to reach out to them through the distance-learning mode. Self-instructional material modules need to be developed for capacity building of PRI members. Such modules could include: introduction to disasters; risk assessment and vulnerability analysis; disaster preparedness; disaster response; disaster medicine; and rehabilitation and reconstruction. There is also need for developing audio-video programmes for enabling PRI members to visualise response and recovery measures and put them to effective use.

Assessing Abilities for Disaster Risk Reduction: Discovering Possibilities and Exploring Options, Chapter 15, is based on a project undertaken by the International Centre for Entrepreneurship and Career Development (ICECD) and the ITDG in two villages of two districts of Gujarat, India. Under the project, detailed activity plans were made for rain water harvesting and livelihood programmes. The chapter highlights the approaches and strategies for project implementation. These include the participatory approach which has a demonstration effect, and measures for sustainable development and gender balance. The strategies incorporated in the project were—identifying specific needs of the people; creating awareness for water harvesting and livelihood alternatives; making people psychologically ready and technically capable; training and capacity building of men and women; ensuring community participation; and demonstrating the impact of sustainable water and enterprise development interventions. The activities and the achievements of the project are also mentioned. Certain useful observations which need to be taken care of for reducing disaster risk are made. Various issues which emerged during the implementation of the project are summarised under three sub-heads: disaster management; rehabilitation, displacement, and development; and management of displacement and the rehabilitation process.

Chapter 16, *Disaster Risk Reduction through Disaster-Resistant Construction Techniques*, looks at the combination of extreme environmental changes and rapid social, economic and institutional transitions that have triggered off a spate of disasters in the last few decades. The pernicious effects of disasters cannot be wished away, their negative consequences, can however, be reduced through effective disaster risk reduction methods. One way to do it is through effective disaster-resistant construction. The chapter lists the factors that should to be kept in mind when planning disaster-resistant housing for earthquakes, floods and cyclones. These factors are: location and siting; layout and designing; and construction techniques. After analysing these factors at length, the chapter states that disaster-resistant construction must go hand in hand with rehabilitation and development activities in the affected areas. The rehabilitation package should entail housing and infrastructure re-development, social rehabilitation programmes and economic rehabilitation schemes. An effective disaster risk reduction plan must consist of schemes for health services, resurrection of educational activities and rehabilitation of the disadvantaged sections, especially women, children and the elderly. Disaster risks could be effectively reduced if the development of infrastructure, agricultural rehabilitation, growth of alternative employment opportunities, livelihood options and viable disaster-resistant construction methods are given due emphasis in development schemes. However, if disaster-resistant construction, a very pertinent disaster risk reduction method, is looked on in isolation from the broader economic, social and cultural requirements of the target groups, it becomes redundant. All risk reduction strategies draw sustenance from the development-disaster interface. Unless the aftermath of disasters is turned into a development opportunity with due focus on the requirements of target groups, availability of infrastructure, mobilisation of resources and agencies involved, disaster risk reduction through disaster-resistant construction will only remain rhetorical exercises.

Chapter 17, *Community Participation in Disaster Mitigation and Risk Reduction (with Special Reference to Cyclone)* begins by stating that the community is the worst affected in any disaster. A disaster derails the normal life patterns of the victims and their families. The social and economic consequences of disasters are of great magnitude and it takes a very long time for the concerned community to get back to normally. For disaster risk reduction, it is imminent that the community adopts a culture of prevention, preparedness, response and recovery. This can be achieved mainly by ensuring the participation of the community in all measures initiated by the organised stakeholders. There is need for creation and sustenance of community awareness on disaster preparedness. An analysis is made of community risk perception to conclude that risk perception is situational and differs from individual to individual at the community level. The major characteristics of risk perception are: perception varies from person to person; people take risk because of the non-availability of choice; people make decisions only at the eleventh hour, when disaster strikes; people leave with their valuables; and, a community never thinks of a disaster, which may occur in the next five years. The chapter also refers to the various important areas of community responsibility.

Chapter 18 which is titled *Armed Forces in Disaster Risk Reduction*, looks at the involvement of armed forces in civilian functions as the last resort for the government. When the military begins its actions, civilian functions cease. Too much reliance on the armed forces weakens the civilian democratic set-up. The involvement of the armed forces is generally uneconomical as it diverts the army from its primary role of defending the country from external aggression. The constitution and the legal framework provide for the armed forces to render assistance during disasters/calamities when the situation is beyond the capabilities of the existing civil administration. Disasters are events, which are beyond the coping capacity of local people. Hence, during a disaster it becomes imperative that the armed forces be involved. Although, at present no specific disaster legislation exists within the country, the assistance of the armed forces is provided to the civil government for various internal duties during peace and war. The civil authorities seek the assistance of the armed forces for: maintenance of law and order; maintenance of essential services; assistance during natural and other calamities such as earthquakes, floods, famine and fires; and other types of assistance, including assistance in development projects. The important issues that need to be kept in mind when the armed forces are working in the aid of civil authorities during disaster situations are: cooperation; planning and preparation; humanitarianism; and economy. The armed forces have always been involved in assisting people in distress. It is important that at the planning stage all agencies, including the armed forces, should be involved in formulating a joint disaster management plan. Although various agencies are important during different types of disasters, but they all rely on the armed forces to play a major role in one way or the other. Exchange of information, cooperation and understanding each others needs and peculiarities will go a long way in better disaster management that contributes towards effective disaster risk reduction.

India, with its wide range of climatic and topographic conditions, is subject to various types of natural disasters. Chapter 19, *Pilot Demonstration Project for Drought Risk Reduction*, deals with one such disaster in Rajasthan—drought. It takes stock of the various risk factors and alternative livelihood options for risk reduction. After an analysis of the droughts and their impact, the chapter identifies *drought risk reduction and preparedness* as the need of the hour. This can only be taken up through active participation of the community. A pilot project undertaken in Lalwadi is part of an extensive exercise that aims to enhance livelihood options for reduction of drought risk. The project will ultimately contribute to long-term self-reliance among the community members to face future disasters and will serve as a model to be replicated elsewhere. The specific activities undertaken include—mass awareness programmes for community mobilisation, water management and advocacy. An analysis of past conditions makes the concerned stakeholders come up with simplistic but workable and innovative schemes with clearly laid down objectives, approaches and strategies for improving the conditions of the affected communities through drought risk reduction measures.

Landslides pose a severe threat to life and can cause extensive damage to natural resources and economic infrastructure. Often, landslides disrupt livelihoods and thereby pose a strain on development. The focus of Chapter 20, *Timely Intervention for Disaster Risk Reduction*, are the landslides that have been constant occurrences in the central and southwestern regions of Sri Lanka since the 1980s. The chapter is based on a study conducted by ITDG and the National Building Research Organisation in Nawalapitiya, Sri Lanka. The chapter identifies the gaps in institutional measures. The non-availability of disaster control or a management agency at the national level is one of the most notable gaps in the institutional setup. Disaster risk can be mitigated at the stage of planning. Adhocism prevails in the listing of priorities and in managing problems arising from uncertainties related to investigations and analysis. The chapter also lists the various coping mechanisms demonstrated by various communities in the universe of the study. Recent initiatives have revealed a high level of enthusiasm on the part of communities to prevent destruction from landslides. Efforts need to be made for undertaking detailed, data based vulnerability assessment and risk analysis. Community awareness should be further created and sustained. A realistic approach needs to be adopted for strengthening alternative livelihood options available to the community. The capacity and capability of governmental, non-governmental, community-based and social organisations need to be enhanced for reducing disaster risks through timely interventions.

Chapter 21, *Disaster Risk Reduction: Livelihood Options in Drought Situations (A Case Study of Tharparkar, Sindh, Pakistan)*, makes an analysis of the drought situation in Tharparkar. It is based on a study conducted in February 2000, by ITDG on Livelihood Options for Disaster Risk Reduction. The aim of the study was to establish the links between the drought and socio-economic conditions in relation to the livelihoods of communities and to identify approaches that can support and create options/opportunities for them. Major activities included in the project were the construction of rain water harvesting tanks and grafting the indigenous *ber* bush with improved varieties and establishing kitchen gardens. The key strengths of the project that contributed to its success included—the capacities of the partners, low scale of activity, clear framework for community involvement and a favourable geophysical fibre.

In Chapter 22, *Livelihood Strategies in Disaster Risk Reduction in Bangladesh,* a brief but updated scenario of disasters in Bangladesh is presented. Special reference has been made to floods, cyclones, river bank erosions, droughts, and arsenic contamination of ground water. The gender perspective is also examined. The most vulnerable sections during a disaster are women and children. Though they are usually at greater risk than men, they make it possible for the community to cope with disasters. Hence their role is absolutely central to the management of strategies to cope with disasters. The chapter also refers to a study conducted by ITDG on *Livelihood Options for Risk Reduction in Bangladesh.* Although a number of communities and households receive assistance either from the government or NGOs during disasters, the assistance falls short of the required needs. On the basis of a number of research studies, the chapter makes suggestions for long-term planning to reduce risks and ensure survival in disasters. Disaster risk can be reduced by adopting appropriate sustainable livelihood strategies aided with institutional support from the government and other agencies before, during and after the disaster.

Impact of Flood on Gender: A Case Study, Chapter 23, is based on a study conducted in Chitwan, Nepal, to analyse gender issues and concerns in the context of floods. Ways of incorporating gender-based capacities into disaster mitigation and management planning are suggested. A multi-pronged approach to disaster management by specifically incorporating women's issues in disaster preparedness, mitigation and management are recommended.

Chapter 24, called *Towards Risk Reduction—A Perspective of Disaster Management in Nepal: Challenges and Opportunities*, describes the major natural disasters in Nepal. The chapter brings to focus the role of the Ministry of Home Affairs, which is the nodal agency for disaster management in Nepal. An account of the functions and duties of various committees at different levels for effective disaster management is also provided. The problems in disaster management in Nepal and the measures to solve the problem are highlighted. There is need for a concrete, effective, practical and proactive policy for disaster management in Nepal.

Drought conditions in India are analysed in Chapter 25, *Drought Profile, Management and Risk Reduction in India*. It lists the administrative districts frequently affected by droughts, in different Indian states. The damage caused by drought in the past and the various government policies and programmes to deal with drought in India are also mentioned. The role of the India Meteorological Department, the National Remote Sensing Agency, the Central Water Commission, the Ministry of Agriculture, etc., in managing droughts is highlighted as are the commonly adopted drought mitigation strategies. A brief analysis of the droughts of 1982, 1984, 1985, 1986 and 1987 and drought conditions in some of the worst hit states like Rajasthan, Orissa, and Gujarat is done to depict the impact, damage and losses due to drought in India. The chapter stresses drought risk reduction and risk transfer by specifically referring to disaster insurance, Seed Crop Insurance, Kisan Credit Card, etc.

Chapter 26, *Disaster Risk Reduction through Development Orientation in Disaster Management: The Significance of People's Participation in India*, attempts to capture some of the synergies and tensions between development, community participation and disaster management. Participation is a qualitative process, which offers opportunities but also poses constraints and dilemmas on the complex and multi-dimensional development process. The chapter analyses different kinds of participation and highlights the objectives of community participation, and the constraints and contradictions of participatory approaches. A strong recommendation is made to systematically and continuously document how community participation was actually implemented.

Stakeholders Response in Disaster Risk Reduction, Chapter 27, deals with the worldwide efforts for risk reduction. It focuses particularly on the role of stakeholders in disaster reduction in India. The chapter highlights the international initiatives for disaster risk reduction, including the initiatives taken in the International Decade for Natural Disaster Reduction and those taken as part of the International Strategy for Disaster Risk Reduction. The paper mentions the various efforts made towards risk reduction in India over a period of time. The roles played by various agencies including NGOs at the national, state and district levels are highlighted. Various agencies like medical and public health services, civil defence, police, paramilitary forces and defence forces play significant roles in this regard. The chapter also discusses the cultures of prevention, quick response, strategic thinking and prevention for reduction of disaster risks.

REFERENCES

Disaster Risk Reduction Hemispheric Conference, Costa Rica, December 4–6, 2001 (http://www.summit-americas.org/eng/quebec-scummit1.htm).

ISDR—A safer world in the 21st century: Disaster and Risk Reduction (http://www.unisdr.org/unisdr/safer.htm).

ITDG: Disaster Reduction Strategy (http://www.itdg.org/html/disaster_reduction/strategy.htm).

Livelihood Options for Disaster Risk Reduction, ITDG—South Asia, Sri Lanka, 2002.

ProVention Consortium (http://www.proventionconsortium.org/objectives.htm).

Sapir, Debarati Guha, *Disasters in South Asia* (http://www.punjabilok.com/india_disaster-rep/introduction/disaster_southasia.htm).

Smith, K., *Environmental Hazards, Assessing Risks and Reducing Disasters*, Routledge, 1996.

Van Oppen, Charles, *The role of Insurance in disaster reduction,* 2001 (http://www.bghrc.com).

World Bank Mission: Disaster Management Faculty (http://www.worldbank.org/dmf/mission.htm).

World Bank paper, *What is Risk Management?* (http://www.worldbank.org/html/fpd/dmf/risk_management).

2

Bringing together Disaster and Development—Concepts and Practice, Some Experience from South Asia

— **Madhavi Malalgoda Ariyabandu**

INTRODUCTION

The global scenario, in relation to disasters, is dismal. Statistics indicate that the impact of natural and man-made hazards on life and livelihoods is increasing. Globalisation is aggravating poverty and vulnerability. We are grappling with increasing levels of poverty, a growing population, and limited and depleting natural resources, problems of governance and rapid urbanisation. These phenomena increase the numbers of disasters and their levels of impact. The following figures are an indication of the gravity of the situation.

During the last decade, annually, over 59,000 people were reported killed and over 195,890,000 people were affected the world-over by natural and industrial hazards. The alarming trend of the increasing impact of disasters is evident from the fact that there was a more than 700 per cent increase in the numbers of people affected in 1999 (41,244,335), in comparison to 1997 (4,698,656), in South Asia alone. (World Disaster Report, 2000.)

Further, recent studies by the United Nations on global warming point an alarming future scenario. There are predictions that global warming can bring about drastic climatic changes, resulting in droughts, floods, cyclones and volcanic eruptions in places, which have not experienced such phenomena earlier. It is also predicted that such a scenario might occur as early as 2025 and millions of people will be affected.

This chapter argues that disaster risk is a part of the dynamic forces at play in the process of development. Disasters need to be seen in the context of where they take place and within what complex and dynamic physical, socio-economic, institutional and political forces. Thus, no development plan is complete unless it addresses contributory factors to disaster risk. Ultimately, the objective of development and disaster management needs to be the same, that is, reducing socio-economic vulnerability.

Presented in this chapter are some of the concepts and definitions that appear in the current discussion on the questions of disasters and development. It also demonstrates the application and non-application of the same at the institutional and community levels. This analysis is primarily based on the case studies carried out by ITDG South Asia under the project *Livelihood Options for Disaster Risk Reduction in South Asia* (2000), and the discussion at the *Future of Mitigation, South Asian Disasters,* Duryog Nivaran Policy Forum held in New Delhi in 1999.

DEVELOPMENT PLANS AND DISASTER MANAGEMENT: INSTITUTIONAL ARRANGEMENTS

Development and *disasters* have been considered separately by the specific institutions, professionals and staff assigned to deal with the disciplines within the current institutional arrangements. In most South Asian countries, the subject of development is handled by departments and ministries with titles such as Rural and Urban Development, Development Planning and Implementation, Ministry of Agriculture, Irrigation Development Department and so on. There is a different set of departments and ministries to deal with disasters and related issues; in Bangladesh it is the Ministry of Disaster Management and Relief, Department of Agriculture and Cooperation under the Ministry of Agriculture in India (this department is now under the Ministry of Home Affairs, Government of India), Department of Social Services under the Ministry of Social Services in Sri Lanka, Ministry of Home in Nepal, Cabinet Division—Emergency Relief Cell and the Federal Flood Commission in Pakistan.

In addition, during recent times, particularly with the declaration of the International Decade for Natural Disaster Reduction (IDNDR) in 1990, special units have been created in some South Asian countries to specifically focus on disaster related matters. These include National Disaster Management Centres in India and Sri Lanka, and Central and Provincial Disaster Relief Committees in Nepal.

This *arrangement* indicates that no possible linkages between the two disciplines are assumed, foreseen or understood. The two sets of organisations largely operate on their own, without any linkages in terms of content, action planning and investments. Under such an arrangement, while some efforts are made to place development issues within the larger context, disasters are taken to happen in a vacuum, with no relationship with the social, institutional or economic context. Further, it also depicts that the action taken in dealing with disasters is largely one of emergency management and relief with welfare orientation.

Since the independence from the colonial powers in the 1940s until now, state institutional structures in most South Asian countries have remained similar to what has been outlined above. In development parlance hazards and disasters are identified as happenings or events, which hinder the development process. In this setting, disasters are given religious and superstitious connotations and are often termed as *God's fury*. This attitude is termed in literature as the *Dominant Approach* (Duryog Nivaran, 1996). A key characteristic of the dominant approach is making interventions once a disaster occurs, with the objective of returning the situation to as it was before the event.

Under this approach, in the analysis of a particular disaster, the emphasis is placed on physical forces, structures and damages. This nature of problem analysis gives no scope for any linkages with socio-economic forces in society, or to place the disasters in the context in which they take place. Hence, this kind of ideology only allows *after the event action*, i.e. emergency management, relief and rehabilitation. People are placed at a distance and/or removed from planning and action. Departments in charge of development activities identify people as *beneficiaries*, while in disaster terminology they are termed as *victims*.

AN ALTERNATIVE TO DOMINANT APPROACH

Social and economic issues related to disasters have been discussed for over a decade. During this period some important concepts and terms have been introduced and clarified, and researchers and practitioners have elaborated the process, which leads to what is commonly termed as a *disaster event*. Consequently, the linkages of disasters with development processes have been exposed.

Different ideas and vital variables have been brought to bear on the analysis of disasters. The most crucial variables are the social and economic variables. Disasters are thus identified as a process behind which a number of crucial factors are at play. Let us define some key terms.

Hazard is defined as the probability of the occurrence of a dangerous phenomena at a given place within a given period of time. A hazard can be related to numerous causes such as rising water levels, prolonged dry periods, high winds and so on.

Much has been written about *vulnerability* and many authors have taken efforts to describe this reality (Blaikie, 1998; Maskerey, 1989). It is defined as the degree of susceptibility to a hazard, or the lack of capacity to absorb the impact of a hazard and recover from it. Vulnerability is related not only to physical factors, but also to a range of social, economic, cultural and political factors.

Risk is defined as the product of hazard and vulnerability; a statistical probability of damage to a particular element which is *at risk* from a particular source or origin of hazard (Lewis, 1999).

Accordingly, a disaster is an outcome of a hazard impacting on vulnerable populations, which can be presented in the following formula:

Disaster = Hazard × Vulnerability

Thus, hazard by itself is not a disaster unless there are vulnerable populations who don't have the capacity to absorb it and who are unable to cope with it.

This conceptualisation is termed as the *Alternative Perspective* (Duryog Nivaran, 1996). This approach takes disasters as part of the normal development process, and recognises that linkages with society during the normal times are fundamental for understanding disasters. The objective of the interventions is to reduce vulnerability of people, and strengthen their capacity and to work on the social structures that make people vulnerable. Thus, the foundation of the alternative approach to disasters lies in the analysis of the disaster–development linkages.

Disaster–Development Linkages

Why are some groups in society more vulnerable than others? Why is it almost always the same resource poor and powerless groups that get affected by hazards and continue to be prone to disasters and other shocks? The concept of vulnerability helps us to answer these questions. In the context of development, it is the social, cultural, economic and political environment that makes people vulnerable to shocks, disease and other negative forces. Class, caste, ethnicity, gender, disability and age are other factors that affect people's vulnerability.

The same argument can be applied to encompass disaster risk. Being poor and having no choices increase people's vulnerability to disasters and increases the degree of risk to any potential hazard. Economic pressures force many of the poor to live in cheap but hazard prone locations such as flood plains or unstable hillsides, but there are many less visible underlying factors; social and political as well as economic, that affect peoples ability to protect themselves against disasters or to recover from them (Ariyabandu, 1999). Thus, disaster risk is one of the key factors that constitute the overall vulnerability of people, along with resource poverty and powerlessness.

Development in broad, general terms means a positive change. Chambers (1997) presents five key words to explain development: *well-being*, *livelihood*, *capability*, *equity* and *sustainability*, and suggests that each term is linked with the others. It vulnerability is taken as the condition, which explains the dynamics between disasters and development, one can see that socio-economic vulnerability can negatively impact all the above-mentioned five conditions. Disaster–development linkages are most visibly demonstrated in the livelihood analysis of people who live with various hazards. Dealing with hazards, and adjusting to live with them forms a large share of livelihoods and life sustaining activities of communities that live in hazard prone situations. Also visible are the varying degrees and types of vulnerabilities and the related survival possibilities at different degrees of vulnerability in confronting hazards. A series of case studies documented in Bangladesh, India, Sri Lanka, Nepal and Pakistan, on *Livelihood Options for Disaster Risk Reduction* by ITDG South Asia, present disaster–development linkages and the related dynamics.

A closer look at the research findings show that despite new thinking and much discussion on alternative approaches to disasters at various levels, matters largely remain the same, as they were a number of decades ago. Development desks continue with their assigned tasks, while emergency management orientation dominates the disaster scenarios, at the central and the provincial levels. Under such circumstances it often happens that the development process itself becomes a cause of increased vulnerability to disasters. Research studies indicate that while at institutional levels disasters continue to be seen as *events*, which *happen* at particular points of time. But for communities that live with disasters, they are a part of daily survival. The remarkable and striking capabilities and capacities that people possess and apply in the risk management process are evident from the case studies.

For communities preparing to face hazards and risk management are a continuous process linked to their livelihood activities. This contrasts with the *event* or *relief* approach adopted by the institutional structures and offers many lessons for integrating the alternative approach to disasters into the main stream.

FLOODS IN THE DEVELOPMENT CONTEXT

The situation of frequent flooding, as recorded in the case studies on the Kamra and Kot Murad villages in Punjab, Pakistan, is a case in point. Communities in both villages depend on agriculture for a living. Animals occupy an important place in the farming system. Land-size and ownership varies. For the community in Kot Murad a major share of the land cultivated belongs to the feudal landlords and the local population is highly dependent on them.

The threat of flood hazards, the case studies observe, has brought about varying categories of vulnerabilities into the lives of people, as shown in the *Vulnerability matrix* by Anderson and Woodraw (1998). Different categories in this matrix, termed as *physical* and *material*, *social* and *attitudinal* vulnerabilities, when applied to the flood situation reflect the complex and dynamic nature of the inter-linkages. With these two communities, it is observed that the existing physical and material vulnerabilities due to poverty have been exacerbated by perpetual flooding, resulting in economic and material losses, and other social and institutional concerns.

Physical vulnerabilities are seen in the form of destruction of houses, damages to agriculture land and crops, loss of livestock and personal possessions, displacement of families and livestock, poor accessibility due to nearby protective embankments and stagnant flood water pools, and the river getting diverted towards village settlements. Followed by the physical damages, there are economic vulnerabilities. For a number of months of the year floods hinder the community from engaging in any economic activity and mobility largely remains restricted to the village. Social and institutional relationships are largely governed by the feudal pattern, since land ownership lies with the landlords. Consequently, formal social organisations are in a poor state, communities are often unable to access services or voice their collective opinions on the unsatisfactory conditions. Literacy is low while infant mortality rates and population growth rates are high. This indicates isolation from mainstream development.

Analysis of flood impact on the livelihoods of the Kamra community show the flood management process that form a part of their calendar every year. From April to June they engage in early flood preparation activities and during July to September they cope with the floods, securing their lives and belongings and taking care of animals. October to December are spent re-building the damaged houses and attending to the damages to agricultural lands. From April to December major livelihood activities such as, growing wheat and fodder get disturbed because of the floods.

Both communities in their struggle to survive floods apply indigenous techniques to gauge the rains, water levels, and assess the danger levels, and act accordingly to manage their livelihoods. In doing so, a continuous risk management process within the given limited resources and knowledge levels falls into place. In effect, the communities are demonstrating a *process approach* with a strong disaster preparedness component.

Considering the frequency and the scale of flood hazard in this area, the institutional structures responsible for general administration and development require a strong flood preparedness focus. In the absence of the same, the research highlights short- and long-term, regular action required.

At the local level, in terms of general infrastructure, in Kamra there is no access road, and people are cut off from the main road, and therefore have no access to marketing channels to sell their produce. An access road built with a small culvert will not only connect them with the main road and the marketing channels, but will also greatly reduce the threat of regular flooding.

In the long-term, and at a higher-level, structural measures such as construction of eco-friendly small dams in the upper catchments of rivers to store excessive water, land-use regulation in flooded areas, development of protective embankments along river banks, re-locating vulnerable settlements, and developing an integrated water management system to effectively control excessive water during the rainy season will help to control the flood hazard. Research observes that it requires consistent policies and well-informed planning to overcome such enormous, perpetual flood related problems.

The Kamra and Kot Murad study demonstrates the continued application of the dominant approach, where development work takes place external to the issues related to flood hazard. In addressing floods which are a regular occurrence in the area, a *relief-based approach* is taken. There is no evidence of linking flood concerns with long-term development plans. Preventive measures, as noted by Athar (1999), are limited to weather forecasting and monitoring of rain water. The Irrigation Department maintains the related infrastructure such as spurs and embankments to check spill-overs. The Health Department takes care of preventive vaccination and local bodies evacuate people from threatened areas. Thus, it is noted that preventive measures are superficial and insufficient and carried out in a dis-jointed manner.

The study indicates the essential development linkages with the regular occurrences of floods and emphasises the need for a change in the approach. The supply-driven interventions made by the state and large voluntary institutions re-institute the vulnerability of the marginalized communities, while taking official pride in the relief measures in response to disasters. An alternate approach, which champions a demand driven approach to address substantial issues in vulnerability of communities like that of livelihood security in response to floods is in the offing. The livelihood options of the affected communities have to be secured in a long-term context of planning for sustainable development, instead of following the disaster and relief approach (Journalists Resource Center, 2000).

MANAGING DROUGHT HAZARD—IS THE PICTURE ANY DIFFERENT?

A closer look at a different hazard, such as drought, which has a slow onset nature, depicts that irrespective of the nature of the hazard, it is the process contributing to vulnerability that we need to grapple with, and how an event or relief-based attitude contributes to the continuous vicious cycle of poverty and vulnerability.

Excerpts from case studies on the livelihoods of communities living with drought hazards in Hambantota and Putlam in Sri Lanka, and in Mithi, Tharparkar in Pakistan presented below give a picture of isolated approaches that continue to be taken for development planning and drought management. Although drought conditions in the two locations referred to are very different, the main related issues, the root causes, the suffering of people, and how these are managed show remarkable similarities.

Drought results in crop failure and drying up of water resources for agriculture and drinking. This directly affects the food and water security of communities. The dry areas of Sri Lanka receive 800–1200 mm annual rainfall, and the livelihoods of a majority of the communities depend on small holdings agriculture: rain-fed, and irrigated paddy and highland cultivation of vegetables and other crops. Lands can generally be categorised as fertile. There are a large numbers of reservoirs in the villages to catch and preserve rainfall. These were built in ancient times over 1000 years ago. Most of them have fallen into disuse due to maintenance failures. Drought or prolonged dry periods are a regular feature in the dry zone of Sri Lanka.

In comparison to the drought situations in Sri Lanka outlined above, the scenario in the Thar is of an extreme nature. Rainfall patterns across the Thar area are not uniform, neither seasonally nor geographically. In some areas, yearly average rainfall is as low as 100 mm. The case study *Drought in Tharparkar* (Waheed and Sheikh, 2000) narrates the relationship between the regular cycles of drought and livelihoods. After every four to six years a drought period of two to three years sets in.

The main livelihoods of the people in the Thar are agriculture, animal husbandry that includes large numbers of animals, and crafts making. Water scarcity is the main issue is the Tharparkar region. There is no canal or river that flows through the area and the rainfall pattern is unpredictable and irregular. If there are insufficient or no rains, the crops will not grow, and hence such drought causes severe food insecurity. The same applies to animals, which suffer acutely from water shortage. This becomes the major cause of their migration, with their owners, to barrage areas.

In drought years, the wells do not get re-charged and dry up, or the level of water goes down further. Absence of rain does not let *tarais* (rain water harvesting structures) store water and the entire burden of catering to the water needs falls on wells. Animals are used to draw the water. During drought years, when animals migrate, or get weaker, it becomes even more difficult to draw water from the depleted well. Usually this task is then undertaken by the women.

Depending on the degree of dryness and the length of the drought period, people in both locations employ various methods to produce food and to access, save and preserve water. The drought management continuum of communities demonstrates the dynamics of vulnerability, as well as how people's capacities are applied to the highest levels in their struggle for survival. Preparedness is a visibly strong component of community action.

In Sri Lanka, ahead of dry periods, communities carry out maintenance activities in the irrigation canal systems, and repair *bunds* in the small village reservoirs where rain water is harvested. With regard to production, when drought is predicted, farmers delay the sowing of the main food crop paddy and/or switch it with other drought resistant grains such as millet. The combinations of vegetable and root crops cultivated too are changed to cope with the reduced quantities of water available. Thus, a number of on-farm water management measures are taken based on their own experience as part of drought risk management. Once the drought situation aggravates and domestic wells dry up, people dig wells in the reservoir bed to access water for domestic purposes, before they are finally forced to walk long miles daily in search of water.

As a part of household risk management strategies, people try to stretch the quantities of available food to the maximum possible level. They begin with reducing the number of meals, quantity of food, and the nature of food taken. As a last resort they consume grains reserved as seed paddy, which is their livelihood supporting reserve. With an increasing degree of drought, migration and selling of household assets starts. State level drought management usually gets activated at this stage and includes distribution of water and other forms of relief such as dry rations and cash.

FROM RELIEF TO MITIGATION—BRINGING TOGETHER DIVIDED DISCIPLINES

Formal, state level drought management in both countries shows a strong relief focus. At present, the most usual/regular way of dealing with the drought is leaving the communities to cope with it as long as they can, and distributing drought relief when the situation is drastic. Once drought conditions set in, relief programmes are activated. Relief packages include food for work programmes, and transportation and distribution of drinking water brought from elsewhere at high cost. In Pakistan, relief include not only food for the human population, but also fodder for animals.

Each year the national budgets for drought relief are shrinking and people who are forced to live under the same conditions fall deeper into the trap of poverty and their vulnerability is enhanced. Thus, the subsequent cycle of drought has a harsher impact on them. Livelihood management strategies adopted by the communities living in desert conditions in Thar, Pakistan and in the dry zone of Sri Lanka offer many lessons for departments working in isolation to come together. However, neither the development plans, nor drought management institutions pay attention to this in any meaningful way. There is no responsible action to capture the capacities demonstrated by people in managing drought risks into long-term water resources and infrastructure development initiatives.

In both scenarios the situation outlined is familiar to all concerned, and has been occuring in regular cycles for the last 50 years or more. What would be the appropriate development strategies/programmes, which could take drought risk into account, and support livelihoods on a sustainable basis?

One can argue that the relevant departments should take care of water resource development, infrastructure facilities and so on. The question that remains is how much overall development planning is linked to the real needs and constraints of a hazard situation. The mandate of the general development network (agriculture and irrigation authorities) in the centre does not include substantial disaster preparedness elements. Nor have there been any known efforts to invest drought relief in short- or long-term water resource development.

However, there seems to be some hope. There is movement from *relief* to *mitigation*, by bringing together divided disciplines. The latest evidence presented by the Centre of Science and Environment, New Delhi, India (*Down to Earth*, 2001), shows emerging changes in India and indicates there is hope for major change. For the first time in the last 50 years, several state governments are dealing with drought in a different way—moving away from drought relief to drought mitigation. The droughts of 2000 and 2001 have seen Andhra Pradesh, Gujarat, Madhya Pradesh (MP) and Rajasthan undertaking major rain water harvesting programmes—getting people to conserve rain water that falls in their villages.

Reference is also made to Haryana and Maharashtra in India, where applying this approach began decades ago and has brought visible results. Water harvesting in these states started in the 1970s and is now showing the beginning of rural ecological and economic regeneration. Water improves agriculture, improved agriculture improves animal husbandry and once people begin to harvest water they begin to take care of their watershed, which means more trees and forests. The combined incomes from improved agriculture, animal husbandry and tree wealth have the potential to not just alleviate, but literally eradicate rural poverty (*Down to Earth*, 2001).

The above given is a rare situation where an alternative approach to disaster management brings about desirable results. It shows how addressing some of the forces behind the vulnerabilities of the people living with drought hazard has resulted in enabling them to achieve overall development. At the same time it is also a case in point, which demonstrates how general development can take care of hazards and vulnerabilities in the long-term.

CONCLUSION

Despite the call for a different, holistic approach towards disasters, where the socio-economic context is recognised, at the ground level, isolated, relief based, compartmentalised action continues to dominate.

There is ample evidence to show that disaster risk is part of the dynamic forces, which form the development process. Thus, disaster mitigation planning and development planning in effect share common goals. No development plan or intervention can be complete unless there is a clear understanding of the risk element and this is translated into practical interventions. Similarly, disaster mitigation action lies not with the immediate concerns in and around hazards, but in the more distant forces behind hazards and vulnerabilities.

This understanding allows building on the capacities people apply in their daily living to manage and face hazards. This approach stops us from seeing resourceful people as helpless victims, whose lives need to be supported by outsiders. Most importantly, this approach offers the key to leave behind the relief mode based on handouts. It opens many windows by which vulnerabilities can be turned into capacities, which in turn can make a dent on poverty and achieve long-term development.

Ultimately the objective of development and disaster management needs to be the same, aiming at removing socio-economic vulnerability. To achieve this status, disaster management needs a transition from *emergency management* to *risk management*, while development makes *risk reduction* a part and parcel of its thinking and action.

REFERENCES

Anderson, M.B. and P.J. Woodrow, *Rising from the Ashes: Development Strategies in Times of Disaster*, Lynne Rienner Publishers, London, 1998.

Ariyabandu, M.M., *Defeating Disasters, Ideas for Action*, Duryog Nivaran, Intermediate Technology Development Group, Sri Lanka.

Athar, H., *Disaster Management and Mitigation Policies in Pakistan: Present and Future*, Policy Forum on Future of Mitigation, South Asian Disasters, Disaster Mitigation Institute India and Intermediate Technology Development Group, Sri Lanka, 1999.

Blaikie, P., T. Davis, and I. Wisner, *At Risk: Natural Hazards People's Vulnerability and Disaster*, Routledge, London, 1994.

Case Studies of Moazza Kamra and Kort Murad Villages in Jhang District Punjab, Journalist Resource Center, Islamabad, Pakistan, 2000.

Chambers, R., *Whose Reality Counts,* Bath Press, London, 1997.

Down to Earth, Vol. 10, No. 6, Centre of Science and Environment, New Delhi, India, August 15, 2001.

Lewis, J., *Development in Disaster-prone Places: Studies of Vulnerability*, Cromwell, Trowbridge, 1999.

Maskerey, A., *Disaster Mitigation: A Community Based Approach*, Development Guidelines No. 3. OXFAM, 1989.

Rupasena, L.P. and R. Ariyabandu (Eds.), *Study of drought in Hambantota and Putlam in Sri Lanka*, Case study prepared for the ITDG South Asia, 2000.

Waheed, A. and R. Sheikh, *Drought in Tharparkar, Pakistan*, Case study prepared for the ITDG South Asia, 2000.

World Disaster Report, International Committee of the Red Cross and Red Crescent, Oxford Press, 2000.

3

The Principle of Risk Partnership and the Role of Insurance in Risk Mitigation

— **Anselm Smolka**

INTRODUCTION

To reduce the growing losses from natural disasters, a cooperative effort from all involved parties is required. Under *risk partnership* the roles played by the affected persons and entities, the financial sector and the state are described from an international perspective drawn from actual business practice. The potential role of the insurance sector in risk mitigation is addressed specifically and some examples of private/public partnerships are presented.

GROWING LOSSES—NEED FOR ACTION

The data on losses from great natural disasters since 1950, has shown a dramatic increase over the last few decades. The reasons for this development are manifold and encompass:

- The increase in world population and enhanced concentration in large conurbations.
- Social and economic factors such as the development of highly exposed regions and high vulnerability of modern societies and technologies.
- Changes in the natural environment (e.g. global warming and the related regional effects).

As the underlying factors for the observed losses remain unchanged, a further increase of losses from natural disasters is inevitable.

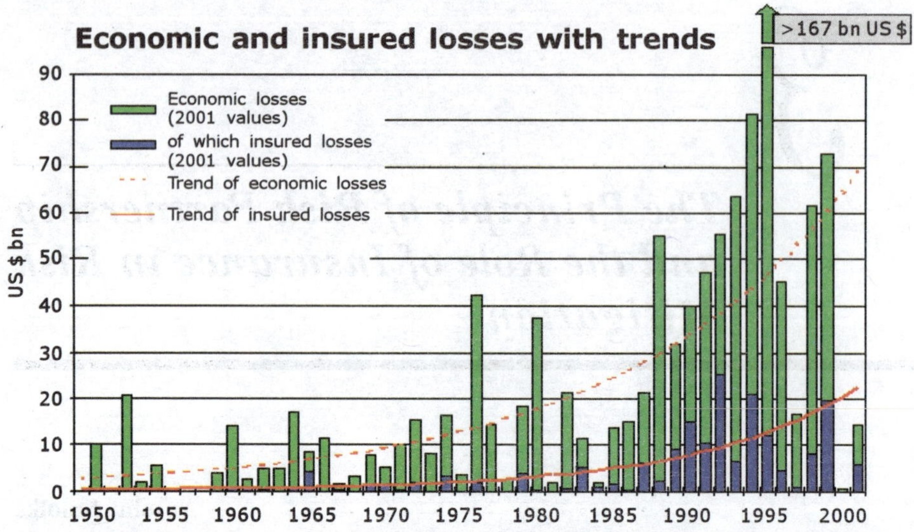

FIGURE 3.1 *Losses from great natural disasters (far exceeding 100 deaths or 100 million US $ in losses), 1950–2001. (Figures are adjusted for inflation.)*

THE PRINCIPLE OF RISK PARTNERSHIP

Coping with future loss burdens represents a formidable challenge, which requires the cooperation of all involved parties, i.e. the potentially affected private individuals and industries with the financial sector and the state. While discussing the role to be played by insurance in natural disaster relief, we can distinguish between

- The insured persons or entities
- Primary insurers
- Reinsures
- Capital markets
- Governments/public authorities

Each of these parties has its own tasks and responsibilities in managing the risks arising from natural disasters. Beyond financing future losses, which is a reaction after the event, more efforts need to be made towards a pro-active strategy to reduce and prevent future losses. Such a strategy is not only a matter of financial resources, but also, and maybe even more so, a result of good and foresighted planning and coordination at all levels, from households and industrial companies to public institutions and authorities. The following are the tasks of the involved parties:

1. **The insured persons or entities.** Householders and business owners can do a lot to reduce the risk to their property by proper maintenance and security of sensitive items like equipment, electronic installations and machinery. In industrial

businesses, emergency planning can help to prevent or minimise losses from future disasters. Finally, a certain portion of the financial risk has to be borne by the insured entities to keep alive the interest in loss reduction. Typical forms of participation are; deductibles, preferably expressed as a percentage of the sum insured and/or co-insurance, i.e. a percentage participation in each and every loss.

2. **Primary insurers.** Primary insurers have to provide and secure capacity by:
   - Charging technically adequate rates.
   - Applying appropriate underwriting guidelines.
   - Accumulation control and portfolio management.
   - Establishing reserves for natural perils.
   - Limiting their liability according to their financial strength, that is, reinsurance protection.
3. **Reinsurers.** They are often the main risk carriers in the matter of natural disaster losses, making proper risk management a primary task which includes:
   - Balancing the risk over time and regions.
   - Technical support to the clients in rating considerations and assessments of probable maximum losses (PMLs).
   - Controlling and limiting liabilities (setting cession/occurrence limits, budgeting, retrocession).
4. **Capital markets.** They have entered the scene only some years ago. This type of alternative risk transfer (ART) must be seen as a supplement rather than a competition to reinsurance. Their function is mainly to provide additional capacity for top-ranking losses.
5. **The state.** In the context of insurance the state has to act as an insurer of the last resort for very rare, extraordinary losses and/or uninsurable risks. The main task of the state is however risk management and risk reduction by:
   - Designing and enforcing land-use and building regulations.
   - Securing the serviceability of critical facilities and infrastructure.
   - Developing emergency plans that precisely define the responsibilities and the coordination of the authorities involved.
   - Granting tax exemption for catastrophe reserves of private insurers.

Within this context, the role of the insurance sector has been well-established and tested. In contrast, the capital markets have still to prove their willingness to provide reliable and continuous services when investors may have lost their money after large disasters. Furthermore, it is worth mentioning that the entirety of ART programmes have been in place so far only for highly developed countries. The complexity of the programmes, investor attitudes and the usually high prices require mature insurance markets. The state should create an environment where the greatest possible use of private resources for disaster recovery is combined with the availability of protection for as many people as possible. Linking the availability of such protection to the observance of building regulations can provide an efficient mechanism for the enforcement of codes, especially for new constructions. Mechanisms aimed at code compliance may serve to encourage rehabilitation measures as well. Another important role in the *mitigation cycle* could be played by mortgage banks requiring natural disaster insurance as a precondition of the loan.

EARTHQUAKE INSURANCE—A TOOL FOR RISK MITIGATION?

The use of earthquake insurance as a motivating tool within the framework of loss mitigation programmes has been increasingly discussed in recent years. However, actual implementation of this concept is much below its potential. The reasons are manifold. There is often a lack of knowledge among the public at large, about insurance mechanisms or an idealistic perception of the function of insurance. In the insurance sector, competition and a short-term financial perspective do not create a favourable environment for actively promoting prevention and mitigation measures, as the timescale for a possible positive outcome tends to be too long. A unique project in this direction is the community classification scheme of the insurance-sponsored Institute of Home and Business Safety (IHBS) in the USA where, to promote loss prevention, communities are classified according to code compliance.

The classic example of successful loss prevention in property insurance is the inspection of insured objects by fire engineers employed by insurance companies to make recommendations on enhanced fire protection. The level of fire protection is a well-established criterion for rating and PML assessment. As far as earthquake risk is concerned, private firms, as a consequence of the shrinkage of insurance capacity, took similar initiatives after the Northridge earthquake in California. On the basis of risk management surveys, earthquake protection was improved and the insurance coverage bought was adjusted to the minimum demand or given up altogether in favour of direct investment in loss prevention.

Nevertheless, in natural hazards insurance, and especially in earthquake insurance, other features that foster loss reduction are widely used. These features are risk-adjusted premiums and participation by the insured party.

1. **Risk-adjusted premiums.** Tariff schemes reflecting the actual risk level commensurate with the location and the constructional characteristics of the insured object are increasingly being used globally. But the correct application of such schemes presents a problem, and in actual practice, rates are mostly still dictated by pure competition. Sometimes, for instance, rebates are given for alleged compliance with anti-seismic building regulations. Often, however, code compliance has not been checked and, although stated, it does not exist in reality. Therefore, this element can be counter-productive and even unjustified for old generation codes wherein the principal goal is avoiding the loss of life rather than reducing monetary loss.

2. **Self-participation.** This can be of three types:
   - Deductibles, expressed as a percentage of the sum insured or as a flat amount. Typical deductibles in earthquake-prone countries start at 2% and go up to 15% in highly exposed regions like California. But even in regions of moderate seismicity, deductibles of 10% are used, if insurance penetration and, consequently, potential catastrophic losses are high (e.g. in Israel). Insurance payments start only in excess of the deductible.
   - (Proportional) co-insurance, again expressed as a percentage of the sum insured. Under this arrangement, the insured party carries a fixed proportion of each and every loss. Typical values range from 10 to 25% and reach a level of 70–85% in Tokyo Bay.
   - First loss co-insurance/liability limits, expressed as a percentage or a flat amount. Here, the insurer pays from the ground up or after a deductible up to a certain limit.

All these elements can be combined and are accompanied by corresponding premium rebates. The greatest incentive to take loss prevention and reduction measures is given by proportional co-insurance of at least 10% or by deductibles of 5% or more, as the insured party has to carry a substantial portion of any loss on its own. The effectiveness of these elements depends to a critical degree on the actual spread of insurance. In this sense the following distinctions can be made between *free* insurance markets and countries where earthquake coverage is obligatory or semi-obligatory:

1. In an unregulated market that is completely exposed to competition, it is a delicate task to find the right balance between tariff elements geared to loss prevention and what will be acceptable to the consumer so that a sufficient spread of insurance is achieved or maintained. A common reaction in such cases is the *zero option*, i.e. no insurance and no loss reduction.

This option is neither in the interest of the public, which ultimately has to pay for future losses without having set aside reserves beforehand, nor of the insurance industry, which wants to generate business. As a matter of fact, insurance conditions that are unattractive, or in extreme cases completely unaffordable, result in a situation where globally, less than 10% of the people have any earthquake insurance at all. Such low market penetration makes attempts to foster loss prevention by means of insurance almost futile.

2. A much better environment for using insurance as a direct incentive or as an indirect contributor to loss reduction programmes is provided by insurance markets where the coverage is either mandatory or at least widespread. Attempts to educate and raise consumer awareness by means of brochures and videos reach many more people and consequently have a greater chance of success in free markets with low insurance penetration. In this context, mortgage banks can play an efficient role in fostering high market penetration without the support of legal measures, by requiring disaster insurance as a precondition for the loan, as practiced in Israel and Colombia.

These measures can be successful only if used properly. If, for example, full coverage without substantial deductibles is granted under mandatory schemes, the goal of loss prevention is missed again. A portion of the premiums collected under such schemes can be invested in loss reduction programmes or in relevant research. The New Zealand government's Earthquake Commission or the Swiss Earthquake Insurance Pool provide examples of such a policy.

CONCLUSION

The foregoing discussion has identified several levers for mitigating losses from natural disasters. The challenge is to knit these components into a secure and tight network of risk reduction measures.

At present, various solutions are in use for the splitting of responsibilities between the involved parties. In New Zealand, for instance, the Earthquake Commission (EQC) provides basic insurance coverage for every household, up to an annually adjusted, actual building value. Additional coverage for the replacement cost, as well as insurance for commercial and industrial risks, and for business interruption can be obtained from the private market. In Japan too, residential risks are covered to a large extent by the state-run Japan Reinsurance Company, whereas large businesses buy insurance in the private market, and the corresponding reinsurance is supplemented to a small extent by ART instruments like CAT bonds. The California Earthquake Authority (CEA) also introduced the concept of *basic coverage* in the aftermath of the

Northridge earthquake in 1994, when earthquake insurance was difficult to obtain for homeowners. An often-used concept in natural disaster insurance schemes is the *insurance pool*, which involves every company participating in disaster losses in proportion with its market share in premiums. This concept ensures that companies avoid being too badly hit or even going bankrupt because of a disproportionately high loss burden from specific events. Such pools now in Switzerland, France and Spain and are sometimes supported by state reinsurance. They are also being considered in several other countries in Europe, Latin America and Asia. The most recent such example is the Turkish Catastrophe Insurance Pool (TCIP). Losses exceeding the capacity of the above-mentioned programmes usually fall under government responsibility.

The involvement of capital markets in natural disaster coverage is still minor and limited to very few, well-developed markets. This may have something to do with a very cautious attitude on the part of potential investors, the reasons being a lack of confidence in risk modelling tools for regions outside the USA and Japan on the one hand, and the general socio-economic stability of developing and emerging markets on the other. In the context of the buyer, i.e. the insurer, the high price of such transactions, as compared to conventional reinsurance, cannot be forgotten.

Efficient incentives for code enforcement are almost completely missing. The Turkish Catastrophe Insurance Pool (TCIP), which started operations in late 2000 and has since sold several million policies to middle-class homeowners, represents an innovative concept for a less developed marketplace where code enforcement is to be linked to the availability of insurance protection and/or governmental disaster assistance. This scheme also illustrates the potential role of international organisations like the World Bank in designing and backing new, proactive strategies for risk prevention and reduction. Another example of such a project is the Natural Disaster Insurance Scheme in Honduras, which addresses low-cost housing, public infrastructure and crop insurance. This project, however, is still in the planning stage and its feasibility has not yet been proven.

REFERENCES

Gülkan, P., *Rebuilding the Sea of Marmara Region*, 2001.

Kunreuther, H., *Mitigation and Financial Risk Management for Natural Hazards*, Proceedings 6th Int. Conf. Seismic Zonation, Palm Springs, 2000.

Lessons for the US and Emerging Economies, Proc. Wharton-World Bank Conference on Challenges in Managing Catastrophic Risks, January 9–10, 2001.

Recent Structural Revisions in Turkey to Mitigate Disasters, Proc. Wharton-World Bank Conference on Challenges in Managing Catastrophic Risks.

4

Risk Reduction through Managing Disasters and Crises

—Earl James Goodyear

INTRODUCTION

In every culture and as far back as history records, the tragedy of natural and man-made disasters is embedded in human memory. No country in the world is free from the risk of natural hazards. People trapped in poverty, living in unplanned settlements and exposed to environmental and industrial risks, and those living in nations caught in internal and external conflict are more vulnerable to the threat of man-made hazards. It is important to recognise that the risk of significant loss of lives, livelihoods, critical infrastructure and development investments is too important to be consigned to a disaster management plan embracing only emergency relief measures. Each nation must decide what measures are necessary, feasible and affordable to embrace acceptable levels of protection against natural, social, economic and environmental catastrophes. Many nations in the world now subscribe to processes that embrace a wide diversity of measures to protect persons and properties while lessening the scale of disasters. Risk reduction can take place through preparedness and mitigation initiatives.

The International Federation of the Red Cross and the Red Crescent Society, have, in their ten-year *Strategy 2010*, set the agenda for disaster preparedness through the 176 Red Cross and Red Crescent Societies around the globe. This agenda defines action to predict and, where possible prevent disasters, reduce their impact as well as respond to and cope with their consequences at local, national and international levels. The key to this strategy lies in initiatives to reduce the vulnerability of households and communities in disaster-prone areas and improving their ability to cope with the effects of hazardous events.

The European Coordination Programme, established in 1997, by the United Nations Office for the Coordination of Humanitarian

Affairs (OCHA) was created to respond to requests from European states for improving the effectiveness and coordination of international humanitarian assistance in the region. In June 2000 ministers and representatives of 52 countries from Europe and the Commonwealth of Independent States, met in Fribourg, Switzerland, to affirm the policy guidance, political support and commitment necessary to improve coordination and cooperation in international emergency humanitarian assistance through the European region.

Both models used in this chapter, identify the need for dramatic changes as evidenced in the increasing demand for effective disaster prevention, mitigation and response. Stakeholders are obliged to improve the coordination between national, regional and international actors, eliminate obstacles to the effective delivery of humanitarian aid, strengthen civilian emergency management capacity and encourage bilateral humanitarian assistance by neighboring and partner states in emergency situations.

REDUCING RISK

Disaster episodes in recent years clearly demonstrate the increasing loss of lives, livelihoods and assets because of natural catastrophes and their disproportionate impact on low-income populations. In the year 1999–2000, international losses exceeded US $ 90 billion when 707 large-loss events were documented. The number of major catastrophic events over the past ten years has increased three-fold. In Africa, it is anticipated that by 2020, half of the continent's population will be living in urban areas. Mega-cities of ten million people or more present a challenge to disaster prevention experts. Urban catastrophes can take the shape of natural disasters, from hazards such as tropical storms, earthquakes, floods and fires to human-induced hazards such as air and water pollution and industrial accidents. While many disasters cannot be prevented, their impact can be reduced by an understanding of who is vulnerable and by creating effective risk reduction programmes before the next disaster.

Creating a *culture of prevention* is essential to address everyday hazards and the consequences of a disaster episode. A city with good sewers, drainage and waste collection systems is better equipped to cope with flooding. Well-designed, sited and constructed housing greatly reduces the risk of physical hazards from earthquakes, flood and tropical storms. Effective military, fire, police and emergency health services can serve as an effective first line of response for rapid and effective disaster interventions.

Reducing risk to critical hazards facing a general population is no longer the sole responsibility of a national government. Humanitarian organisations share a similar responsibility, but complementary to that of the governments', towards disaster preparedness. Many humanitarian organisations have developed long-term assistance strategies that support processes to create local competence that will become sustainable over time. Some organisations strive to reach new standards of excellence in offering disaster relief, technical assistance, training, food aid, material resources and management in combinations appropriate to local needs and priorities. And some organisations also advocate public policies and programmes that support these needs.

POLICY FOR DISASTER PREPAREDNESS

Central to the work of the 176 National Societies of the International Federation of the Red Cross and Red Crescent is building organisational capacity to predict and, where possible, prevent disasters, reduce their impact as well as respond to and cope with their consequences at local, national and international levels. This task essentially involves: (a) reducing the vulnerability of households in disaster-prone areas by improving their ability to cope with the effects of disasters, (b) strengthening the capacities of Red Cross/Red Crescent Societies for disaster preparedness and post-disaster response, (c) determining the society's role and mandate within the government's National Disaster Plan and (d) establishing a regional network of societies that will create a strong, collective impact on disaster preparedness and response at the international level.

The current Policy for Disaster Preparedness, adopted by all National Societies at the XII Session of the General Assembly in Geneva (November 1999) includes the following salient features that compliment the strategies formulated for the International Decade for Natural Disaster Reduction and other global stakeholders in disaster and risk reduction:

1. Recognise that disaster preparedness should be one of the primary activities of the International Federation and each National Society. It must be regarded as the most effective way of reducing the impact of small and localised as well as large-scale disasters. The National Societies have a role to play at the branch, national and the international level. These roles shall be complemented by the actions of the Federation at the international level.
2. Recognise the role of the Red Cross/Red Crescent in disaster preparedness as complimentary to that of the government and not as a replacement of state responsibilities. In addition, the National Society should engage in debate with the government on the focus and nature of the National Emergency Plan and encourage the assignment of clear role and responsibilities to the National Society, supported by appropriate legislations.
3. Advocate, where necessary, with governments, donors, non-

governmental organisations and the public, the need for and effectiveness of disaster preparedness. National Societies should contribute to raising awareness of hazards, levels of risk and coping mechanisms adopted by society, and mitigation programmes, such as early warning systems, that may reduce the loss of lives and properties when a disaster strikes.

4. Improve coordination by promoting better cooperation and partnership between National Societies, the International Committee of Red Cross (ICRC), governments, non-governmental organisations and other disaster response agencies at the local, national, regional and international levels.
5. Identify those persons, communities and households most at risk to disaster through assessment and analysis of risks, vulnerabilities and capacities (Vulnerability and Capacity Assessment) as a basis for prioritising location and focus of programme activities.
6. Raise awareness of disaster hazards through public education, encouraging vulnerable people to take preventative measures where possible before disaster strikes. Ensure that the knowledge from prediction and early warning systems can be accessed, understood and acted upon by local communities.

For many years, disaster preparedness was viewed by the International Federation as focusing on activities that improved the emergency response capacities of National Societies to respond to natural or man-made disasters. Now, the International Federation, along with many humanitarian organisations, perceives that the present challenge is to create more effective mitigation programmes that enable people in at-risk environments to gain mastery of their own lives and overcome the vulnerabilities that inhibit social and economic development. This challenge is being addressed through processes that enable greater participation by vulnerable groups in the decision-making process of developmental and risk reduction programmes along with communication strategies that effect positive behavioral changes in the decision makers in government, development agencies and the donor community.

The notion to contribute to disaster mitigation is reflective of a world with growing instability. A world where destabilising forces stem from a complex interplay between; population growth, accelerating gaps and disparities caused by globalisation, environmental stress and pollution with worsening competition for natural resources, a differential access to technology, reduction in and privatisation of government's social welfare responsibilities, an erosion in the value of community and a further destabilising of sub-national groups and interests.

THE OUAGADOUGOU DECLARATION

In September 2000, 52 African National Red Cross and Red Crescent

Societies convened the 5th Pan African Conference in Ouagadougou, Burkina Faso. The conference focused on key issues of food security and health for:

- Making food security a strategic priority for this decade, recognising that food insecurity is directly linked to a number of root causes, including poverty, the HIV/AIDS pandemic, the worsening debt crisis and armed conflict.
- Responding to the HIV/AIDS pandemic as an unprecedented humanitarian and development disaster in Africa, by scaling up our responses in terms of advocacy, prevention, care and mitigation.

As a prelude to the Pan African Conference, the International Federation supported the initiation of a yearlong food security study of 15 nations in East, West and South Africa. The results of the study were consolidated into a draft position paper presented to the participating Red Cross and Red Crescent Societies as a foundation for initiating the definition of both strategic directions and action plans for the ensuing decade. The strategic directions adopted by the National Societies included:

1. The initiation of VCA as a diagnostic tool for both, the development of a strategic plan for food security programming (working within existing government food security policies or advocating their creation) and capacity-building within the organisation of each National Society.
2. Systematically incorporating food security planning within all Red Cross/Red Crescent disaster preparedness and response programmes—linking relief and development interventions through internal and external capacity-building initiatives.
3. Participating in early warning systems and advocating for their establishment where they do not exist.
4. Developing food security programmes with a measurable impact through selective targeting, flexible implementation strategies, proactive monitoring and evaluations that measure impact in addition to processes employed in the programme.

The Action Plan, now to be created in each African National Society, would seek to develop more innovative responses to food insecurity. Gender issues, including advocating for women's rights, in particular to issues such as land access, land tenure and credit, would be considered in each food security programme. Reforestation and water conservation strategies would be reviewed in programmes addressing environmental conservation.

To respond to the HIV/AIDS pandemic, the National Societies adopted the following strategy:

1. Proactive advocation for a comprehensive and coordinated

strategy from governments and the public and private sectors for continent-wide support to provide access to treatment and care, to people living with HIV/AIDS in addition to awareness and prevention campaigns.

2. Advocating increased access to affordable drugs and basic health infrastructure necessary to sustain the life and livelihood of people living with HIV/AIDS.
3. Making National Societies more responsive to people living with HIV/AIDS through the development of programmes that protect their rights and dignity, and offer family members training in basic care, first-aid and psycho-social support.

With its continent-wide network of National Societies and its two million volunteers, the Red Cross and Red Crescent movement is uniquely placed to make a difference to the food security and health of vulnerable populations. Their strategy—to move from being service providers to facilitators—by delivering essential food and non-food services in emergencies, to mobilise and coordinate resources and to mobilise communities to address their root problems affecting the attainment of health and food security—shall have a positive impact on the efforts to reduce the suffering of people affected by drought and famine and the HIV/AIDS pandemic in Africa.

FRIBOURG FORUM

The Fribourg Forum, held in June 2000 in Fribourg, Switzerland, was convened by the UN Office for the Coordination of Humanitarian Affairs (OCHA) to bring together ministers and representatives from 52 countries of Europe and the Commonwealth of Independent States, regional, international and non-governmental organisations, including the Red Cross and Red Crescent Movement. The purpose was to obtain the policy guidance, political support and commitment necessary to improve coordination and cooperation in international emergency humanitarian assistance throughout the European region.

As it could be recalled, following the end of the Cold War, several new states were created in the region. In this transition, many states in the region have reduced the level of resources allocated to civil humanitarian emergency management. The downsizing of many civil defense organizations left gaps in the individual emergency response capacity of many states. The region also witnessed a growth in the number of new actors involved in the provision of humanitarian assistance from the broad community of governmental, regional and international organisations and civil society. These stakeholders in disaster preparedness, mitigation and response often work with overlapping mandates and compete for the same financial resources and emergency response assets. This increases the likelihood for ineffectiveness and confusion.

The demand for greater disaster preparedness was driven by four factors: (a) increasing industrialisation of urban and border areas, (b) continuing population growth and urbanisation in vulnerable areas, (c) declining ability to evacuate urban populations and (d) growing expectations of assistance from regional and international organisations. With an increasing frequency of major disasters and emergencies, more people are at risk and with a lower domestic response capacity in many states, there existed an immediate need for international humanitarian assistance to complement national efforts.

The Forum was part of an ongoing initiative to improve the effectiveness and coordination of emergency humanitarian assistance by offering further insights and policy guidance on issues including:

1. Reducing the collision of organisational mandates and responsibilities within and between countries, regional and international organisations operating in disaster response and humanitarian assistance.
2. Removing inappropriate and outdated protocols for the movement of relief goods and rescue and relief personnel within the region.
3. Strengthening and improving capacity and resources for civil emergency management institutions.
4. Improving the coordination of emergency response assistance between border states through the development of bilateral and multi-lateral arrangements between states.

Other concerns raised about humanitarian assistance in the region included:

- Adherence to humanitarian principles and the role of civil society.
- Reinforcement of legal frameworks for humanitarian action and cooperation.
- Continuing need for security and protection of civil populations.
- The role of the United Nations in coordination and response.

The participants created and endorsed the Fribourg Communiqué and the Framework for Action—documents that provide the political support and commitment sought to move forward in addressing regional issues and concerns. The communiqué identified four priority issues in need of immediate assistance: coordinating initiatives, expediting relief, encouraging bilateral response and strengthening civil institutions. Each member state agreed to focus immediate action on:

Coordinating initiatives. The proliferation of actors involved in emergency response and provision of humanitarian assistance necessitates a continuous review of initiatives and roles to ensure the most effective use of available resources. This includes reviewing the mandates of agencies within each government for eliminating unnecessary duplication and ensuring coordination among international initiatives. The importance of planning and pre-crisis coordination was recommended to become part of a framework for action that would include risk prevention analysis. The promotion of bilateral agreements among neighboring countries to establish unified reporting procedures and designing common training mechanisms for upgrading the skills of disaster professionals and first responders.

Expediting relief. Most of the obstacles that still exist to impede the rapid delivery of emergency humanitarian assistance are found in the transitional and recepient states. All states and organisations are encouraged to abide by the existing protocols for the rapid movement of emergency humanitarian assistance, to review national policies, to prepare for and practice the receipt of international aid, to establish uniform professional credentials, and to ensure the safety and security of the personnel involved in humanitarian efforts.

Encouraging bilateral response. In most emergencies, neighboring and partner states are the most immediate source of external assistance. States are encouraged to cooperate by concluding bilateral agreements for mutual assistance in times of emergencies.

Strengthening civil institutions. The task of providing emergency assistance is best performed when led by national civil emergency or crisis management agencies, with support, from regional, international and civil society organisations. States and organisations are encouraged to strengthen their capacities, especially with respect to improving coordination and ensuring adherence to humanitarian principles.

The member states noted that action taken by governments and regional and international organisations could result in major improvements. The maintenance and improvement of disaster response capacity in the region requires a dynamic process of participation, a continuous process of review, dialogue and action to clarify mandates, share best practices and lessons learned. The communiqué, while not a binding document, demonstrates the commitment expressed by participating states and organisations to address these issues in a collegial manner and to move forward with a much-needed intergovernmental dialogue.

The member states further suggested additional steps to be undertaken including:

- Advance accreditation of relief professionals.
- Broader involvement of local and international non-governmental organisations.
- Creation of a coordinating mechanism for heavy airlift transport facilities in the region.
- Tax exemption agreements for emergency relief organisations.

- Procedures for emergency meetings between border-states prior to and during crisis episodes.
- The sharing of national and international assets to respond to an acute emergency.
- The free entry of telecommunications equipment in emergency zones.

THE NEED FOR ACTION

Through the ever-expanding media attention on Africa, we can see the plight of countless individuals who daily demonstrate incredible resilience and ingenuity in coping with disasters and fighting for survival. Despite these efforts, millions of people are experiencing crises that exceed their immediate capacities to cope. Governments continue to fund humanitarian assistance to alleviate some of this suffering. The UN agencies, non-government organisations—both international and national, respond.

These responses should not be viewed as charity but as an expression of the international commitment to assist and protect civilians caught in conflict and people responding to a disaster episode. The four Geneva Conventions and the two attendant protocols are the centre of the International Humanitarian Law that sets out that all civilian non-combatants in conflict are to be treated humanely in all circumstances. The 185 signatory states to the Geneva Convention have accepted legal responsibility to protect and assist those in need. The Universal Declaration of Human Rights, states the right of those affected by natural disasters to food, water and shelter. And, the 1951 Refugee Conventions define the obligation of states and warring parties to provide refugees with humanitarian assistance, or allow it to be provided to all those in need.

While the right to assistance is universal, the reality of the provision is varied and inequitable. Not all those who require assistance receive it. Each year, thousands of neglected and marginalised people in forgotten emergencies face critical food insecurity and receive little or no assistance. We merely have to look at the Democratic Republic of Congo, Angola and Burundi for examples of humanitarian crisis looming on the horizon. The underlying causes of violent conflicts and disasters—intensified poverty and scarcity of resources, economic marginalisation within and between countries and exclusive and divisive politics—seem set to continue, if not worsen.

ENHANCING COOPERATION AND COORDINATION MECHANISMS

To improve the cooperation and coordination mechanisms that address human suffering has been the intent of both the International Red Cross and Red Crescent Societies in Africa and the member states to the Fribourg Communiqué. They have also reaffirmed their commitment to international humanitarian law and the principles guiding humanitarian action that form the basis for any emergency response.

Several objectives and measures that may be relevant in the development of a crisis management pact can be identified from these examples.

1. To reaffirm and to ensure respect for international humanitarian law and the principles guiding emergency humanitarian assistance in crises. To achieve this objective, it is necessary to ensure adherence by states, regional and international organisations to the principles guiding emergency humanitarian assistance in crises and recognition of the need to differentiate between emergency humanitarian assistance provided under civilian responsibility and activities carried out for political and/or military objectives. Principles defining the use of the military in support of humanitarian assistance programmes may need to be defined.
2. To support improved coordination at all levels of the state through coherent policies and complementary structures. States have the primary responsibility for providing emergency humanitarian assistance when a crisis occurs. National crisis management organisations are best placed to understand the nature of hazardous events, to prepare to rapidly offer assistance to them and to provide the at-risk populations with appropriate response measures that include:

- Initiate an inventory assessment of governmental, non-governmental, military and civil society stakeholder agencies in disaster preparedness and response to identify strengths and gaps in operational capacity to respond to hazards and risks.
- Review and revise state policies for improved coordination and cooperation in the provision of or receipt of international emergency humanitarian aid.
- Create activities that develop, promote and initiate harmonised training programmes and methodologies among participating states in a crisis management pact.
- Develop a framework to monitor and evaluate emergency humanitarian assistance programmes. The results of such actions are the key to the evolution of internationally agreed methodologies which are more efficient and effective.
- Review of the Crisis Management framework to ensure that each organisation has a clearly defined role and responsibilities that match their human and material resource

capacity. This review is critical to ensure the appropriate use of humanitarian aid.

3. To improve cooperation and coordination between neighbouring and partner states in crisis. When a crisis occurs, neighbouring and partner states represent the most accessible and rapid source of emergency humanitarian assistance. The provision and delivery of relief personnel, support equipment and relief supplies, however, is often delayed by the absence of bilateral agreements as well as because of governmental decision-making processes and procedures. Thus, it is necessary to consider:
   - Draft and initiate bilateral agreements between neighbouring and partner states to facilitate the reception and delivery of emergency humanitarian assistance and to identify the resources that will be made available in response to specific hazards.
   - Develop a schedule to periodically review agreements in order to share lessons learned from disaster episodes and incorporate the best practices.
   - Develop internal procedures and protocols to facilitate customs and cross-border instruments.
   - Establish a series of international, regional and bilateral training curricula, including exchange programmes and plan exercises to test procedures and protocols.

4. To improve cooperation and coordination in crisis prevention and response by civil society humanitarian agencies. Civil society organisations can compliment national and bilateral capacities in the provision of emergency humanitarian assistance as well as contribute to the capacity-building and risk prevention of vulnerable rural and urban communities. The key to greater harmonisation and effectiveness in disaster management is the creation of an appropriate division of labour and sharing of responsibilities among all the key stakeholders. As such, it is appropriate to consider the following:
   - Recognition of the role of civil society organisations to support preparedness, mitigation and response initiatives through an assessment of respective organisational capacities to support complementary activities.
   - Development of coherent protocols that ensure national policies are observed by civil society organisations in order to avoid unnecessary duplication and colliding initiatives at the time of a crisis.
   - To share information on national and international agency capabilities and develop a strategy through which individuals and organisations demonstrating excellence in preparedness,

mitigation and disaster management can be made available to other states.

5. To provide a secure environment for the conduct and delivery of humanitarian assistance. The protection of the civil population and relief personnel within a country, in the case of an emergency, must be assured by the state. National authorities should provide a secure environment for the conduct and delivery of humanitarian assistance. Security and protection in a crisis environment should be a precondition for cooperation among states. Thus, it is useful to consider:
   - Create a security plan and relevant capacity to ensure the protection of both service providers and relief recipients at the time of a crisis.
   - The use of military capacities, when required to ensure a secure environment, should be programmed to revert to civilian leadership as early as possible.

CONCLUSION

The International Federation of the Red Cross and Red Crescent has adopted policies to affect a pro-active disaster preparedness capacity in all of the current 176 National Societies around the world. The dimensions of disaster preparedness now include both the development of a response capacity in addition to the creation of more effective mitigation programmes that enable populations at risk to gain mastery of their own lives while overcoming the vulnerabilities that inhibit social development.

The Fribourg Forum initiative sought and obtained the political commitment necessary to begin the improvement process of cooperation and coordination of international humanitarian assistance in the broader European region. The Fribourg Communiqué and the Framework for Action affirmed the commitment of all the states and organisations in the region to work together at improving effectiveness and coherence of emergency programmes.

For millions of people, humanitarian assistance offers a vital lifeline. It is an expression of solidarity and humanity. The wisdom of promoting crisis prevention is incontrovertible. Aid alone, however, is not the solution to stem the consequences of violent conflict or natural disaster. Humanitarian assistance is by its nature short-term and cannot substitute for a commitment to crisis prevention. What is required is a commitment to the principle of universality at the heart of humanitarianism, through greater and more equitable provision of assistance and protection. A crisis management pact that empowers states in a commitment to strengthening prevention and response capabilities is necessary to alleviate human suffering.

There is an old Russian saying that, People only cross themselves when they hear the sound of thunder. We are able to identify the hazards and the risks that will affect millions of people in Africa and throughout the world. And now it is time to produce the sound of thunder for all disaster stakeholders to hear and consider measures that can be taken to create a safer world in the 21st century. This is the paramount challenge to cope with natural, social and economic disasters.

5 Disaster Risk Reduction through Livelihood Concerns and Disaster Policy in South Asia

— Amjad Bhatti

INTRODUCTION

If agriculture coordinates between heaven and earth, as believed by many, then South Asia is the credible courier between the two. This region has a distinct agrarian past, which teaches us to live intimately with nature. Climatic hostilities taught people to invent a pattern of adjustable relationships, as ancient life was largely based on personally perceived and collectively shared resource bases.

Agriculture, being a vital source of livelihood, was essentially dependent on the natural seasons. Disasters were believed to be an inseparable part of nature and a series of undesirable fluctuations in climate could endanger food security. People looked upon disasters as a part of rural-agrarian life. And they tried to cope with undesirable climate by offering sacrifices to gods, and by inventing various small-scale devices to ensure the security of their lives, lands and livelihoods. They made their rustic disaster policy, with which they were able to manage the risk, damage and stress of disasters.

Today, South Asia is populated by over a billion people, a quarter of the world's population, and eight out of ten people live in places officially classified as *rural*, surrounded by agriculture. A much smaller proportion works on the land and non-agricultural employment is growing rapidly, but a substantial majority still depends on agriculture for its livelihood. Agriculture includes farming and also animal husbandry, pastoral activities, fishing and harvesting the forest.

Disaster, owing to geo-physical peculiarities, has become an intensive and recurring social experience for most South Asian communities living in hazardous sites and at-risk locations. The 'Natural-Technological Syndrome' or *Na-Tech Syndrome* as it is called, caused by rapid growth and development results in bigger environmental threats, and the risks are multiplied given the lack of disaster resilient infrastructure coupled with varied layers of vulnerabilities among the marginalised majority of this region. But, quite alarmingly, the present disaster policy lacks the elements of integration at the conceptual and functional levels, although it uses modern technology.

This chapter attempts to address the definitional issues of disaster in relation to livelihoods, identifies the contours of disaster policy in South Asian countries and suggests advocacy options mainly drawn from the ITDG research project called Livelihood Options for Disaster Risk Reduction in South Asia (2000–2001).

DEFINITIONAL ISSUES

Disasters are popularly understood as aberrations in the course of nature, which are presumed to be out of the domain of human control. Floods, earthquakes, droughts, thunderstorms, etc., were considered to be pre-ordained acts of nature or the wrath of God and humans could do little to domesticate them. "Accidents are always there, we simply meet them" was how disasters were interpreted. Nature was seen to be controlled by divine forces. This subsequently culminated in faith-based rationality to understand disasters. However, there were other interpretations of disasters also. Some of these were suggestive and focussed on the functional aspects of the issue as history indicates that practical strategies were also devised to cope with such natural calamities. Hence, there was not just one approach towards disasters, rather diverse perspectives were applied to rationalise occurrences like natural disasters in the traditional societies of South Asia.

Yet, the dominant argument remained reductionist in nature and looked at disasters as isolated events with no linkages to general conditions in society. Recent studies differentiate between *hazards* and *disasters*. These studies suggest that hazards might have their origin in nature but these turn into disasters through a societal process. Climatic hazards like floods, cyclones, earthquakes, dry spells and volcanic eruptions are geological or meteorological risks but these turn into disasters when the structural and non-structural infrastructure is too shabby to withstand these risks.

Vulnerability is defined as a set of prevailing or consequential conditions, which adversely affect the ability to prevent, mitigate, prepare for and respond to events. Vulnerability has been divided into three main categories: (a) physical/material vulnerability, (b) social/organisational vulnerability and (c) attitudinal/motivational vulnerability. Same categories of vulnerability apply to capacities. *Capacities* refer to resources and means, skills and knowledge.

Thus, natural events such as earthquakes, typhoons and volcanic eruptions are just hazards, not disasters, but when they hit vulnerable areas with vulnerable populations, they become disasters. According to this approach, managing disasters necessarily entails managing vulnerabilities.

A comprehensive approach to disaster management, thus, involves five basic phases: *prevention, preparedness, mitigation, emergency response* and *recovery*. Preparedness involves measures taken in anticipation of a disaster to ensure that appropriate and effective action is taken in the aftermath. Mitigation covers measures, which can be taken to minimise the destructive and disruptive effects of hazards and thus lessen the magnitude of a disaster. Prevention covers activities designed to impede the occurrence of a disaster. Emergency response includes essential services and activities that are undertaken in the aftermath of a disaster to assist disaster victims.

DISASTER POLICY IN SOUTH ASIA

South Asia is one of the most disaster-prone regions where one disaster or the other happens through out the seasonal calendar. According to one estimate, every year on an average 55,000 people are killed, and over 72,000,000 people affected by disasters in South Asian countries. The severe tropical cyclone (02A), which caused heavy damage in south-eastern Sindh on May 18–21, 1999, moved into Rajasthan on May 22, leaving about 7,000,000 people and 4,000,000 acres of crops badly affected in Sindh-Pakistan. During October–November 2000, monsoon flooding and subsequent river erosion caused by water draining from northeast India, Nepal and Bhutan into the Bay of Bengal, triggered severe flooding in the western districts of Bangladesh. On December 25–29, 2000, a cyclone crossed the northern parts of Sri Lanka and on December 27 the tropical cyclone reached the southern tip of India where it damaged houses, fishing boats, paddy and banana crops.

In 2000, parts of India and Pakistan experienced severe drought conditions. A major earthquake occurred in Gujarat, India, about 65 miles north-northeast of Jamnagar, India or about 180 miles southeast of Hyderabad, Pakistan. More than 15,000 people died, 55,000 were injured badly, while 200,000 became homeless. Flash floods in Nulla Leh, Rawalpindi and land slides in Dadar, Pakistan in July 2001, brought horror to communities, which are still striving to cope with the outstanding devastation.

The responsibility for disaster management in South Asia lies with state institutions, which have assumed the role of security providers for citizens. Counter-disaster policies at the mega level are almost the exclusive domain of state functionaries. Centralised institutions dominate the intervention strategies (DN, 1999). South Asian states have made some institutional arrangements while different departments are assigned different roles to manage natural calamities.

States have put in place some institutional structures to address disasters at the national level. Bangladesh has developed a Disaster Management Plan (NDMA). The Ministry of Water Resources (MWR), the Ministry of Disaster Management, and the Ministry of Local Government, Rural Development and Cooperatives (MLGRDC) have formulated the components of disaster management in Bangladesh.

In India, a high powered committee (HPC) was set up by the government in August 1999, for the preparation of a disaster management plan. The HPC has after interaction with a wide cross section of the government system at all levels and voluntary agencies evolved a comprehensive plan of action.

Government policies for disaster management in Nepal were initiated with the enactment of the Natural Calamity Act in 1982. The Act made provisions for the formation of Disaster Relief Committees at the central, regional, district and local levels. The subsequent amendments of the Act in 1986, 1989, and 1992 incorporated emerging issues in disaster management and gave more operational mandate to the committees. The government has also specified objectives, policy measures, programmes and priorities to prevent disasters and minimise their adverse effects on the society (MOH, 1994).

In Pakistan the Emergency Relief Cell was established in 1977 under the cabinet division. It collects information about disasters, monitors losses and provides cash compensation to the affected. It also manages warehouses at different stations and maintains helicopters for rescue operations and VIP missions. At the local and provincial levels, already existing administrative structures are mobilised when emergency strikes in various hazardous areas. The Pakistan Meteorology Department operates 73 Met observatories all over the country. They perform the task of weather forecasting. The Federal Flood Commission and the flood departments are entrusted with the task of monitoring and managing floods in the country.

The Ministry of Social Services in Sri Lanka deals with policy formulation and the Department of Social Services implements the post-disaster relief and rehabilitation measures. The department has a network of 300 divisional secretariats in the 25 districts of Sri Lanka. There is a structure for the close coordination of the activities of the district secretariats, divisional secretariats, and other relevant agencies including NGOs. On a recommendation made by the cabinet of ministers, a National Disaster Management Center (NDMC) was established in 1996. It prepared the Sri Lanka Disaster Counter-measures Act which has to be approved by the Sri Lankan parliament. The main functions of the NDMC are: preparation of a national disaster management plan, monitoring the implementation activities by respective organisations and establishment of a databank and an information system.

In some countries like India, Sri Lanka and Bangladesh the policy discussion on disaster mitigation is gradually improving at the conceptual level. However, there are still some crucial gaps in the disaster policies of the member countries of this region. This indicates the ill-placed focus of the disaster policies, which leave priority issues unattended and make the whole exercise of disaster management somewhat cumbersome. Some of these policy gaps are outlined below.

Firstly, the state perception of disasters in South Asia is, for the greater part, ill-informed. This leads to an inconclusive policy and inadequate planning with regard to disasters. For instance, states are over-occupied with the structural dimensions of disasters, giving sole attention to the construction of embankments, installation of machinery and banking upon heavy physical infrastructure. Very little attention is paid to the non-structural aspects of disasters by shrugging off the alternative notions of risk reduction through investments in human resources.

Secondly, the applied disaster policy carries certain strategic biases that are aimed at protecting locations and infrastructure of *greater economic importance* at the cost of people in certain cases. The top-down approach remains the central bias in formulating a disaster policy. The lack of political will and transparency in execution are crucial hurdles in smooth implementation.

Thirdly, these policies are invariably dominated by the relief approach instead of prevention and preparedness perspectives. The focus is on post-event response while pre-event linkages are ignored.

Fourthly, there is no exclusive institutional arrangement in some countries to address disasters on a permanent basis. Rather, there are whimsical arrangements which deal with disasters only when they strike. As in Nepal and Pakistan there are no specific disaster management institutions, which can sustainably carry out disaster management activities.

Fifthly, the clash of interests among various state institutions that are responsible for disaster management hampers the proper flow of information and creates hurdles in institutional coordination. Lack of horizontal and vertical coordination, for instance, among flood relief agencies is one of the major gaps restricting the actual implementation of flood management plans in Pakistan (Kashif, 2001). This gap creates questions of jurisdiction; whether disaster is a provincial subject or a federal subject.

Sixthly, the prevailing apathy towards risk assessment during mainstream development planning results in new forms of disasters in certain cases. The non-recognition of arsenic contamination as a disaster in Bangladesh, for instance, exposes the perceptual biases of the state in relation to risk assessment.

Finally, policy institutions lack the necessary training, logistic and financial support and most of the disaster policies in the region remain ad-hoc, incoherent and inconsistent.

DISASTERS AND LIVELIHOOD OPTIONS

There is a dearth of literature on the social linkages of disaster in South

Asia; however, some recent efforts have been made to identify the knowledge gaps in this discipline. One such recent effort is the ITDG research project on Livelihood Options for Disaster Risk Reduction in South Asia carried out simultaneously in Bangladesh, India, Pakistan, Nepal and Sri Lanka. The research project was initiated to:

- Explore the impact on livelihoods and assess the needs and opportunities that result from disasters on livelihoods across the South Asian region.
- Identify practical options which can enhance livelihood support in a disaster situation.
- Pilot test and demonstrate options for wider dissemination.

Ten case studies on droughts, landslides, floods, cyclones, land erosion and arsenic contamination conducted in all the five South Asian countries found out that disaster-prone communities are living at virtually subsistence levels and very limited livelihood options and opportunities are available to the at-risk communities in these South Asian countries. The available major livelihood options in the research areas include: agriculture, animal husbandry and agricultural, wage labour. Most of the disaster-prone population still depends on agriculture as a major source of earning. The continuous onslaught of nature's fury has not only led to the loss of lives but has also led to economic losses. For example, the saltpan workers in Kandla did not have any livelihood option left with them after the Gujarat cyclone. In Kot Murad, a flood prone village in Pakistan where farming is the main livelihood, 87 per cent of the total households are landless. Livelihood support measures become more important especially in a disaster scenario. The findings from the case studies in Sri Lanka indicate that the main source of income for 72 per cent of the population in Andarawewa and 56 per cent of the population in Mahameddawa is from subsistence agriculture during normal years. Off farm income is the major source of income during periods of drought. During such periods people often have to migrate to other places in search of wage labour. Women suffer the most in such situations as they are unable to leave their families in search of employment. Hence, the total income earning opportunity for a family, as a whole is lower in the drought years compared to that a normal year. Rainfed highland cultivation is the only income source for a majority of the families in villages under study.

Yet, livelihood concerns are practically missing from the disaster policies and plans of member countries. The research argues that besides human lives, livelihood security is the primary concern that comes under severe threat in the case of natural or man-made disasters. This concern becomes graver in the case of countries and communities of South Asia, where economic opportunities are generally very few. The limited economic opportunities make disaster-prone communities more vulnerable and they find no way to exit from the burning hell of miseries.

LIVELIHOOD OPTIONS RESEARCH (LOR) PROJECT IN SOUTH ASIA

This project does the following:

1. Applies participatory research tools and methodologies in study areas by conducting field surveys, focussed group discussions and extensive interviews.
2. Identifies gaps in institutional arrangements that largely neglect the livelihood options for disaster risk reduction and opts for emergency measures instead of preparedness and risk reduction approachs.
3. Criticises existing disaster management approachs that do not include at-risk communities in decision-making processes and exclusively remain a top-down phenomenon thus making early warnings ineffective.
4. Confirms the notions that there is a wide gap in our knowledge on risk and vulnerability of communities in South Asia to disasters. As a result, most interventions are largely insensitive and supply driven.
5. Maintains that there is a lack of analytical concepts and methodologies in understanding and addressing risk and vulnerability in South Asia. It also notes that emerging local knowledge, methodologies, and practices on risk reduction, based on local resources and knowledge remain isolated in the background, and lack support for wider sharing and replication.
6. Appreciates indigenous coping systems and tracks down differential impacts of disasters on class, community and gender in different social settings.
7. Suggests enhancing the knowledge base of various social determinates that multiply the effects of natural hazards on socially weaker and economically marginalised people.
8. Recommends pilot projects for demonstration aimed at reducing the risk of disaster by enhancing and strengthening livelihood options of vulnerable communities living in disaster-prone areas of South Asia.
9. Concludes that natural hazards are not disasters; they become disasters only when at-risk communities are no longer able enough to fight the menace through their physical and social resilience. A strong and durable livelihood base available with at-risk communities can efficiently reduce the risk of disasters.

Messages for Advocacy

A quick overview of LOR studies conducted in five South Asian countries reveals that there is a marked similarity among disaster-prone communities of these countries as the basic structure of facilities and problems is the same. Risk perception of both—communities and state institutions—forms interesting parallels in these different case studies.

At-risk communities of South Asia, with some difference in degrees, are condemned to face the same threats and problems with identical levels of capacities and options. The issue of disaster, as mapped out by the ITDG research project in South Asia, relates the imminent risk of disasters to the general conditions of rural under-development prevalent in these countries.

The study reveals that a majority of disaster-prone rural populations depend on agriculture and livestock as the two major sources of livelihood in all research locations of the five countries. These impoverished conditions and few available choices make them more vulnerable to any natural. The larger message of this regional research project is to take disaster risks and livelihood options as mutually exclusive notions in planning. Both have an inverse relationship: enhancement of one leads to reduction of the other.

Disasters can be controlled and damage can be curtailed by enlarging human choices. The Livelihood Options Research identifies particular issues to be taken up at the regional, policy, and community levels to ensure intersection between micro and macro considerations for a broader debate and larger scale of implementation.

Regional Level

Natural disasters can be a point of closer interaction cooperation among South Asian countries. Severe climatic fluctuations in the last years in the form of droughts, cyclones and earthquakes in geologically associated countries have evidently validated the assumption that disasters have no frontiers. This notion raises another related premise that when disasters have no frontiers, response itself has to be above borders. A strategy should be laid down to facilitate information and experience sharing about disasters between member countries of South Asia. Weather-related information, ongoing disaster planning and policies, and local level mitigation initiatives should be exchanged and replicated to better cope with the menace of disasters at the regional level. Cross-country exposure visits and training should be designed to enlarge the understating of disasters at the public and policy level by learning from each other.

Policy Level

LOR suggests a timely shift of paradigm in applied disaster policy from disaster management to risk reduction and it insists on neutralising the hierarchical biases of the top-down approach in dealing with disasters. Disaster mitigation and risk assessment should be incorporated into mainstream development planning ensuring hazard free development at the macro- and micro-level.

Research findings recommend the establishment of disaster-specific institutions where they do not exist. In place institutions should ensure horizontal and vertical coordination within and between responsible organisations. The crux of the theme is to understand the social linkages of disaster risks by incorporating livelihood concerns in general disaster planning. It also deliberates upon the need to strengthen available livelihood options and also proposes to diversify the resource base for at-risk communities by providing alternative economic opportunities to them.

The regional research also suggests to supplementing the existing local knowledge base of disaster management rather than replacing it with modern obsessions which are culturally incompatible technology devises. Forecasting data and technological idioms should be contextualised to enable effective and meaningful communication with people at large.

Investments in disaster-resilient infrastructure should be made possible and non-structural aspects of disaster management—training, capacity building and skill development—should be prioritised in policy layout. LOR also emphatically stresses the inclusion of gender concerns in disaster management. It advises that land use regulations should be enacted and effectively enforced and legislative measures should be taken to develop organised disaster management mechanisms at the national level.

Community Level

Time tested coping strategies like rain water harvesting in drought hit areas should be streamlined, farming of disaster resilient crops and changes in cropping patterns should be practiced to reduce the risk of disasters at the community level. Sharing of experiential knowledge about disasters should be encouraged and skills should be enhanced to maximise the economic opportunities at the local level. Informal insurance and support systems should be encouraged and strengthened, and modern knowledge should be acquired to manage disasters more efficiently.

Based on the findings of the LOR project, livelihood-sensitive disaster management micro projects at the community level are being implemented in Bangladesh, India and Pakistan. These projects are intended to demonstrate policy options for disaster management to be replicated, scaled-up and disseminated further. Related government institutions and development organisations can study these demonstration pilot projects to assess their viability of replication. These projects include:

- Alternative strategies for community-based flood preparedness in Pakistan
- Drought preparedness strategies in one village of Tharparkar, Sindh, Pakistan

- Drought risk reduction strategies in Rajasthan, India
- Alternative strategies for community-based flood preparedness in Bangladesh
- Drought preparedness and landslide risk reduction in Sri Lanka.

Donors/Intermediary Organisations

Donor organisations and other disaster-related non-governmental organisations, those that are working in the fields of relief distribution and emergency management, can incorporate vulnerability and livelihood issues into the ongoing disaster planning. They can help communities to diversify their occupational base through skill development and micro credit schemes. Training can be designed and organised by NGOs on watershed management, community-based early warning systems and disaster management systems.

Media

Media can play an important role in informing public opinion and influencing policy on disaster management. It also has a crucial role in transmitting and decoding early warning symbols and forecasting data for risk-prone people. LOR studies provide a good insight for journalists to enhance their understating of concepts and methodologies of disaster reporting. Additionally, policy gaps in disaster management at the national and regional levels can be identified and analysed in editorials, reports and features by applying the frames of livelihood options and concepts of disaster risk reduction.

A long-term training plan for journalists on the concepts and methodologies of disaster reporting can be organised at the national and regional levels. Media, if informed well, can efficiently inform and educate constituencies at the other end. The need is to first communicate to the communicators. This can multiply the outreach of advocacy messages of disaster risk reduction.

CONCLUSION

In a nutshell, it can be stated that there is ample need for an integrated and holistic approach to be adopted for effectively managing disasters and thus reducing risks emanating out of these for the ultimate benefit of the society, polity, community, etc.

6 Applications of Information Technology in Disaster Risk Reduction

— N. Vinod Chandra Menon

INTRODUCTION

Information Technology has penetrated all disciplines, thereby making the management of information and knowledge more efficient, cost-effective and virtually real time. The design and development of databases, relational database management systems, management information systems, decision support systems, expert systems, knowledge bases, simulation models, etc., have made each discipline more dynamic and functional. In this chapter, an attempt is made to explore the various areas of applications of information technology (IT), which make disaster management more effective.

The developments of the World Wide Web and the Internet have opened up the possibility of creating websites for specialised institutions, which work in the various domains of disaster management. In advanced countries, the specialised institutions which allow net surfers to browse the contents of their sites, have tapped the fascinating potential of the Internet and the World Wide Web. Even in developing countries, the details of disasters are now available almost instantaneously on the World Wide Web due to the user-friendly tools available for web hosting and uploading content on the net. For each specialised discipline there are *user networks* and *list serves*, which allow exchange of information across the world on a real time basis.

The phenomenal expansion of telecommunication facilities in developing countries in the recent past makes it easier for the integration of IT with telecommunication interfaces. Thus, the facility of IT interfaces with satellite telecommunication provide opportunities for video teleconferencing, setting up site operations centre at the site of a disaster for video teleconferencing with the emergency operations centre at the state headquarters, apart from setting up intranets, extranets, wide area networks, etc.

Another major breakthrough of IT applications in disaster management is the design and development of Geographical Information Systems (GIS), which permit the development of base maps with district and *taluka* boundaries, village locations and their access to critical infrastructure like primary health centres, blood banks, pathological laboratories, police stations, fire brigades, transport depots, etc. This will facilitate more efficient decision-making, policy analysis and problem-solving both during an emergency and during normal times.

APPLICATIONS OF INFORMATION TECHNOLOGY IN DISASTER MANAGEMENT

The phenomenal expansion of the World Wide Web and the Internet in every nook and corner of the world has made access to the Internet from anywhere in the world smooth, fast and efficient. Websites have been created by millions of specialised institutions on the World Wide Web, thereby making searching and browsing the net an integral part of the new professional's daily routine.

The Yokohama Declaration was a very significant outcome of the World Conference on Natural Disaster Reduction from May 23 to 27, 1994, at Yokohama, Japan. The declaration made the following observation: Disaster prevention, mitigation and preparedness are better than disaster response in achieving the goals and objectives of the International Decade for Natural Disaster Reduction (IDNDR). Disaster response alone is not sufficient, as it yields only temporary results at a very high cost. This has been further demonstrated by the recent focus on response to complex emergencies, which, although compelling, should not divert from pursuing a comprehensive approach. Prevention contributes to lasting improvements in safety and is essential to integrated disaster management.

On the issue of applications of technology in disaster management, the Yokohama Declaration exhorted that regional and international cooperation will significantly enhance our ability to achieve real progress in mitigating disasters through the transfer of technology and the sharing of information and joint disaster prevention and mitigation activities. Bilateral and multilateral assistance and financial resources should be mobilised to support these efforts.

Further, according to the Yokohama Declaration, the information, knowledge and some of the technology necessary to reduce the effects of natural disasters can be available in many cases at low cost and should be applied. Appropriate technology and data, with the corresponding training, should be made available to all freely and in a timely manner, particularly to developing countries.

In the spirit of the Yokohama Declaration, the International Decade for Natural Disaster Reduction witnessed the development of several cross-cultural initiatives across countries with the active support and participation of institutions and disaster management professionals from the developed world. Two such major initiatives are the Risk Assessment Tools for Diagnosis of Urban Areas against Seismic Disasters (RADIUS) Project and the Asian Urban Disaster Mitigation Programme (AUDMP) Project.

DATABASES

A very comprehensive definition of disaster management treats it as the process by which the uncertainties that exist in potentially hazardous situations can be minimised and public safety maximised. The objective is to limit the human, material and economic cost of an emergency or disaster, and this is achieved through a range of strategies from hazard management and prevention to speedy restoration of the affected community (Granger, 1993). Comprehensive emergency management is often viewed as the two Ps and two Rs: *Prevention, Preparedness, Response* and *Recovery.* At each stage of this emergency management process, decision-making plays a very critical role. The availability of accurate, authentic and reliable data is an essential ingredient in the decision-making process. Thus, it has become imperative to accord the highest priority to the design and development of institutional mechanisms, which can streamline the collection, compilation, analysis, and interpretation of data and information pertaining to vulnerable areas and vulnerable communities. This is evident in the assertion that emergency management is 10 per cent telecommunications, 20 per cent operations and 70 per cent information. Information, like people and money, is a resource and the only resource that makes possible the coordination of vital services during an emergency (Everson, 1986).

It is necessary to ensure that a risk assessment and vulnerability analysis of the disaster prone areas and disaster prone communities is carried out through administering structured questionnaires supplemented with participatory rapid appraisal techniques. This will yield the creation of dynamic databases of the disaster prone areas and the disaster-prone communities. It should also be possible to document the coping strategies of communities which are frequently exposed to disasters: like the settlement of an urban slum on the banks of a river or canal, which leads to stress situations during monsoons.

The Relief Commissioner's office has to be a clearing house of data, which collects, analyses, compiles, stores and communicates facts and figures for faster decision-making. Even when resource constrains are imposing serious challenges in humanitarian efforts, "the plunging cost and soaring scope of computer hardware, information software and telecommunications increasingly offers, disaster professionals in the developing world—and, ultimately, disaster victims themselves—the chance of taking far greater control of all aspects of crisis management, from relief operations to preparedness programmes" (World Disasters Report, 1997).

Relational Data Base Management Systems (RDBMS)

The databases can be compiled using Relational Data Base Management Systems (RDBMS) so that it will be possible to subject this database to queries for more informed decision-making. It is also possible to interface such an RDBMS to a Geographic Information System (GIS) of the area, which will act as a front-end so that scenario analysis results can be simulated to see the options on the GIS. Such systems can also support forecasting and predictive models, especially if time series data sets are available for such areas and communities.

As the World Disasters Report, 1997, observes, each disaster stakeholder has very different information needs and priorities, as well as having their own influence on its management. Low understanding of what each stakeholder requires is one reason why systematic organisation of disaster management information has not kept pace with advances in the agency use of information and communications technology. The report further observes that much predictive data for natural disasters—from meteorological satellites for the mapping of seismic activity—is available, as is information on man made and technological hazards, and both are being used by emergency management professionals. However the humanitarian community needs an enormously diverse range of information, such as: availability and movement of relief supplies, population displacement, capacity of local infrastructure such as ports, hospitals and airfields; disease surveillance including base line epidemiological information, specialist, relief expertise available; data on other agencies involved; national and local facts and figures on the economic, social and political situation; and mapping and other geographical information vary widely during the different phases in disaster management. For example, information required at the disaster preparedness phase includes: country specific information such as demographic and geographic data; emergency plans and national development plans; reports from other agencies; in-country capability including financial, material and human resources; and documented information from previous disasters, such as recovery programmes.

The database can form a very valuable resource, especially when it is properly archived with the facility for retrieval for specific purposes through well-designed query interfaces. One of the successful illustrations of an integrated approach in building such comprehensive databases is the establishment of an Integrated Information Centre for the Ministry of National Resettlement, Rehabilitation and Reconstruction in Sierra Leone which coordinates resettlement of nearly 1.2 million internally displaced people and refugees.

Management Information Systems (MIS)

The Management Information System for disaster management brings together data and information so that policy analysis and problem-solving exercises can be carried out for more informed decision-making. All aspects of disaster management like search and rescue, evacuation, preparedness, prevention, rehabilitation and recovery can be carried out using the MIS approach. This approach also facilitates the backward and forward linkages, which are very critical for the success of disaster management efforts. It is possible to analyse the damage caused by disasters and the impact of disasters by designing and developing comprehensive MIS for disaster prone areas.

Decision Support Systems (DSS)

Decision Support Systems are value added extensions of the MIS, which facilitate more informed decision-making, problem-solving and policy analysis. In a specialised field like disaster management, DSS can play a very significant role for more effective, efficient and faster decision-making. These can also provide linkages to predictive and forecasting models for carrying out simulation modelling and scenario analysis.

Knowledge Bases

There are thousands of websites on the Internet, which provide information on various specialised fields of disaster management. A knowledge base on earthquakes will bring together all scientific and technical information on earthquakes, including case studies on some of the major earthquakes. If such knowledge bases are available with a glossary of terms, frequently asked questions, multi media content, etc. on a few powerful servers in different specialised institutions in different parts of the world, anybody from anywhere in the world can log on to the Internet and visit the site where the server is located for making specific queries. There can be bulletin boards, chat rooms, list serves, etc. with auto-robots responding to specific queries. These knowledge bases can be periodically updated.

Expert Systems

In disaster management, it is possible to design and develop expert systems, which can evaluate and analyse options and arrive at faster decisions where the system will play the role of the domain expert and suggest ways of effective disaster management interventions. Professionals in various fields have used such expert systems for the last several years. For instance, medical professionals are using MYCIN as a very useful expert system. On the Internet it is possible to identify various other expert systems, which can be used in disaster management.

Simulation Modeling and Scenario Analysis

If good databases are available, it will be possible to design and develop powerful simulation models, which can simulate the ground reality and offer policy options by evaluating various alternative scenarios. GIS can be used as effective front-end tools for further enhancing the policy analysis options. These simulation models can be used as planning tools before designing a response strategy for disaster management interventions in the event of an alert or early warning. These can also be used as decision-making tools for using real time operational control to supplement desktop planning efforts. They also have a very significant role in training and in debriefing. Simulation models, forecasting and predictive models can assist the planner by providing an opportunity for evaluating alternatives for more effective decision-making in the area of disaster management.

GEOGRAPHIC INFORMATION SYSTEMS (GIS)

According to Dale and McLaughlin (1988), *Geographic Information System* is a system of capturing, storing, checking, integrating, analysing and displaying data about the earth that is spatially referenced. It is normally taken to include a spatially referenced database and appropriate applications software. In the words of Cowen (1988), GIS is decision support systems involving the integration of spatial reference data in a problem-solving environment. GIS can be used for risk monitoring, risk assessment and vulnerability analysis, risk mapping, risk modeling and public awareness and dissemination. These systems apply remote sensing, GIS photogrammetry, topographical surveys, urban survey and town planning, geology, hydrology, geomorphology, traffic and transport engineering, land use pattern, rainfall pattern, drainage pattern, etc. as thematic overlays. A Disaster Management Information System (DMIS) has been prepared for the 43,000 villages and 32 districts of Maharashtra on a 1:2,50,000 scale with multiple thematic overlays using ARCINFO and work is in progress to develop the system on a 1:50,000 scale for all the districts.

HAZARD MAP AND VULNERABILITY ATLAS

As a follow up of the World Conference on Natural Disaster Reduction at Oklahoma in 1994, the Building Materials and Technology Promotion

Council, Ministry of Urban Development, Government of India decided to prepare a Vulnerability Atlas of India. This Atlas, the first edition of which was published in 1997, provides state-wise hazard maps for earthquakes, cyclones and floods. The zoning maps on macro level for earthquakes, cyclones and floods are available for the country as a whole. The National Atlas and Thematic Mapping Organisation (NATMO) of India had prepared a Natural Hazard Map of India in 1991 on a 1:6 M scale. A revised version on a 1:4 M scale is expected to be released shortly.

INTRANETS AND EXTRANETS

The information technology revolution has facilitated the setting up of *local area networks* (LANs) within organisations, which link together computers scattered across different departments and different locations. *Wide Area Networks* (WANs) link together computers across geographical locations covering a wider area. The government of Maharashtra has set up an Intranet linking 40 nodes consisting of the Mantralaya at Mumbai, Yashwantrao Chavan Academy of Development Administration (YASHADA) at Pune, six divisional commissioners and 32 district collectors for audio-video and data interchange. The district collectorates are further linked through their servers to their field offices through dial up modems. This is a satellite-based network of VSATs unlike the fibre optic-based network commissioned in Andhra Pradesh for linking government offices.

LIST SERVES

Special Interest Groups (SIGs) can be set up on the Internet for disaster management professionals for exchanging data, information and experiences in handling disasters. The Relief Web set up by the United Nations Department for Humanitarian Affairs, the Global Disaster Network (GLODISNET) set up by the United Nations University, Tokyo, the EQNET (Network of Earthquake Management Institutions), Response.Net and VITA are some of the significant initiatives in information management using the information technology applications of the Internet. The United States Geological Survey (USGS) has established several e-mail lists that will provide information on a regular basis to subscribers on relevant areas. These list serves are organised by topics such as: water-pr; geologic-hazards-pr; biological-pr; mapping-pr; products-pr; lecture-pr; etc. These USGS list serves can be subscribed to by sending an e-mail to (*listproc@listserver.usgs.gov*).

DISTANCE EDUCATION AND TRAINING

Distance education and training in various areas of disaster management is being increasingly carried out through the Internet. Some useful sites are *www.ucl.ac.uk*; *www.udel.edu*; *www.hoshi.cic.sfu.ca*; *www.colorado.edu/hazards*; *www.sfsu.edu*; *www.iris.edu*; *www.dir.ucar.edu*; *www.es.mq.educ.au*; *www.jcu.edu.au*; *www.paho.org*; *www.millersv.edu* etc. The Indira Gandhi National Open University (IGNOU), New Delhi, India, has started a Certificate Programme in Disaster Management through distance education and is in the process of launching a Post Graduate Diploma in Disaster Management through distance education in collaboration with YASHADA, Pune.

AWARENESS AND DISSEMINATION

Mitigation, prevention and preparedness campaigns can be carried out through radio, television, cable television channels, folk theatre, street plays, etc. Community participation and stakeholder awareness on risk reduction are integral components of a disaster mitigation strategy. It is necessary to explore ways of integrating information technology and telecommunication for creating greater awareness among disaster-prone communities. It is also imperative to document the best practices and coping strategies of disaster-prone communities so that these can be disseminated to create greater awareness among disaster-prone communities elsewhere through the Internet, intranets and extranets.

EARLY WARNING SYSTEMS

Early Warning Systems have been installed at Area Cyclone Warning Centres (ACWCs) at Calcutta, Chennai and Mumbai and at Cyclone Warning Centres (CWCs) at Bhubaneswar, Visakhapatnam and Ahmedabad. The Indian Meteorological Department (IMD) has commissioned a satellite based communication system called Cyclone Warning Dissemination System for dissemination of cyclone warnings to the cyclone-prone coastal areas. IMD also has a Limited Area Analysis and Forecast System based on the optimal interpretation method and a Limited Area Primitive Equation model to provide numerical guidance. It also receives numerical guidance from the European Centre for Medium Range Weather Forecasting. The IMD also monitors seismological changes through a network of 36 seismic monitoring stations and also collects, compiles and interprets rainfall data from various locations. These are used for mitigation efforts and for carrying out scenario analysis on the receipt of warning of an impending disaster.

VIDEO TELECONFERENCING

The Latur earthquake of September 30, 1993 was one of the most severe tragedies witnessed by Maharashtra in the recent past. The Government of Maharashtra prepared a very comprehensive Maharashtra Earthquake Emergency Rehabilitation Programme (MEERP) with the support of international agencies like the World Bank, the United Nations Development Programme and the Department for International Development (former ODA of UK). As a part of this exercise, a network of control rooms have been set up linking the Emergency Operations Centre (Eoc) in the *mantralaya*, Mumbai with a Standby Control Room at the Centre for Disaster Management, YASHADA, Pune, Divisional Commissioners' Control Rooms in each of the six divisional headquarters and District Control Rooms in each of the 32 districts of Maharashtra. These control rooms can facilitate video teleconferencing across these 40 locations and the system can be scaled upto 200 nodes. This facility can be used during normal times for review meetings, monitoring as well as for distance education. This system supports electronic mail, fax, satellite hotlines, audio-video and data exchange over satellite across these nodes.

INFORMATION TECHNOLOGY AND MULTI MEDIA INTEGRATION

It is possible to put together content in various media formats like audio, video clips and scanned picture files for disseminating information through the Internet or through intranets so that a larger network of people can be contacted. The information technology and multi media integration offers immense potential for major breakthroughs in reaching out to local communities with content in regional languages, which is relevant to their background and socio-economic and demographic profiles. This process is increasingly becoming multi-disciplinary and offers challenges for preparing modules for training, dissemination and awareness creation.

NET-BASED APPLICATIONS

The Internet has become the most revolutionary breakthrough in information technology. With the recent advances in software programming, applications programming and in the emerging frontier areas of net-based applications of virtual reality, the java family of developments in JINI, JNI, graphical user interfaces, java applets and extended markup languages, the Internet has become the platform for professionals to explore ways for creative expression of their talent. The Virtual Emergency Management Information System (VEMIS) set up at the Simon Fraser University in Canada illustrates the possibilities of net-based applications in emergency management.

CONCLUSION

The potential of integration of various applications like remote sensing, global positioning systems, databases, management information systems, decision support systems, expert systems, artificial intelligence, simulation modeling, scenario analysis and distance education holds the promise of exploring more effective ways of decision-making, problem-solving and policy analysis in the various fields of disaster management. Appropriate use of such applications could have substantial impact before during and after disasters, thus reducing disaster risk qualitatively.

REFERENCES

Cowen, D.J., *GIS versus CAD versus DBMS: What are the Differences?* Photogrammetric Engineering and Remote Sensing, 5f (10), pp. 1367–1370. American Society of Photogrammetric Engineering and Remote Sensing, Falls Church, Virginia, 1998.

Dale, P.F. and J.D. McLaughlin, *Land Information Management: An Introduction with Special Reference to Cadastral Problems in Third World Countries*, Clarendon Press, Oxford, 1988.

Everson, P.R., "Emergency Planning Decision Support Systems" in Marston, S.A. (Ed.), *Terminal Disasters—Computer Applications in Emergency Management*, University of Colorado, Institute of Behavioural Science, pp. 11–27, 1986.

Granger, K.J. and R.W. Johnson, *Hazard Management: Better Information for the 21st Century*, EMA, Australia, 1994.

Granger, K.J., "Submission by the Australian Urban and Regional Information Systems Association, Inc. to the Senate Inquiry into Major Disasters and Emergencies". Senate Hansard Record of Inquiry Evidence, Australia, 1993.

IFRC, *World Disasters Report*, International Federation of Red Cross and Red Crescent Societies, Geneva, 1997.

7

Trigger Mechanism—The Concept for Emergency Response Plan for Disaster Risk Reduction

— S.K. Swamy

INTRODUCTION

Man feels that he has conquered nature. Nature, on its part, expresses its fury and takes its toll every now and then reminding mankind to be cautious. There are a number of factors, which contribute to the natural disasters striking different parts of the world. These can be traced to the process of evolution of the planet earth, the impact of indiscriminate developmental activities on nature and environment, consumption of natural resources without regeneration for meeting the endless demands of man, explosion and concentration of population and the changing patterns of social life. The effects of the modern culture are evident as global warming, excessive generation of heat energy and continuous depletion of the protective atmospheric layer. Whereas natural disasters are the manifestations of nature trying to reinstate equilibrium, man made disasters are the fall out of the top sided developmental activities man sponsors in pursuit of his disastrous ambition to reign supreme in this planet.

CATEGORISATION OF DISASTERS

Disasters can be broadly categorised as:

(i) Those which are neither anticipated nor expected.
(ii) Those which are anticipated but not expected.
(iii) Those which are anticipated and also expected.

In case of disasters, which can only be prophesied and cannot be visualised, there is not much that can be done. These will continue to be the catastrophic events that elude the comprehension of man. However, in case of disasters, which are anticipated but not expected, precautionary measures play an important role. Most of the man made disasters come under this category. The process of development does create hazards but effective steps reduce the associated risk level to zero. It is the Natural disasters, which belong to the last category of disasters that are anticipated and also expected. Preparedness plays a very crucial role in reducing the risks and minimising the losses such disasters inflict.

MAJOR HURDLES IN DISASTER MANAGEMENT

Whenever a disaster, natural or man made, strikes, the disaster managers struggle to mitigate its effects of human life and property. This is on two counts. (i) The first is attitudinal insensitivity. The management is distanced from the location of the event physically and from the affected people emotionally as a disaster usually does not involve its own stakes. There is also public apathy towards disasters and the risks involved pose a big challenge to mobilising the community towards preparedness. The most vulnerable sections in disasters are the poor. This section of the society attaches very little value to life, which in turn, lowers its desire to survive against odds and fight for itself. People in this category generally expect providential help when disasters strike. They lack self reliance and the capacity for individual response. This problem is aggravated further where the writ of the law does not run and it makes the disaster management a very difficult task.

(ii) The other problem is that the present disaster management plans, though very exhaustive and detailed, do not have a spontaneous functional mechanism for the response to be enacted in time. The response managers are required to swing into action as soon as a the disaster strikes. Generally, in such situations, the managers start organising, planning and activating the mitigation process whereas the need of the hour would be to start the mitigation process on the ground, with virtually no time available for activities like organising and planning. Time is the essence of disaster management, the objective being to effect immediate relief and rescue to save the maximum possible human lives and to mitigate human miseries within the crucial first 48 to 72 hours after a disaster has struck. What is required to be done thereafter is actually part of long-term rehabilitation and reconstruction programmes.

THE TRIGGER MECHANISM

In view of the above major drawback, the trigger mechanism envisages that on receiving signals of a disaster happening or likely to happen, all the resources and activities required for the mitigation process are energised and activated simultaneously without loss of any time and the management of the event is visible on the ground. The primary objective of the trigger mechanism is to undertake immediate rescue and relief operations and stabilise the mitigation process as quickly as possible.

The Trigger mechanism requires the planners:

1. **To identify disasters and the probability of their occurence.** The response managers and planners should identify the disasters to which the area of their concern is prone to and also be aware of the likelihood of such a happening. This would mean identification of the hazards and the vulnerability of area to the hazards. In other words, a profile of the district or the state or the area of concern is required to be prepared, which will include vulnerability assessment, records of previous disasters, details of, the geographical features of the area and the impact and intensity as also the damages inflicted by disasters in the past. Taking into account these parameters, the developmental process and other noticeable changes in the area, an assessment of the likely intensity and the impact of the probable disasters of the future may be carried out. This would enable the response managers to define the parameters for the response.

2. **To evolve an effective signal/warning mechanism.** After identifying the likely disasters and their probability, the next step would be to evolve an effective signal/warning mechanism and to put it in place. This mechanism may avail the techniques being used in the existing signal warning systems of the India Meteorological Department and the Central Water Commission and also enlist the services of the All India Radio. The TV, local sirens and other traditional means of communication may also be used to alert the community and the field level machinery immediately in the event of a likely disaster.

3. **To identify the activities.** Disaster management is always a coordinated effort. A number of activities need to be undertaken as the response in the event of a disaster. The activities involved in disaster response generally would be: (i) co-ordination, command and control, (ii) rapid assessment of damage, (iii) restoration of power, (iv) restoration of Telecommunication, (v) restoration of surface transport, (vi) deployment of search and rescue teams, (vii) deployment of medical and para-medical teams, (viii) supply of food material, (ix) supply of medicines, (x) arrangements for drinking water, (xi) setting up of temporary shelters, (xii) sanitation and hygiene, (xiii) transportation of men and material, (xiv) identification of resources i.e. men, material and money, (xv) earmarking of resources i.e. men, material and money, (xvi) maintenance of law and order, etc.

4. **To identify the sub-activities.** Having listed down the likely activities, which would be needed in specific terms, it would be necessary to determine the sub-activities under each category of main activity identified. In fact, when it comes to actual performance on the ground, each activity will generally involve three or four sub-activities. For example, for search and rescue, one needs to organise the search and rescue team, the necessary equipment, means of transportation, the dog squad and the like. Similarly, for deployment of medical assistance, besides the doctors and the para-medical staff, other arrangements like medicines and medical equipments would be needed. For effective response all the sub-activities need to be listed under the respective main activities.

5. **To define the level of response.** The level of response should be in view of the magnitude of the disaster in terms of intensity and impact. Lower the magnitude of a disaster, lower the level of activity required. In other words, the arrangements needed would be less elaborate. If the magnitude of the disaster were very severe then the same response activity would be required to be undertaken on a much larger scale. Therefore, it is necessary to grade the levels of activities keeping in view of the assessment of intensity and the impact of probable disasters carried out as already described. This will enable to undertake activities as per planned level of response.

6. **To specify authorities.** Every activity is to be assigned to a specific authority. Since a number of departments/organisations would be involved in the response programme, details like name, designation, telephone/fax numbers, e-mail addresses, etc. of the authorities specified for the different activities should be listed down. As an alternative, the similar details of the second-in-command authority should also be listed for each activity. As already discussed, a number of sub-activities need to be undertaken to effect every main activity. The sub-authorities entrusted with these sub-activities should also be specified. As far as possible, these should be from within the organisation so that proper command and control is exercised. Details like the name, designation, telephone/fax numbers, e-mail addresses, etc. of the sub-authorities and the second-in-command for each sub-activity must also be listed down.

7. **To determine the response time.** In an emergency response, one should not forget that time is the essence. Therefore, each specified authority should work out the response time needed for operationalising the activity and the sub-activity it is responsible for. This response time will be graded as lesser time will be required for operationalising an activity and sub-activity

at a lower scale and obviously more time would be needed if the activity is to be operationalised at a higher scale. Thus, each authority and sub-authority should very clearly know its response time with reference to the scale of activity it is to operationalise as part of the response.

8. **To work out individual activity plans.** After identifying and defining the activities and sub-activities and the scale at which these will be required to be operationalised with reference to the pre-determined response time, every authority and sub-authority has to work out the individual activity required of it in view of the assessment of probable disasters and their likely intensity and impact determined earlier. These plans will lay down a functional mechanism of coordination in both, horizontal and vertical manner. These plans will require clear identification of the resources i.e. men, money, material and equipments needed for each activity and the sub-activities under it. This is needed for the distribution of the resources during an actual operation as well as for earmarking the resources. The identification and earmarking of resources should be both in terms of quantity and availability and also functionality.

9. **To have Quick Response Teams (QRTs).** By now it is clear what is required of each authority and sub-authority in the context of the activities and sub-activities they are associated with. The job to be undertaken is known in advance and the process of planning and organisation has been completed. Therefore, at this stage, a Quick Response Team may be constituted for each activity. The team members should know one another and should also know their specific task as an individual activity besides the overall task assigned to the team. The specified authority and sub-authorities should have complete details of the Quick Response Team members to ensure their availability and proper communication at the time of response.

10. **To undergo preparedness drills.** The Quick Response Teams should undergo preparedness drills and carry out mock exercises so as to pre-empt the possibility of any functional hiccups or restrains which are likely to occur while implementing their plan on ground. On the basis of field experience or the lessons learnt from previous disasters, the team should update their functional methodology to be more effective both in terms of time and delivery.

11. **To provide appropriate delegations.** The performance of the Quick Response Teams will depend on mobilising the resources at their command. It would, therefore, be necessary to ensure

the necessary administrative, financial, and legislative provisions in advance so that the response mechanism becomes functionally viable and sustainable. The provisions for the different authorities and sub-authorities should be well defined and clearly laid down. In fact, the Standard Operating Procedures (SOPs) should be laid down in advance so that one knows what to respond to and how to respond.

12. **To have alternative plans.** Even in an emergency response to a disaster, one has to aim for zero level of failure. It is therefore recommended that alternative plans and contingency measures also be thought of in advance so that the response mechanism does not get stuck up for any reason till the mitigation process is stablised.

CONSTITUTION OF TRIGGER MECHANISM

The Trigger mechanism has two basic components. These are the Trigger authority and the coordination, command and control activity. The Trigger authority would generally be the authority in charge of the disaster response by the unit and may be the District Magistrate, the Relief Commissioner, the Central Relief Commissioner or any other officer designated for the purpose. The flow chart for Trigger mechanism is mentioned in Figure 7.1 and the Trigger mechanism network in Figure 7.2.

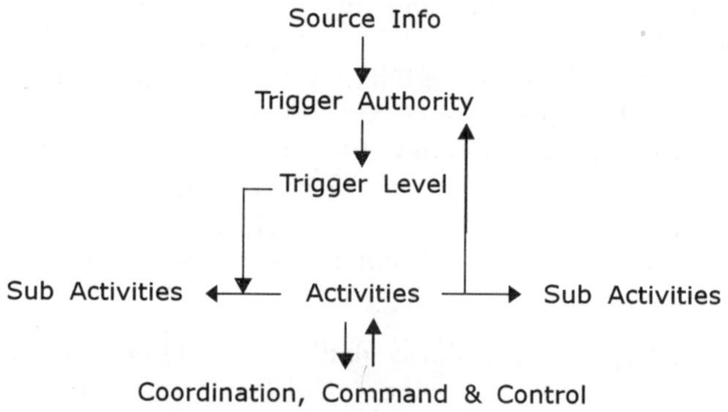

FIGURE 7.1 *Flow chart for trigger mechanism.*

The Trigger mechanism in fact is a preparedness plan in which all the participating managers know in advance the task assigned to them and the manner in which they have prepared themselves to respond. As such, the organisation and planning would have already been taken care of. The resources including manpower, material and equipment are identified. The performers have adequate delegation of financial and administrative powers and have the mandate for accomplishing the task.

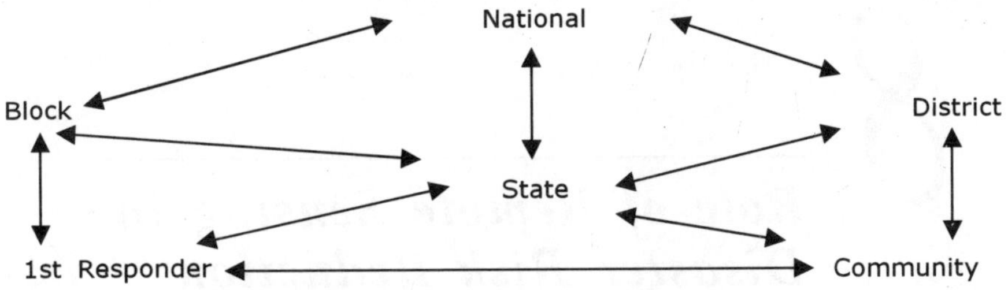

FIGURE 7.2 *Trigger mechanism network.*

The success of the Trigger mechanism depends on the vision and perception of the planners. They are required to anticipate the likely activities with reference to the nature of the disaster and its impact. In fact the Trigger mechanism can also be called the 'Operating Standard Procedure' since the implementation of the efforts on ground are well laid down in advance. All the major activities, which are generally common to all types of disasters will require sub-division and preparation of sub-action plans by each specified authority. They will be required to list all the requirements and their availability within the prescribed response time. Trigger mechanism would thus be a horizontal network of interaction between the different responders.

To sum up, the Trigger mechanism is an emergency quick response mechanism like an ignition switch, which when energised spontaneously sets the vehicle of management into motion on the road of disaster mitigation process to play a prominent role in disaster risk reduction.

8

Role of Remote Sensing in Disaster Risk Reduction

— Manoj K. Arora

"Two thirds of the world's population lives in developing countries and approximately 90 per cent of all natural disasters and 95 per cent of disaster-related deaths occur in developing countries."

(Anonymous, 1998, Page 6)

(http://www.oosa.unvienna.org/unisp-3/docs/backgroundpapers/)

INTRODUCTION

Natural hazards or disasters can strike any part of the world at any time, often without any warning and causing colossal loss to life and property. Though, we may not be able to interfere with God's schemes of natural devastation, we can absorb their impact by taking suitable measures to decrease the damage disasters can cause. This requires devising reliable methods of hazard prediction and mitigation. Remote sensing and Geographical Information System (GIS) are becoming essential components of disaster management programme since they can play a significant part in the early detection and response to natural disasters particularly cyclones, drought, floods and forest fires.

In developing countries, every year thousands of people lose their lives and property worth billions of rupees is damaged due to the occurrence of a number of natural disasters. These include cyclones, floods, drought, volcanic eruptions, landslides and earthquakes. The great earthquakes of the past (e.g., the years 1897, 1905, 1934 and 1950) in the Himalayas and the recent ones in Bhuj (2001), Chamoli (1999) and Uttarkashi (1991), the last two occurring in the International Decade for Natural Disaster Reduction (IDNR, 1990–2000), are alarmingly convincing examples of natural disasters inflicting heavy loss on life and damage to property. We must also not forget the damage caused by the cyclones in Orissa, land subsidence in Jharia Coalfields in Bihar, drought in Madhya Pradesh and the frequent landslides in the Himalayas. It has been estimated that, on an average, annually, the damage caused by landslides in the Himalayan range amounts to more than one billion US dollars and that there are more than 200 deaths every year, which works out to 30 per cent of such types of losses occurring world-wide (Naithani, 1999). It is thus imperative and urgent to seek reasons and remedies for these devastating disasters.

Although natural disasters are regular phenomena since ancient times and one never knows when these will strike, their frequency appears to have greatly increased in recent times. This may partly be due to the environmental degradation caused by deforestation, unplanned land use and, of course, the population explosion, especially in our region. Thus, though it may not be possible to eliminate natural disasters completely, creating proper awareness among the people about the destructive potential disasters are capable of, can reduce the sufferings caused by disasters. This requires adequate disaster management and planning by developing suitable warning systems and being prepared for any eventuality. Fortunately, we are living in the era of Information Technology (IT) when the dissemination of information from one place to another has virtually become real time. The IT tools can be, and are being, sufficiently exploited for disaster management programmes (Gupta, 2000). It is worth noting that improved technologies may not be able to prevent disasters, but will lead to a better understanding of the causes underlying disasters which will help us in our preparedness to combat them.

Remote sensing and Geographical Information System technologies are now, the leading IT tools that can be effectively utilised to reduce the degree of damage due to disasters in a region. For example, the remotely sensed data can be readily applied to prepare hazard maps showing the proneness of different regions to disasters and the anticipated severity of possible disasters. These maps may thus assist in reducing the disaster risk by informing the people in advance about likely hazards so that preventive measures may be taken. Further, since hazards and/or disasters (see next section) are the result of a combination of natural and human induced environmental factors, a collective analysis of these using a GIS would significantly help in devising suitable methods for disaster management, monitoring and assessment programmes.

In this chapter, the role of remote sensing and GIS for disaster risk management is presented. The chapter is divided into many sections. This, the first section has laid the foundation for the discussion envisaged. In the next section, the various issues related to natural disasters are deliberated. The third section explains the basics of remote sensing describing various data collecting devices and their analysis. A brief overview of the concepts of GIS and its applicability as a multi-disciplinary approach for dealing with problems such as disaster management are presented in the fourth section. The role of these technologies in various phases of disaster risk management are clarified in the next section with regard to some disasters occurring commonly in Asian regions. A case study carried out by the author, relating to landslide hazard zonation in Bhagirathi valley using remote sensing and GIS is presented in the sixth section. Finally, the conclusions drawn from the paper.

The following table 8.1 shows a list of abbreviations used in this chapter.

TABLE 8.1 Abbreviations of Sensors

Acronym	*Expansion*
VIS	Visible
AVHRR	Advanced Very High Resolution Radiometer
MSS	Multi Spectral Scanner
TM	Thematic Mapper
ETM	Enhanced Thematic Mapper
PAN	Panchromatic
HRV	High Resolution Visible
LISS	Linear Imaging Self Scanning
WiFs	Wide Field Sensor
SAR	Synthetic Aperture Radar

NATURAL HAZARD AND/OR DISASTER?

A *hazard* is a phenomenon that puts humans in a potentially dangerous and disadvantageous position (Wadge, et al., 1993). Hazards, in fact, are the natural or human induced processes or events with the potential to inflict heavy loss to human lives and affect their welfare. These may be drastic geophysical events, biological processes and major technological accidents that pose unforeseen risk to mankind (Anonymous, 1998). Floods, cyclones, earthquakes, tornadoes, landslides, volcanic eruptions, drought, wildfires, avalanches, etc. to name a few, are among these.

A hazard becomes a *disaster* when it results in a sufficiently large scale loss of life and property. Literally, hazard is a danger or risk whereas disaster is a tragedy or calamity. Thus, a hazard may turnout to be a disaster depending upon its severity. A disaster management programme in fact warrants for our preparedness to contain the hazard before it becomes a disaster. Therefore, both the words hazard and disaster may be treated as synonyms in our context.

Disasters may be categorised as rapid or slow, depending upon whether it is short-lived or long in space and time (Wu and Usher, 2001). Earthquakes, for instance, are usually short-lived (rapid) but their impact is intense. Drought, on the other hand, is a slow phenomenon though its impact may also be no less severe. In fact, no clear distinction can be made between slow and rapid when describing a disaster. For example, a flood may either be classified as slow or as rapid depending upon the conditions under which it occurred. Nevertheless, no matter what a disaster is, its prediction, monitoring and its impact after it has occurred needs to be assessed for taking remedial measures for the future.

Prediction involves collecting the necessary information to determine the possibility of a hazard so as to initiate the preventive and mitigation activities. Monitoring implies tracking or recording a disaster as it happens to provide timely information to help those responsible for performing the relief and recovery activities. Assessment is the post event activity of collecting information that can be used to estimate the extent of damage caused by a disaster. It may also be of assistance in finding out what led to the occurrence of the disaster. Thus, a disaster management programme may be divided into the following phases (Anonymous, 1998):

(i) Pre-disaster planning phase
(ii) Disaster preparedness or early warning phase
(iii) Monitoring phase
(iv) Emergency response or damage assessment phase
(v) Recovery and relief phase.

In the pre-disaster planning phase, the emphasis is placed on planning activities related to minimising the possibility of risks by gathering information on the characteristics of hazards, their location, the probability of their occurrence and the degree of severity, as well as the susceptibility of life and property to such hazards. The disaster-preparedness phase involves initiation of actions like hazard forecasting, warning and prediction. A range of activities such as assessment of the extent of damage, the relief and remedial measures called for etc. are performed before, during and after the occurrence of a disaster in the emergency response phase. Disaster recovery and relief phase may include actions required to re-establish the community and its infrastructure.

Remote sensing and GIS can play a crucial role in furnishing important information in each of these phases.

REMOTE SENSING: AN INSIGHT

Remote sensing may be defined as the science of extracting information about the earth surface from images acquired at a distance. Although, aerial photography also fits into this definition, the name remote sensing was first coined in the early 1960's. The type of remote sensing that we shall focus here is concerned with the observation of the earth by means of reflected or emitted electromagnetic energy. While aerial photography deals with the energy in the visible region of Electro Magnetic

Spectrum (EMS), remote sensing goes beyond this and includes the infrared (IR) and microwave (MW) regions also. The EMS is a continuum of energy ranging from kilometres to nanometres in wavelength categorised into a number of bands. The wavelength bands commonly applied in remote sensing of the earth's surface range from visible (VIS) to microwave (MW) regions. The visible range (i.e., 0.4 μm to 0.7 μm) is described as the photographic remote sensing band. Wavelengths longer than visible are subdivided into IR and MW bands. The IR band (i.e., 0.7 μm to 1 mm) is not a uniform region. Short or near IR (i.e., 0.7 μm to 0.9 μm) behaves like the visible range and can be detected by special photographic film. IR region with a wavelength up to 3 μm is primarily of solar origin and is reflected by the earth's surface. Beyond 3 μm, IR radiation emitted by the earth's surface can be sensed in the form of heat emissions. Therefore, the longer IR bands are also called thermal IR bands. Although the atmosphere absorbs most of the emissions, there is a transmission window between 8 μm to 14 μm, which allows a satellite sensor above the atmosphere to detect the thermal emittance from the earth. The region of EMS with wavelengths between 1 mm and 300 cm is called the MW band.

A typical remote sensing system has four basic components namely, a source of energy, its interaction with the earth's surface, interaction of reflected and/or emitted energy with the atmosphere and a sensor to capture the reflected and/or emitted energy. The energy may come from a natural source such as the Sun or an artificial source such as the microwave Radar (RAdio Detection And Ranging). The sensor, for example, a camera, captures the energy that has interacted with the atmosphere and the earth. The sensors to produce digital and photographic images record the difference in the amount of energy reflected by various earth surface features. Radiation in the MW region can penetrate clouds and thus is a useful region for remote sensing purposes. The microwave sensors record even small amounts of radiation emitted by the earth's surface. However, the remote sensing images produced are of very poor spatial resolution. Therefore, some other devices known as active sensors that can generate and detect their own source of energy are used in this region. One of the active microwave sensors is Synthetic Aperture Radar (SAR).

The sensors are placed on a platform such as a balloon, aircraft or satellite launched in space, the latter two being the most common remote sensing platforms. There are a number of sensors operating from these two platforms in various wavelength regions suitable for remote sensing purposes. We shall focus on the satellite based sensors. Depending upon the orbits, the satellites may be classified as geo-stationary and sun-synchronous. The geo-stationary satellites appear stationary with respect to the motion of the earth and are situated at very high altitudes of around 36,000 km, and observe a single region of the earth's surface throughout the day. These are largely used for meteorological and communications purposes and have a major role to play in disaster management, particularly in the pre-planning and warning phase, as we shall see later. The satellites such as INSAT, MATEOSAT, GOES, etc. belong to this category. The sun-synchronous satellites are placed in the orbit in such a way that they maintain a constant angular relationship with the solar beam. Due to this, the satellite will always pass over a particular place at the same local solar time. Having a constant local solar time will produce similar seasonal illumination and shadowing in the images of a given area. This aids in the interpretation of the images for retrieving earth surface information for mapping purposes. The spatial resolution of sensors on-board these satellites are very good which makes them useful in various phases of disaster management. The earth resource satellites such as LANDSAT, SPOT, IRS, RADARSAT, IKONOS, etc. belong to this category.

The data from the satellite sensors are downloaded in digital form at the earth receiving stations located around the globe. Here, the data can be converted into photographic form, if desired. For the interpretation of photographic data, the visual interpretation technique is applied while digital data are interpreted using digital image processing (DIP). The former involves a human analyst as the interpreter while the latter involves the use of a computer. The various objects are discerned from one another by examining the relative differences in their brightness or colour. In visual interpretation, seven basic image interpretation elements namely tone; texture, shape, size, shadow, site or association and pattern are used either individually or in combination. However, with the availability of huge amount of digital data, the present trend is to utilise DIP to extract accurate information in an efficient and cost-effective manner. It permits a range of image analysis tasks that may not be accomplished just by visual interpretation technique. The principal advantages of DIP are versatility, repeatability and presentation of data with precision. Not only is the analysis procedure fast, but the digitally derived products from remote sensing data can also be directly entered into a GIS for further analysis. There are a number of image processing operations that may be conducted. These include image enhancement, image classification, image fusion, image change detection, etc. Image classification is generally treated as the equivalent of visual image interpretation to derive information and in particular, to produce a variety of maps from remote sensing data. These maps become useful entities for any disaster planning and management programme. The change detection algorithms may be adopted for disaster monitoring jobs.

GEOGRAPHICAL INFORMATION SYSTEM (GIS): A PRIMER

Disaster management is a complex process and generally requires participation from many institutions. Moreover, due to the complexity of the current man-made systems and the unexpected nature of the global environmental situations, a large amount of information from different sources has to be included in the disaster management process. A well-organised disaster management team must therefore incorporate not only resources obtained by remote sensing but also other ancillary data such as topography, climate, land cover, socio-economic data, etc., pertinent to reducing the risk of disaster. This can expediently be done using a GIS, whereby the spatial and non-spatial data from different sources, at varied scales, accuracies and formats can be effectively analysed in a cost-effective manner for modeling, mapping and the spatial decision support required later. GIS based decision support systems may be used for enhancing preparedness as well as in responding to actual emergencies.

By definition, a GIS is a powerful set of tools for collecting, storing, retrieving, transforming and displaying real world spatial data for a particular set of purposes. In fact, it is a computer-based system that is used to manipulate the spatial and non-spatial data in digital form. The software part of a GIS system consists of five basic elements namely, data input, data storage and data base management, data output and presentation, data transformation and the user-interface. The input module is used to capture data from available maps, remote sensing sensors and from statistical and field surveys using digital equipment such as Total Stations and Global Positioning Systems (GPS). The job of the data base management and storage module is to allow for proper organisation and for linking the various forms of data in an efficient manner. The data output is concerned with the presentation of the results in suitable forms such as maps, reports, tables and figures. The data transformation encompasses the analysis algorithms to manipulate the data as desired by the users in the form of queries asked by them. In fact, this module may be treated as the heart of a GIS. A user interface and inference mechanism is required to answer the queries from the users in a simple way.

Thus, a GIS system allows the user to link different map layers and databases (*see* Figure 8.1) to perform different analyses on those in order to derive relevant and useful information for the job in hand. The input of remote sensing data in almost real time makes GIS an indispensable tool for disaster risk management. Further, the advent of GPS systems has added another dimension to GIS, remarkably enhancing their capability. The GPS is a satellite-based radio navigation system that can provide three-dimensional position and time information in one go. Besides acting as real time data acquisition devices for GIS, these are very useful as stand alone systems for a wide range of activities related to disaster management. For example, in earthquake studies, the GPS may be used to measure the deformations caused by an earthquake to the order of a few millimetres. These deformations in conjunction with seismological and other data may readily be analysed in GIS to locate the source of an earthquake.

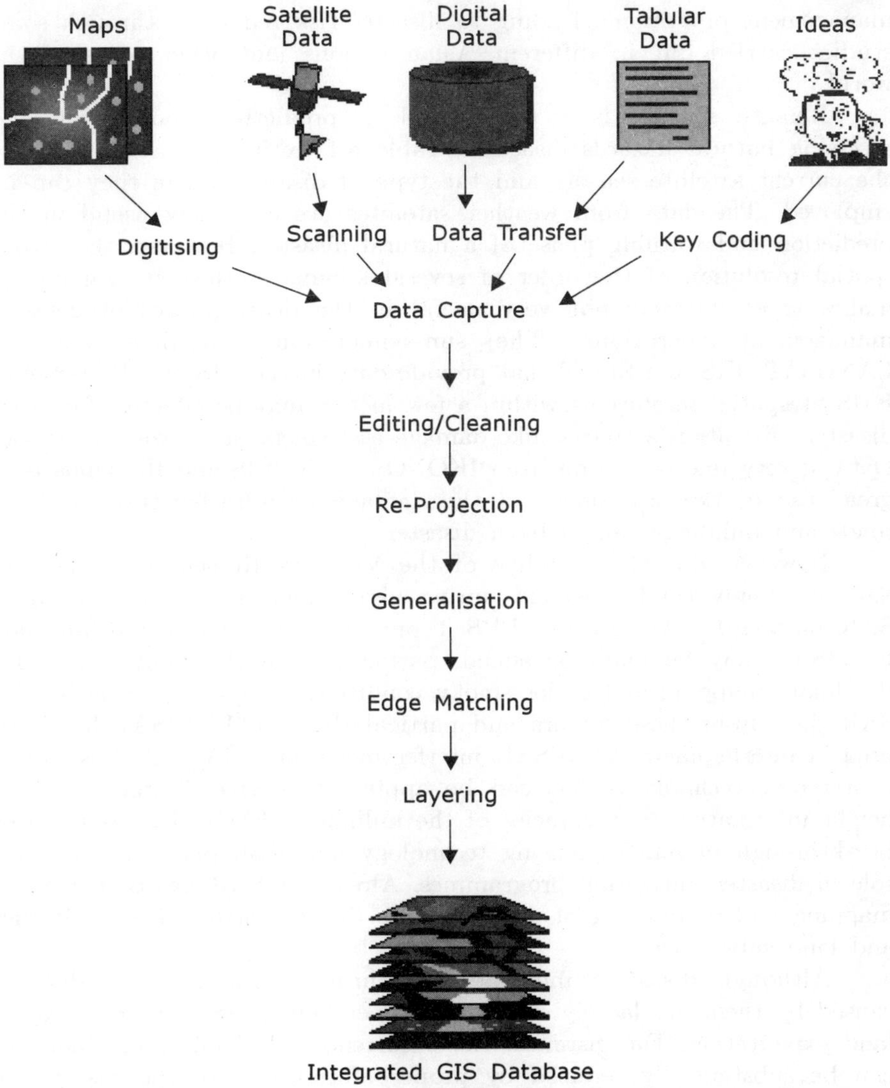

FIGURE 8.1 *The general processing steps in any GIS operation.* (*Source*: Heywood, et al., 1998.)

To summarise, it may be said that remote sensing, GIS and GPS whether individually or in combination, are extremely useful tools on which a disaster risk management programme may be based.

CONTRIBUTION OF REMOTE SENSING AND GIS TO DISASTER RISK REDUCTION

Following our brief review of the basic concepts of remote sensing and GIS, let us focus on how these assist in the various phases of disaster management programme leading to disaster risk reduction through some studies carried out in different Asian regions and other parts of the world.

Remote sensing is a useful tool in predicting, monitoring and assessing natural hazards/disasters. Table 8.1 exhibits a list of some of the current satellite sensors and the type of disaster where they can be employed. The data from weather satellites are extremely useful in the prediction and warning phase of a natural disaster. However, the coarse spatial resolution of the order of several kilometres they are capable of makes these datasets not very useful in the other phases of disaster management operations. The sun-synchronous satellites such as LANDSAT, IRS and SPOT that provide data in the VIS and IR range of EMS at spatial resolutions within a few metres may be effective for other disaster mitigation activities like damage assessment and hazard zonation. Data of very fine resolution from IKONOS in the VIS and IR bands is of great use in the assessment of the damage to infrastructures such as roads and buildings caused by a disaster.

However, due the inability of the VIS and IR sensors to acquire data in cloudy conditions and during night-time, the Radar data from SAR on-board RADARSAT, ERS 1 and 2, and the recently launched ENVISAT may be more beneficial, particularly in the rainy season for floodplain mapping and under smoky conditions during fire hazards. The SAR data from these sensors and particularly from ENVISAT has been used in interferometry also. SAR interferometry or InSAR, as it is called, is a recent technology that can be applied to derive 3-dimensional or height information to accuracies of the millimetre levels. This is a major breakthrough in remote sensing technology and shall play an important role in disaster mitigation programmes. Already, InSAR has been used in mapping and monitoring of displacements due to earthquakes, landslides and land subsidence.

Although disasters may be a natural phenomena, the damage caused by them are largely due to improper human activity and reckless land exploitation. For instance, the occurrences of floods and landslides can be substantially reduced by proper aforestation of the susceptible region. Thus, accurate mapping of land use and land cover patterns at regular intervals becomes a pre-requisite to effect preventive measures to avoid a disaster and to assess the damages after it has occurred. Remote sensing data at spatial resolutions ranging from 10 m to 100 m available at varied temporal resolutions have been successfully utilised to produce land use land cover maps at a desired accuracy of at least 85% (Arora, 2002). The remotely sensed derived land use land cover maps are frequently input to a GIS for several disaster related studies, for example landslide hazard zonation in the Himalayas (Saha, et al., 2002) and flood risk management in Bangladesh (Rahman, et al., 1991).

Now, let clarify the role of remote sensing and GIS in some commonly occurring hazards.

Cyclones and Tornadoes. The weather satellites (geo-stationary) such as INSAT and GOES provide almost continuous data on day-to-day basis for the prediction of weather and monitoring climatic changes. This is beneficial in providing information for issueing early warning and specifying the geographical location of hazards such as severe thunderstorms and tornadoes, thereby assisting the emergency response teams for timely evacuation of the people in the hazard-prone areas. Tornadoes are frequent in USA and one of them, at 260 miles per hour touched the Maryland town as recently as on April 28, 2002. Due to the availability of real time pictures from GOES satellite, the residential and commercial areas were evacuated in time resulting in minimum loss of life.

In an other application, the visual interpretation of SPOT images before and after the cyclone in the Cook Islands sufficiently demonstrated the potential of the remote sensing datasets to identify the large scale changes in coastal and shallow marine areas produced by the cyclones (Loubersac, et al., 1991). The large-scale damage due to cyclone visible in SPOT images was helpful in planning emergency aid and reconstruction.

Floods. From the early civilisations till now, people have preferred to live by the rivers as rivers offers many advantages such as fertile soil, access to water, food and energy, and transportation routes, etc. However, living by the rivers has some disadvantages also, of which the most obvious one is the threat of floods. Floods are a recurring phenomenon especially in less developed countries, where factors such as population density, absence of strict regulations, and lack of early warning systems contribute to flood risks. Every year, floods play havoc in Bangladesh and many parts of India. Remote sensing and GIS tools can be used in flood analysis to determine and delineate the floodplain. Further, these may also assist in mapping and monitoring flood inundated areas, assessing damages due to floods and in flood hazard zoning. The flood plain management and zonation would make for alerting the risk-prone residents and commercial establishments in advance about the areas that are threatened by the floods.

In 1988, Bangladesh experienced the worst flood in its living memory that caused terrible loss of life and property. AVHRR images obtained before and after the flood were classified to generate the flood maps of the region. These maps were compared to assess the damage due to floods. The data from these maps along with those from other maps (namely soil, geology, land use and contour) were analysed in GIS to prepare the flood risk map indicating the zones of high to low flood risk (Rahman, et al., 1991).

Forest Fires. Forest fires are another natural disaster where remote sensing and GIS can be employed successfully. The detection and monitoring of fires in the quickest possible time helps the emergency management agencies to prevent large-scale damage to life and property.

Also, the residents in an area can be notified in advance so that they may have sufficient time to vacate the places.

In 1983, in Eastern Kalimantan, Indonasia, about 3.5 million hectares of tropical rain forest was damaged by what was the biggest fire of the 20th century. To monitor and detect fires at global scales to prevent such heavy loss, the images from thermal band of NOAA AVHRR and Landsat TM sensors could have been used. These data were effectively used in May 1987 in the northern part of China when a forest fire happened there. These datasets provided timely grasp of the fire condition including precise location of fire scenes and the spreading direction of the flames. The information obtained from the remote sensing data was of immense value to the emergency fire extinguishing departments (Changda, et al., 1991). It was the coarse resolution data that helped in the detection and monitoring of the forest fires. The fine resolution data from IKONOS may be utilised for damage control and post fire assessment to help the emergency response teams.

Earthquakes. The economic and social effects of earthquakes can be reduced through a comprehensive assessment of seismic hazard and risk that can be used in increasing public awareness, planning land use in seismically sensitive areas and formulating seismically sound building construction codes. Seismic hazard describes the potential for dangerous, earth quake-related natural phenomena such as ground shaking, fault rupture, landslides and liquefaction. Remote sensing data and GIS may be used to plot seismic hazard maps to assess the exact nature of risks.

In fact, GIS is being used in various areas of earthquake assessment. For instance, a GIS can be used to correlate the relationships among aftershock epicenters to understand what happened underground. In a study by Gupta, et al. (1994), GIS was used to integrate geological, topographical and seismic information to generate an isoseismic map of the Uttarkashi region. The results obtained from the analysis showed close approximation to the damage pattern identified after the Uttarkashi earthquake. Moreover, remote sensing and GIS based mapping of known faults may assist the people in apprising them about their living in close proximity to the fault.

Landslides. Among the various natural hazards, landslides are probably the most widespread and damaging, particularly in the Himalayas. The high susceptibility to landslides of the Himalayan terrain is mainly due to a complex geological setting combined with contemporary crustal movements, varying slopes and relief, heavy rainfall, along with ever-increasing human interference in the ecosystem. To effect a quick and safe mitigation measure, strategic planning, identification of landslide prone areas and landslide hazard zonation are mandatory. Remote sensing coupled with GIS can be easily applied to plot hazard zonation maps indicating the probability of occurrence of a hazard in an area. Hazard maps constitute an important source of information for the parameters required in various phases of disaster management. For example, landslide hazard zonation maps help in providing information on the extent of slopes that are likely to be affected and the rate of mass movement of debris in the event of occurrence of a landslide. A study carried out by the author on preparation of landslide hazard zonation map using remote sensing and GIS in Himalayas is presented briefly in the next section.

In a different study, the multi-temporal images from Landsat TM and MSS were used to establish relationships between the land cover and the occurrence of landslides in Southern Thailand in 1988 (Rosenqvist et al., 1991). It was shown how the risk of landslide increased when the natural forest was converted into some other kind of land use demonstrating the impact of land cover on the landslide hazard.

Thus, it can be seen that remote sensing together with GIS tools is being and can be utilised in a big way for managing a variety of natural disasters. The application of these can be found in various stages of a disaster mitigation programme. However, in emergency situations, the temporal resolution of the dataset plays an important role in meeting any eventuality. The temporal resolutions of the current satellites carrying either optical or microwave sensors are not good enough for evolving an emergency response system (*see* Table 8.2). Although, NOAA AVHRR and SPOT 4 VEGETATION sensors provide data on daily basis, their spatial resolution is poor. Probably, in the near future, data with better spatial and temporal resolutions will be available. Nevertheless, these datasets are quite useful for monitoring all types of disasters, whether floods or fires.

CASE STUDY

A Landslide Hazard Zonation (LHZ) of Bhagirathi valley using remote sensing and GIS was performed recently by the author and his team. A brief overview of this study is presented below. The details can be found in Saha, et al. (2002). Landslide Hazard Zonation (LHZ) is a process of ranking different parts of an area according to the degrees of actual or potential hazard from landslides. The LHZ maps assist planners in selecting favourable sites for development schemes involving the construction of buildings, dams and roads. Even if the hazardous areas cannot be avoided altogether, their recognition in the initial stages of planning may help adopt suitable precautionary measures. Thus, these maps help identify and delineate hazard-prone areas so that environmental regeneration programmes can be initiated adopting suitable mitigation strategies.

For the landslide hazard zonation of Bhagirathi Valley in Himalayas, a number of GIS data layers were prepared using IRS 1C LISS data and other existing maps. These include thrust fault, photo-lineament, lithology, land use and land cover, drainage, slope and relative

TABLE 8.2 Applications of some current satellite sensors for disaster management

Satellite	*Sensor*	*Spatial Resolution*	*Type of disaster*
METEOSAT	VIS	2.5 km, day and night observations	prediction/monitoring of cyclones, tornadoes and volcanic eruptions, floods, avalanches
NOAA	AVHRR	1.1 km, twice a day	land cover, detection and monitoring of fires, drought, volcanic eruptions
LANDSAT	MSS TM ETM	30 and 80 m	land use, flood extent, drought, landslide, fire
SPOT	PAN HRV	10 and 20 m	3-dimensional mapping, flood extent, damage assessment, crop identification, drought, landslide
IRS	PAN LISS WiFs	6, 23 and 188 m	3-dimensional mapping, flood damage assessment, drought, fire, landslide
ERS	SAR	All weather, 25–500 m	3-dimensional mapping, flood extent, damage assessment, night coverage, earthquake, fire, landslide
RADARSAT	SAR	All weather, 10–100 m	3-dimensional mapping, flood extent, damage assessment, night coverage, earthquake, fire, landslide
IKONOS	PAN MSS	1 and 4 m	High resolution mapping, infrastructure identification, terrain analysis, property damage assessment

relief maps (*see* Figure 8.2). The various data layers were arranged in a hierarchical order of importance, and a weightage number (from 0 to 9) was given to each map layer. Similarly each class within a layer was given an ordinal rating from 0 to 9 according to their importance. Each data layer was then multiplied by the corresponding weightage and then added up to yield the landslide hazard index for each cell of the layer. These landslide hazard index values were categorised into five zones namely 'very low hazard', 'low hazard', 'moderate hazard', 'high hazard' and 'very high hazard' to produce the landslide hazard zonation map of the area (*see* Figure 8.3).

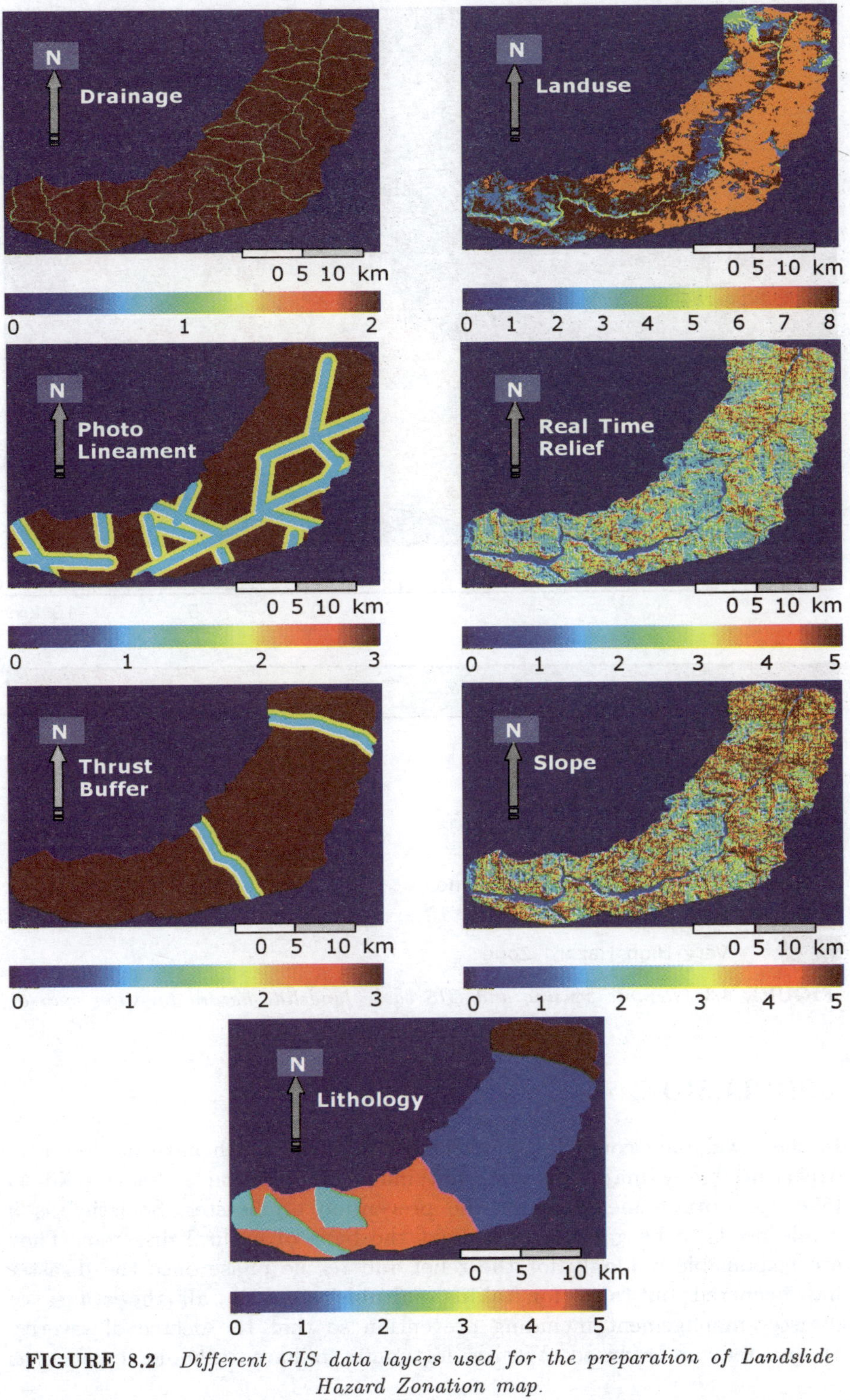

FIGURE 8.2 *Different GIS data layers used for the preparation of Landslide Hazard Zonation map.*

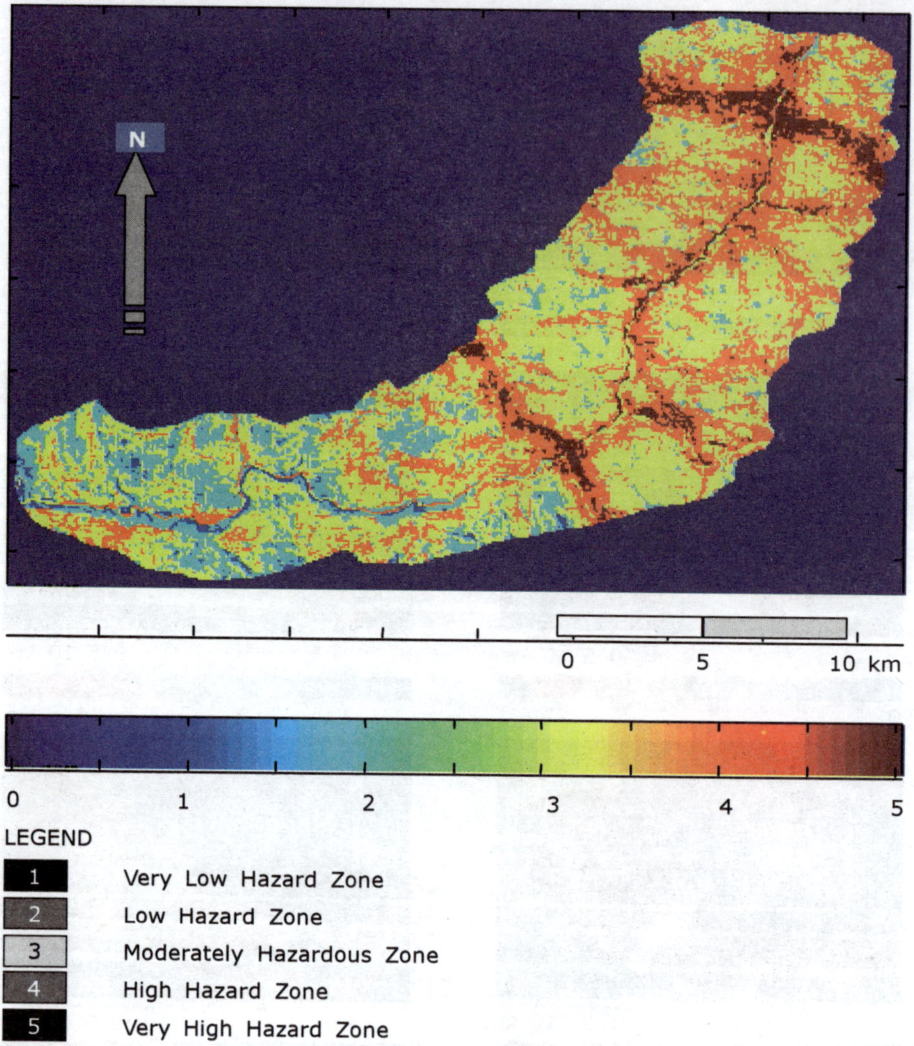

FIGURE 8.3 *Remote sensing and GIS based landslide hazard zonation map of Bhagirathi Valley.*

CONCLUSION

In the developing countries, particularly in those which have not recently experienced any major disaster, not much of attention is being paid to take appropriate measures for the prevention of disaster. Societies as a whole needs to be galvanised to fight the bane of natural disasters. They are responsible not only for the relief and rescue phase once the disaster has occurred but also for taking suitable steps in all the stages of disaster management including prevention so that the degree of severity of disasters are reduced. We are fortunate to have technologies such as remote sensing and GIS at our disposal and must utilise the power of these to face any challenge. These technologies have sufficiently demonstrated their potential in tackling natural disasters (as can be seen from some of the examples presented here). However, there is lot more to be done before these can be implemented at the operational level.

Further, the type of natural hazard may vary from one geographical region to the other. It is therefore imperative that many countries should involve themselves in and contribute to developing disaster risk management methodologies adaptable to the local requirements. The methodologies developed by different countries may be shared with the international community through the Internet for better understanding of natural disasters. While remote sensing acts as a valuable source of data for disaster management, the ancillary data from other sources such as topography, population, socio-economic data, etc. are equally important. The benefits of GIS should be seriously considered to facilitate its rapid integration with any decision making process. In future, when more and more satellite data at improved spatial and temporal resolutions become available to the community, remote sensing and GIS will certainly improve their utility for disaster management and shall form an integrated component of any disaster risk management programme.

REFERENCES

Anonymous, *Disaster Prediction, Warning and Mitigation*, Background paper 2, 1998 (http://www.oosa.unvienna.org/unisp-3/docs/backgroundpapers/).

Arora, M.K., *Land Cover Classification from Remote Sensing Data*, *GIS@development*, March, 2002, 6, pp. 1–3.

Burrough, P.A., *Principles of Geographical Information Systems for Land Resources Assessment*, Clarendon Press, New York, 1986.

Curran, P.J., *Principles of Remote Sensing*, Addison Wesley Longman, New York, 1990.

Gupta, A., *Information Technology and Natural Disaster Management in India*, 2000 (http://www.iscgm.org/more-iscgm/forum2000/).

Gupta, R.P., A.K. Saraf, P. Saxena and R. Chander, *IRS Detection of Surface Effects of the Uttarkashi Earthquake of 20 October*, 1991. Himalaya, International Journal of Remote Sensing, 15(11), pp. 2153–2156, 1994.

Heywood, I., S. Cornelius, and S. Carver, *An Introduction to Geographical Information Systems*, Addison Wesley Longman, New York, 1998.

Kalluri, S., R. Plante, M. Mohamed, R. Bergman, and K. Carr, *Remote Sensing Applications for Operational Decision Making at Local Scales: Current Status and Future Opportunities for Agriculture and Disaster Management Applications*, 2001 (http://earth-outlook.east.hitc.com:1500/).

Loubersac, L., A.L. Dahl, and L.D. Ozouville, Comparison of two SPOT images acquired before and after a cyclone, in *Applications of Remote Sensing in Asia and Oceania—Environmental Change Monitoring* (Ed. S. Murai), 1991, Asian Association on Remote Sensing: Japan, pp. 269–273.

Miguel Ayanz, J.S., M.M. Verstraete, and B. Pinty, *Using current and future Remote Sensing Systems in Natural Hazard Management*, in (Eds.), 2000, *Observing Land from Space: Science, Customers and Technology* (Eds. M.M. Verstraete, M. Menenti, and J. Peltoniemi), Kluwer Academic Publishers, Netherland.

Naithani, A.K., *The Himalayan Landslides*, Employment News, XXIII(47), 20–26 Feb., pp. 1–2, 1999.

Rahman, S.M., *Flood Risk Mapping in Bangladesh*, in *Applications of Remote Sensing in Asia and Oceania—Environmental Change Monitoring* (Ed. S. Murai), 1991, Asian Association on Remote Sensing, Japan, pp. 263–268.

Rosenqvist, A., S. Vibulsresth, S. Viseshsin, and S. Murai, *Landslide disaster in Southern Thailand*, in *Applications of Remote Sensing in Asia and Oceania—Environmental Change Monitoring* (Ed.: S. Murai), 1991, Asian Association on Remote Sensing, Japan, pp. 274–276.

Saha, A.K., R.P. Gupta, and M.K. Arora, *GIS-based Landslide Hazard Zonation in a Part of the Himalayas*, 2002, International Journal of Remote Sensing, 23(2), pp. 357–369.

Trnka, J., and A. Sivertun, *GIS in emergency management as a Core Information System and Related New Requirements to Emergency Telecommunication*, 2001 (http://www.emtel.etsi.org).

Wadge, G., A. Wislocki, J. Pearson, and J.B. Whitow, *Mapping Natural Hazards with Spatial Data Modeling Systems*, 1993, in *Geographical Information Handling—Research and Applications* (Ed. P.M. Mather), Wiley, Chichester, UK, pp. 239–250.

Wu, T. and J.M. Usher, *Application of Remote Sensing for the Prediction, Monitoring, and Assessment of Hazards and Disasters that Impact Transportation*, 2001 (http://www.rstc.msstate.edu/publications/99-01).

9

Disaster Risk Reduction by Education, Information and Public Awareness

— **George N. Ritchie**

INTRODUCTION

Disasters such as (drought, floods, environmental degradation, etc.) occur as part of natural events, affecting people, their livelihoods, economies and infrastructure. Without people there would be no disasters only natural events of different types and degrees of severity. Disaster threats and combating them form the focus of this chapter.

In corresponding equations, development implies benefits to and participation by society in both human and economic terms, for it to be sustainable and successful. To be sustainable, development must manage and preserve the environment in which it takes place so that the benefits can be enjoyed by future generations.

Because of the crucial role people can and should play in disaster reduction, whether as government or non-government organisations, private business sectors or the community itself, it is of the great importance that they are aware of, and are trained and helped in their efforts to fulfill their responsibilities. This necessitates organising programmes of education and training in the public service sector and in NGOs, to create and develop among individuals and organisations analytical, planning and programme management skills which are essential elements of all disaster reduction plans.

To ensure the success of disaster reduction plans, the community must be encouraged and helped to play their complementary and fundamental role which calls for increasing their self-reliance and self-help both, in preparedness and in response when disasters strike. This may be achieved through programmes of education, public awareness, information and training conducted at the community level.

Education is an important aspect of the development process

particularly as it has a direct bearing on health, hygiene and sanitation which are already important considerations for programmes mounted by such organisations as the UNICEF. Awareness of these factors and of the importance of clean drinking water is essential at the family level for controlling the outbreak of diseases and infections, which frequently accompany disasters.

The importance of environmental conservation and management can be impressed upon the community as part of such awareness and training programmes encouraging it to contribute towards reducing environmental degradation which is a significant cause of the threat posed by drought, desertification and floods which plague much of the developing world. Although people and communities are the first victims of disaster their contribution through self-help, community action and self-reliance is the first step in disaster reduction programmes.

In completing this loop we see that development strategies and projects must not only consider the existing threats of natural disaster but also examine how development may engender new threats to the environment in which it takes place.

Take for example the development of pastoral farming in the Sahel region of Africa during the last 50 years of the twentieth century. As the consequence of improved stock and animal husbandry, as well as relatively good rainfall, herd sizes grew in 1950s and 1960s. To satisfy the increased demand for water in an area where surface water is perennially in short supply, ground water sources were tapped using tube well technology. But these developments and measures were not accompanied by increased consumption of animal products (beef, hides, etc.) or by improved management of pasture. When drought occurred, as could have been foreseen, herds concentrated at the wellheads and these areas were soon grazed out.

Similar examples of environmental degradation caused by over-stocking, poor development planning, inefficient environmental management and failure to involve people in the development plans which will affect generations to come, are occurring in many areas of the world now.

Drought is frequently experienced in many countries but floods too are occurring more frequently. Their severity and duration have also increased. Over-grazing of hillsides and deforestation deplete the vegetation cover. This causes the run-off during rains unimpeded, causing erosion and loss of topsoil which is deposited in river channels. This, in turn causes river channels to silt up, flood more easily and the floodwater drains only slowly. The loss of topsoil will affect agricultural productivity for centuries to come.

Topsoil, vegetation and trees are assets, which people and their agricultural advisers (extension officers) must learn to preserve. For too long the focus has been on maximising production rather than protecting these as capital assets.

There are four main actors in this situation:

- Politicians in governments
- The news and information media
- Government administrators and professionals
- People in the community

All must be helped to meet their separate responsibilities because, unless each can address its own responsibilities with regard to disaster reduction, development strategies and projects will continue to be set back by repeated disasters. Conversely, sustained economic and human development will create robust societies and resilient economies capable of absorbing the effects of natural disasters.

The USA in any year experiences a range of natural disasters such as drought, flood, hurricane, etc., but due to its economic strength, rich human and technical resources and its well-organised disaster preparedness and response plans, it survives and develops. In the last 20 years, we have seen India make similar progress, directed towards self-sufficiency in disaster management. Development of such self-sufficiency at the national level in combating disasters reflects self-reliance at all levels of society which is the most important element of disaster reduction and preparedness planning.

THE PROBLEM

In primal societies, the responsibility for the safety of people during disasters rested with the chiefs and the leaders. In modern societies, governments have assumed this responsibility, through international organisations, NGOs and the UN agencies. The external support to countries in disaster over the last 50 years has predominantly been in the provision of food, medicines, medical aid and other forms of disaster relief. Indeed when the UN created a special agency for disaster affairs in 1972, it was to be the 'UN Disaster Relief' Office, which was recently renamed the 'Department of Humanitarian Affairs'. The focus has thus always been on relief rather than on developing sovereign responsibilities and capabilities for disaster reduction, preparedness and disaster management.

To some degree at least, the UN policy has relieved the necessity for national self-sufficiency in disaster response and preparedness and has affected the policies of other international and bilateral donors, including the NGOs.

A great reservoir of various forms of support—human, material and financial—has been grown in the developed countries in response to the needs of disaster-prone developing countries. This has manifested itself in the commitment to work with and amongst the affected populations, mostly the poor and disadvantaged, when disasters strike. The need for support to disaster-prone developing countries is unlikely to diminish for many years to come and disaster relief will continue to be an essential element of development aid. However, it is important that ways are created and developed which will ensure that these disaster relief activities seek to reduce and eliminate dependencies, which can and do arise.

A very large proportion of the world's population lives in disaster-prone developing countries, which lie on the borderline between hunger and extreme food shortage. When a disaster strikes, many find their livelihood and access to food cut off. They become dependent upon others for survival. The food relief operations of the national, international and UN agencies become lifelines. The problem which all must address is how to ensure that these lifelines are only temporary means of pulling victims to safety and do not become long-term life support systems from which they cannot be detached.

Because disasters disrupt various economies, many need to be first freed from the worry of feeding themselves and their families, following which, the need the issue of revitalising and restoring to normal, the economy upon which their livelihoods depend may be addressed.

In Mozambique, the internal security situation led to an almost total collapse of normal production and agricultural activities for many years. The normal economy was replaced by a 'food aid' economy, which was dependent upon, and driven by, the donors' aid programmes, which replaced all normal economic activity, such as production and many commercial activities. National politics and policies also became involved in the aid programme and it demanded a very great degree of political will and an urgent need for peace to be established throughout the country, to end the dependency of all sectors of society upon the food aid programme.

Similar situations have arisen to a greater or lesser degree elsewhere in Africa and also in other parts of the world. Because much of the developing world depends on agricultural economies, the effects of flood and drought are of particular significance, involving large sectors of the population, government departments and to some extent, the government itself. Collapse or stalling of the agricultural economy can jeopardise not only the national economy but also political stability and the general health and well-being of the people.

As the rural populations increasingly drift from the agricultural economy towards the opportunities and promise offered by urban areas, the importance of agricultural production increases. Fewer agricultural workers seek to maintain or to increase production to meet the growing demands, whilst harvest prices remain controlled at below market prices. A precariously balanced economy can be badly upset by a disaster which can dangerously affect national stability and can lead to indiscriminate flow of agricultural labour to the urban areas.

THE WAY FORWARD

We have already mentioned that the responsibility for effecting improved standards of disaster preparedness, better disaster management and the development of self-help and self-reliance in communities rests with four sections:

- The Government at all levels
- News and information media
- Administrators and professionals
- The people, NGOs, and community-based organisations and networks

Research shows that in rapid on-set disasters (cyclone, earthquake, flood, etc.) whereas the event may cause destruction and death, it is the collapse of the government administration under load, which brings deprivation and distress to the victims. However, in the case of slow-onset disasters (famine, environmental degradation, diseases, pestilence, refugee movement, drought, infection, etc.) whether these are man-made or natural, the only factor responsible for creating the disaster is failure to implement effective programmes for promoting environmental management, health and hygiene, reliable agricultural practice and social development.

Educating the masses, staff training, public information and awareness programmes are necessary to devise effective programmes.

NATIONAL POLICY AND RESPONSIBILITIES

Disaster preparedness, relief management, post disaster planning and recovery are all sovereign responsibilities of national governments. In meeting these responsibilities, governments may seek financial, material and advisory support from external agencies such as the UN, the Commonwealth Secretariat, international and national NGOs and from bilateral aid programmes. But the Government is responsible for determining a national policy and for creating and demonstrating the political will without which little will be achieved.

In creating the climate for positive action and thinking, the first step the Government has to take is eliminating national and individual dependency in countering disaster threats, and in meeting the requirements for relief and recovery.

The responsibilities of the political fabric will include:

- Implementation of national disaster preparedness and response plans.
- Development of early warning systems (EWS) based upon monitoring and surveys at community levels.

- Defining the specific responsibilities of the Government and the local authority in planning, response and recovery.
- Identification of community and individual responsibilities in disaster preparedness and response.

It has been found that EWS, which are aggregated at the national level frequently, often fail to identify slow-onset disasters, particularly drought, food insecurity, crop failure and environmental degradation that threaten local communities and geographic areas. They also tend to remove the responsibility of government departments at the district level and end up as mere exercises in statistical gathering and analysis. EWS, although important in their purpose of drawing response from national and international agencies to imminent situations must not be seen as a system that eliminates the responsibilities at the local (say district) levels of government.

In the arid Turkana district of Kenya, an EWS was to be established under a bilateral aid programme based upon the local knowledge and understanding of a team of young men, themselves from that district. These men knew the language, culture and customs of the pastoral herders and were trusted by them. This enabled them to collect important data on the state of herds, the stability of the livelihood of the pastoralists the health and nutritional state of the people, the prevailing environmental conditions and the probable changes.

However, problems arose when this data and information was passed to the district office staff. These were not, in the main, from the Turkana district and had grown up, been educated and had their university education and training in the capital. In consequence they were not immediately confident about the value and the usefulness of the information they were provided, which not compatible with the data gathering systems developed to meet the national EWS. Similar problems have occurred elsewhere and indicate the need for educational and training programmes, which ensure that administrators can the information available to the best and can develop the interface, which translates local knowledge into higher-level statistical data.

In establishing and demonstrating its policies and political will in relation to disaster preparedness, governments should employ all available public information sources, educational and research organisations, as well as news media systems. These should be staffed by professional educationalists who are given the freedom to apply their professional skills in the most effective way.

It is a common plaint throughout the developing world that very often, when technical departments of government became involved in public information and warning programmes (e.g. meteorology departments), the language used is technical jargon (the language of the subject), which is not understood by the target audience. It is for this reason that, as part of disaster reduction plans, public information and awareness programmes are conducted to develop the professional skills of media staff at the national and also at the local levels so that the challenges are met in the most effective way.

Radio, television and newspaper reporting has an urgency and immediacy which render them unsuitable for the development of public awareness through educational and information programmes. Such programmes demand skilful development, presentation and repetition more akin to the practices of the teacher or evangelist than to news reporting. It is therefore important that public information and awareness programmes that use the news media systems, are meticulously developed and are provided with the resources which ensure their effectiveness as elements of national and local disaster reduction programmes. These resources include clear definition of the objectives and expert programme designers who are educationalists rather than reporters besides ensured availability of programme time.

Administrative Responsibilities and Action

To be effective in their purpose of creating effective standards of disaster preparedness, government policies must be translated into action. Although the primary responsibility for this must lies with the Government, it is essential that all concerned agencies, such as the UN and national and international NGOs be brought in for support. Institutes of public administration (IPA) should be required to and encouraged to develop programmes of public service training pertinent to disaster reduction and preparedness training and to open these programmes to the UN, NGO staff, and also, most importantly, to community representatives. The IPA should also be encouraged to establish travelling instructional teams whose responsibilities would include training local authority (at the district level) as part of their disaster preparedness planning, disaster management strategy and post-disaster recovery duties and responsibilities. Such training and instructional programmes should be multi-sectoral at all levels and at the district and community levels must include community leaders and representatives. This will ensure that the people are fully involved in national disaster reduction activities.

The most appropriate and rapidly available relief material is that which is located nearest to the point of need. This applies to both human and material resources and implies the importance of developing disaster preparedness plans from the community level upwards. Study of cultural attitudes, local resources, materials and markets can be included in district and national disaster preparedness plans to curtail expenses, and to avoid unnecessary movement of foreign 'experts', equipment and material around the world. However, the availability of local experts and appropriate local equipment depends upon the local awareness of the disaster threat, the local disaster response plan and of the responsibilities and tasks of the local community.

Community Responsibilities

The most effective environmental management begins at the field level and is practised by individuals as members of a community respecting their individual and community responsibilities. Environmental management involves nurturing young trees, adopting fire wood gathering in place of tree felling, soil conservation by terracing and contour walling, water conservation and harvesting using small dams and tap and pump maintenance and control. It also means maintaining herd size at optimum levels by creating off-take outlets of meat and hides, stall feeding when necessary to limit pasture wastage, control of goat grazing by herding, protecting saplings and young trees, preventing and controlling pasture and forest fires, developing firewood plantations and encouraging re-forestation.

For these to succeed, communities should work together towards acheiving common objectives, observing both individual and community constraints on the resources upon which all are dependent for their livelihoods.

Sanctions must enforce observation of the constraints imposed either by law or agreed upon jointly within and between communities. These responsibilities and duties fall upon the local government and the communities to which they relate, but good laws are those whose objectives are clearly apparent to those whom they affect and are willingly observed by the majority. One of the principal responsibilities and tasks of the local government, community leaders including politicians of all colours, religious leaders of all denominations and NGOs both national and international, is to lead the campaign against disasters such as drought and food shortage and their fall out such as famine and disease in all areas of their country prone to and affected by these.

The past has seen societies and communities crumble altogether under the affects of prolonged droughts and their consequences. One of the most serious effects of ineffective disaster management is that it results in a loss of robustness and self-reliance among communities and leads them to depend upon government and external aid agencies for disaster management. Self-confidence and self-reliance do come from material assets but more from knowledge and understanding.

Droughts and food shortages are not new to the developing countries of the world. What is new is the growth of population, and the consequent increasing demands upon the environment for food, water, fodder and firewood which, in turn calls for increased external assistance for food, nutrition, technology and medical aid. Changes in political systems, livelihoods, lifestyles, economies, education and communications has also led in many areas to a loss of knowledge of the customs practised by and the methods employed by the previous generations to cope up with and to survive disasters.

It is therefore important that those with responsibilities, at whatever level of government, in societies and communities, give due regards to changes and avail the new opportunities offered by the improved means of communication, systems of education, technologies, information media as also external assistance, to address the problems created by drought and environmental mismanagement.

It may already be too late, but 'coping mechanisms' have been developed, generally by 'outsiders', in almost all societies to address the recurrent problems that affect livelihoods and environment. They are local methods for dealing with local problems. In western societies research programmes are carried out and projects initiated to draw up knowledge of how things were done before the new technologies took over. Similar research at community levels in Africa would certainly succeed in revealing some of the 'coping mechanisms' used in the past during periods of drought and crop failure.

The new technologies can also help. It has been estimated that as much as 30% of harvested grain in Africa is lost due to poor transport and storage. Disinfestation methods, improved low-technology storage systems, pest control, improved seed strains can all be used, but success comes only from public awareness of the problem, information and training regarding the remedies imparted to all pertinent sections of the community and participation of external aid agencies and NGOs. Schools and school children constitute an important channel through which understanding of the problem and possible solutions can be conveyed. Comparatively, it is probably much more useful if the agricultural extension officer spends a day every month, teaching school children about soil conservation and improvement than it will be for him to learn about improved chemical pest control to improve production.

What is required is a clear statement of policy and responsibility for the development of community-based projects, which will focus upon environmental management, conservation of forest wealth, ground cover, top soil and water resources. Community-based projects cannot be expected to address or arrest drought as a natural event though they can effectively deploy that major national resource, the people, in countering it.

IMPLEMENTATION POLICY AND PRACTICE

Development of an effective disaster reduction programme, particularly for drought, flood and environmental management is dependent upon close liaison and coordination between government departments, its various ministries, the private and public sectors and the community at all levels.

The theory has always been that disaster preparedness planning must be developed from 'bottom up', guided by national policy, which is developed from 'top down'. Such a process ensures that the national disaster preparedness plan becomes a 'jigsaw', a mosaic in which the specific and particular needs of the community elements are identified and linked into district, provincial and the national plans. The 'bottom up' process of disaster preparedness planning ensures the consideration of needs for mutual support between communities, districts, provinces and so on. This process also reveals gaps and deficiencies in resource availabilities, both human and material, which need to be met from external sources, including international aid. This provides an effective way of turning theory into practice, but in many ways is dependent upon effective integration of community needs and responsibilities, which in turn calls for public information and awareness programmes.

When drought, floods and environmental degradation are the principal threats, it becomes important to establish community involvement as an on-going element of disaster reduction policies and practice. Environmental mismanagement, deforestation, over-grazing and human interference with the natural balance in the environment, all lead to the disaster threats and events mentioned at the beginning of this paragraph. The ancestors of those now affected by these disasters understood and respected this balance. It therefore becomes important for the success of a disaster reduction programme, to reduce the tendency to of the community depend on external help created by previous disaster relief and food aid programmes, to create the spirit of self-help and self-reliance in the community and to convince the people, that the support of a wide range of respected authorities is available to them at all levels. By involving religious, political and community leaders in public awareness and information programmes, the people in the area will gain confidence in the disaster reduction programme and its aims.

It is important that the public sector pursues government disaster reduction policies. State farms, government forestry departments, the armed forces, water conservancy, irrigation and other government departments must set examples, which communities and individuals can follow.

The private sector, particularly large-scale enterprises such as multinational companies must be encouraged or, if necessary, coerced to participate in the programmes.

Experience shows that industrial operations can have a major impact on the environment and the people in developing countries. Examples of the good effects of scrupulously planned industrial operations are to be seen in the United Kingdom where the coal mining industry could effect major reinstatement of the environment in areas seriously scarred by the previous practice opencast mining by enabling the disposal of huge mining waste dumps as fill for road constructions. These areas have also been sensitively landscaped to merge with the environment.

In another example, elsewhere in England, an oil extraction site in a coastal area, designated to be of outstanding natural beauty is the subject of major environmental safeguard and management concern so that its beauty and the integrity of the surroundings remain in tact.

Examples of the ill-effects of reckless industrial operations can be seen in the oil extraction site of a multi-national company in Nigeria and in a multi-national chemical plant in Bhopal, India.

It is therefore of extreme importance that multi-national operations are brought within the purview of national disaster reduction and environmental management plans and actions.

EXISTING PUBLIC AWARENESS PROGRAMMES

In developing countries, the UN sponsors a range of programmes while at the same time, international NGOs aim at social and community development. Because many of these are themselves relevant to disaster reduction it is both economical and effective to graft public awareness and information plans for concerning disaster on to such existing programmes rather than to launch specifically disaster oriented, but possibly confusing new ones, directed at the community level.

Examples of relevant existing programmes are, the primary health care schemes, woman and child care programmes and famine warning programmes, which have already been effectively developed and implemented by the UNICEF and *Save the Children Fund*. In these programmes, home hygiene, sanitation and the importance of clean and safe drinking water are important aspects. In disaster situations, when victims may be crowded together in camps, water supply and normal sanitary practices may be interrupted and hygiene, sanitation and care over drinking water may assume a vital role in controlling the spread of diseases. These aspects of disaster reduction are best addressed as extensions of normal practice rather than as new practices relevant to disaster management only.

Community-based Programmes

Agriculture extension officers work very much within the farming community but with responsibilities, which relate more to the maximisation of production than to disaster reduction and environmental management. Environmental conservation and management are best practised at the field level and responsibility and action for this lies primarily with the farmers themselves. They themselves can ensure the preservation of the top soil and vegetation which are the principal capital assets upon which they depend. The agricultural extension officers' responsibilities and programmes should be extended if necessary, to educate the farmers on the aspects of good farming practice and their duties could include conducting public awareness and information programmes aimed at equipping the farmers with enough knowledge of environmental management and conservation.

In the agricultural sector in India, Food for Work (FFW) now practised as, Income Generation Projects (IGP) have been important elements of disaster reduction, disaster preparedness and relief for over 100 years. These are built into disaster preparedness plans at the district and community level, with provision of resources kept ready for their early implementation when a disaster strikes. Experience over the years has shown how important it is to ensure that FFW/IGP plans are made in close consultation with the communities to which they apply.

FFW/IGP must be an early subject of public awareness and information programmes and must involve community leaders so as to take into account the community's assessment of local projects and their expectations from them. Once mounted it is equally important that the awareness and information programmes are sustained in a manner that ensures that the community understands what their responsibilities are, the importance of their participation, the role of cash or food in work programmes and the responsibilities of the Government and any NGO involved.

Whereas FFW and IGP are intended mainly as relief measures when disasters strike, there are a range of community programmes pertinent to drought and flood control and to environmental management and conservation which can be stimulated by public awareness and information programmes. Though such programmes may fall within different areas of responsibility of the department of Agriculture, all are likely to require financial and technical support, which must come from other departments of government and from NGOs. Such programmes, demanding a wide range of community participation, will include a range of erosion control measures, hillside terracing for top soil conservation, water harvesting using small dams, forestation and tree conservation.

Community leaders must don the primary role in implementing the awareness programmes which are essential to create community understanding of the threats which the action programmes are designed to meet. They must not only participate actively but also set examples that will motivate the community to involve in the work programmes. For this, the community leaders will need the support of the Government as well as the NGOs involved in such programmes.

Role of Schools and School Children

Schools, schoolteachers and pupils must be included in the national and local programmes for disaster reduction and environmental management. This is probably one of the most cost effective ways of creating change and effecting improvement in farming practice and environmental conservation. In addition to those mounted through the agricultural extension programmes, educational and literacy programmes may be undertaken to encourage good agricultural practice and environmental management involving both, the young and the old.

Classroom learning programmes, projects and experiments based upon environmental conservation practices can also be started in schools and as part of adult learning programmes.

Local weather and environmental monitoring programmes can be incorporated in schools as part of higher levels of secondary education to promote among the young, knowledge and understanding of the physical environment in which they live.

The Mass Media

In addition to these programmes taking place in the more formal environment of school and adult learning, all channels and means of communication can be availed to impart relevant information to all sections of the people. The press, radio and television can be employed to propagate programmes of public awareness and information oriented towards disaster reduction, provided that the finance required to support and sustain them is available.

The news media is characterised by an immediacy and ephemeralness, which renders it unsuitable for promoting sustained programmes related to any subject though it is effective as an advertising medium because of its wide audience, ability to attract attention and create an impact. These factors must be considered while judging its suitability for disaster management programmes.

For any disaster reduction public awareness and information programme is to be mounted by any of the means of mass communication, it is essential that either the Government uses its authority or has the finance necessary to ensure a sustained programme that will convey its message correctly to the target audience.

The design of educational, public awareness and information programmes requires knowledge and abilities, which are professional and quite different from those of either news reporters or journalists. This subject is dealt with and described in the research report, *The Effective use of Radio for Mitigation of Drought in the Sahel* by Cranfield Disaster Preparedness Centre.

Simple messages, particularly those related to family and individual responsibilities and actions, can be the subject of effective poster campaigns. The messages must be identified by the authority for disaster reduction and then passed on to professional illustrators for the design and formulation of appealing posters.

Display of the posters is of equal importance. They should be displayed widely, particularly in community centres, churches, temples and mosques, schools, medical posts, shops and markets. Action must also be taken to prevent their deterioration and damage. They should be renewed periodically, if possible in new colours and format so that their impact is renewed.

In all such programmes, including even the simplest weather reports, jargon and technical language must be avoided. Scientists and technicians are frequently poor communicators. They often exhibit 'snobbery' in their use of technical terms and language. It is therefore important that although the message may originate from the technical department it should be translated into everyday language by the communicators. Local dialects and vernacular language should be included in the spoken and written means of communicating the message. Whenever possible illustrations should be used to explain techniques and methods since they are an effective means of communication. Of course, practical training and teaching represents the most effective way of transferring knowledge and skills.

In addition, and importantly, complementary to the formal means of stimulating public awareness and of communicating information at community level, the wide range of traditional methods of transmitting practical knowledge must be availed. Different societies and cultures have created and established their own ways of doing this and it is sensible and useful to incorporate such methodologies in the public awareness and information programmes for disaster reduction. This will also create the opportunity to make the best of existing programmes related to disaster management, such as the UNICEF programmes and the SCF health and hygiene programmes.

The traditional programmes use songs, dances, plays, jingles and slogans to convey their messages. In many ways and throughout the developing world, which we are considering, such traditional modes are being overtaken by more modern ways and the wisdom of the past, particularly in relation to environmental management, is being lost. Nevertheless these traditional ways retain an authority, credibility and authenticity, which continue to be relevant.

The effectiveness of such methods in disaster reduction programmes must be assessed by *those within the culture* who can judge them authentically. This is an area, which probably is best assigned to the local NGOs.

Where the 'pop culture' has invoked widespread local interest, it may well be advisible to recruit local 'pop stars' to promulgate public awareness plans for disaster reduction and to convey relevant information. This blending of new methods with traditional ones also provide local songwriters and musicians with a challenging and interesting opportunity to market their talents and creativity.

How well all these are done, developed and maintained demands earnest commitment from all quarters, active involvement of the national and local governments in collaboration with the people, participation of protagonists of all kinds and unflagging support of the NGOs.

Clearly all disaster management programmes call for time, men, material and money. However, the environment conservation effected and the amount of damage averted will more than compensate for the expenses. One penny spent now in disaster reduction activity or in environmental conservation of soil, water, trees or vegetation will pay its return to the tune of thousand times in the form of reduced relief needs and increased agricultural production in the years to come.

10

Partnership in Health and Disaster Management for Risk Reduction in South Asia

— **Dolly Mathew**

INTRODUCTION

The countries of South Asian Region are committed to the principle and practice of inter country cooperation. Quite recently there has been increased initiation of technical cooperation among the South Asian countries, particularly among the developing countries. With the Alma Ata declaration of *Health for All* and the Yokohama call for a *Safer World*, pertinent to disaster prevention, preparedness, and mitigation, it becomes necessary for the South Asian countries to examine the areas of technical cooperation in health and disaster management in the light of these declarations.

Since most of these countries belong to the category of developing nations, disaster management measures such as superforecasting and communication networks, enforcement of strict codes for seismic and fire proof construction, provision of quick medical relief and a well laid down contingency plan are either not available with them or have not been implemented properly. As such, the population remains extremely vulnerable to the adverse health consequences from natural disasters (Noji, 1997).

Since the overall public health effects of disasters are common to most of the catastrophic natural events threatening these countries, they can work collectively in areas of health and disasters. It thus becomes pertinent to identify common areas of technical cooperation among the South Asian countries within the framework of health and disaster management, for reducing the impact of disasters on public health.

Before exploring the possible areas of technical cooperation, it will be worthwhile to examine briefly the health and disaster situation in the

South Asian countries and also the effects and consequences of disasters on public health.

Health Situation in South Asia

The World Health Report, 1996, of the WHO, has shown that despite an overall improvement in the socio-economic status and the increased life expectancy, the communicable diseases and old diseases like cholera and tuberculosis still dominate the disease pattern, while malaria, plague and kalaazar which were on the verge of eradication have reappeared. HIV/AIDS has now become the most menacing health problem in South Asian countries.

The factors which contribute to the high prevalence of communicable diseases include poverty, malnutrition, ignorance, insanitary environment and lack of safe drinking water. Population growth and rapid urbanisation with overcrowding, poor housing and environmental deterioration have worsened the situation and are responsible for the emergence of new diseases and the re-emergence of communicable and non-communicable diseases which were nearly eradicated.

In the South Asian countries, the public health infrastructure need to be strengthened by implementing a good public health policy, a better system for health care and training of the persons involved. There is urgent need for volunteers to be selected and trained as active partners in health development programmes. There is also the need to strengthen the district health system and establish primary health centres. Participatory learning in health education and training is also required.

Disaster Situation in South Asia

According to the *Disaster History, Significant Data on Major Disasters World-wide, 1998*, India had 1551.8 million people affected by natural disasters, whereas in Bangladesh, Pakistan and Sri Lanka, the number was 214.0 million, 14.6 million and 14.5 million, respectively. Likewise, the total number of disasters taking place in South Asian countries is very high. India has roughly 216 disasters per year while, Bangladesh and Pakistan each, experience 100 disasters every year.

In South Asian countries, disaster management policies are generally weak. There are no up to date disaster vulnerability indices for the micro regions and agro-climate zones for different kinds of disasters, nor is there any objective system for rapid disaster damage and assessment or even trained policy teams. There is no proactive approach with systematic, nuanced, multifaceted and long termed policies built on regional and sectoral linkages. Ecological policies which have a direct bearing on disasters are still not on the scene. Another crippling factor is the absence of the research backup. There is need to achieve an effective policy formulation based on information (Kabra, 1999).

A country like India could have developed a lot of expertise in disaster management considering the number of disasters it faces, but the fact is that there is a diaspora of expertise, which is diffused and uncoordinated, with each one unmindful of the other's work and experience. This is the area where the Government, the planners and the implementers are all challenged and stalled. Experiences are not institutionalised. Those in the forefront in the disaster management lack orientation and training (Mukhopadhyay, 1999).

In short, the prevailing situation exposes the South Asian countries constantly to the risk of adverse health consequences arising from disasters.

Public Health Consequences of Disasters

Disasters, in this region usually unleash a plethora of health problems in the form of epidemics such as cholera and gastro-enteritis. During disasters, the existing system of sanitation is thrown out of gear. People depend on makeshift arrangement. Open defecation is practised. Water gets contaminated. Potable water becomes scarce. There is overcrowding in shelters leading to unhygienic conditions and vector breeding. Medical aid and relief strategies also fail to cater to the large number of affected people. Disasters have affected the health of the South Asian community adversely in many ways:

(a) They have caused large number of deaths, injuries and illness by creating situations beyond the therapeutic capacities of health services and calling for external assistance.
(b) The health infrastructure like hospitals were destroyed and the routine health services got disrupted increasing the casualities further.
(c) Disasters gave rise to a spate of epidemics and communicable diseases.
(d) Disasters had psychological effects on the victims leading to neurosis, trauma and depression.
(e) Disasters like drought and cyclones caused food shortage and severe nutritional deficiency (Noji, 1997).

The effectiveness of the public health response depends on anticipating these problems and developing adequate and appropriate mechanism and strategies to meet them. There is need to build up an information and knowledge base. It should be experienced-based and properly documented (Mukhopadhyay, 1999). Since the overall public health effects of disasters are common to most catastrophic events striking the South Asian countries, it is necessary for these countries to work in convergence and complementarity with each other to develop an effective public health response system to disasters.

PARTNERSHIP IN HEALTH AND DISASTER MANAGEMENT

Partnership in health and emergency assistance in South Asia can provide a broad inter-governmental framework for building a cooperative approach so that the precondition for success can be met and the health of the people protected and improved. The partnership should ensure that the government activities in the region are consistent, coordinated and collaborative. The broad objectives of partnership can be:

(a) To improve collaboration in the national public health and disaster management effort.

(b) To improve coordination and sustainability in public health strategies towards disaster management.

(c) To strengthen the public health infrastructure and its capacity (World Health Forum, 1998).

These will help formalise arrangements for cross-border deployment of expertise and resources and use of information and telecommunication technology for public health activities in the region. This chapter suggests and discusses setting up a *Regional Health and Disaster Management Centre and a Regional Health Disaster Information Network* to facilitate partnership activities in the region.

We will be discussing the role of the Regional Health and Disaster Management Centre followed by the use of Regional Health Disaster Information Network for health management during disasters.

REGIONAL HEALTH AND DISASTER MANAGEMENT CENTRE

The Regional Health and Disaster Management Centre will mainly concentrate in chalking out a Regional Plan of Action for health and disasters. The plan will aim to promote technical cooperation in coordinating the responses to emergencies and in integrating emergency preparedness and the response to health development programmes. The overall objective of the plan is to prevent health hazards and to reduce the adverse effects of disasters on health and health services by strengthening national capacities for disaster preparedness and response. The Plan of Action will focus on promotional and developmental activities rather than on mounting health relief operations and on integrating and coordinating managerial approach rather than on vertical action in emergencies (World Health Report, 1992).

The Plan of Action will emphasise upon:

(a) Use of epidemiological methods and techniques in disaster management.

(b) Setting up of collaborating and cooperative centres for promoting partnerships.
(c) Education, training and research.
(d) Information sharing via information technology (as explained under Regional Health Disaster Information Network).
(e) Drawing upon the member countries to draw disaster management plans.

Epidemiological Methods and Techniques

Widely accepted standard case definitions of disaster related morbidity and mortality could be developed and disseminated. Standard reporting forms and procedures can be developed which can be easily modified for use in various settings and can be used to generate questionnaires as well as to analyse data. The assessment indicators and surveillance methods in disaster settings can be developed keeping in mind the four attributes—simple to use, timely, collectable under adverse field conditions and useful (World Disasters Report, 1996). This will help in developing accurate and appropriate disaster epidemiological knowledge regarding the causes of death, injuries and illness during disasters and this knowledge, in turn can be used to prepare plans to meet disasters in future.

Collaborating Centres

Collaborating centres with the resources, infrastructure and the mandate to collaborate in areas of health programmes and disaster management can be set up. The South Asian countries can have collaborating centres (as on the lines of WHO collaborating centres), as bases for partnerships in health and emergency assistance. These collaborating centres can operate in the following forms:

Health and Disaster Management Collaborating Centre

The responsibilities of this centre may include:

(i) Investigating medico-sanitary issues in the aftermath of different disasters and preparing and implementing methodologies for the planning of health emergency and relief actions.
(ii) Assessing the effectiveness of existing health service structures in meeting emergency situations and preparing recommendations relevant to disaster medicine service and health management of the population in times of emergency.
(iii) Undertaking mutually agreed research activities and the collection and evaluation of technical and scientific information related to disaster preparedness and management.

(iv) Assist in developing national and international training courses on the medical aspects of disaster preparedness and management.

(v) Looking into the health related issues of people displaced by disasters. Developing a comprehensive database of the impact of natural disasters on community health and the effectiveness of the ways in which these have been dealt with in the past. It will also involve preparation of training material relevant to the health management of the displaced population. Epidemiological research and evaluation aiming to improve the existing strategies to manage the health of displaced people may also be taken up.

(vi) Tendering advice and participating in the awareness programmes in the areas of public health implications of exposure to radiation. A nuclear emergency preparedness both region-wide and nation-wide should be designed. Garnering information from past examples of exposure to radiation may be undertaken to manage existing and future cases of exposure to radiation.

(vii) Ensuring the incorporation of nursing care as part of disaster preparedness, and developing national networks of nursing and midwifery personnel for disaster preparedness and for imparting primary health care.

Health and Military Collaborating Centre

The primary responsibility of this centre may be to effect collaboration among the military of the concerned countries, to identify the areas of health hazard and to forge cooperation in different emergency situations as well as develop guidelines and recommendations on contents, strategy, tactics, mechanism and managerial approaches for mutual and humanitarian intervention in times of emergency. The centre will investigate how humanitarian cooperation and military resources can be integrated into a comprehensive emergency preparedness and response strategy in general and for health issues in particular for disaster management.

Health and other Sectors Collaborating Centre

To effectively integrate health and disaster management, the health sector will interact with other sectors such as environment, education and housing. Partnership with international agencies (Rotary, Lions Club, etc.), non-governmental organisations and between the Government and the community will foster participatory relationships that would not only make the implementation of disaster management plans more effective, but would also enable local problem-solving (World Health Report, 1996).

The centre will seek to collaborate with fire fighting authorities for response in fire disasters and with industries for the implementation of eco-friendly practices and for the prevention of industrial disasters. In collaboration with the concerned sectors, it will prepare guidelines, manuals and training tools for field use and also use computerised teaching hypertext material and tele-medical technology for imparting distant learning and for distant management of disasters.

Cooperative Centres

The Plan of Action also entails the formation of cooperative centres in health. Such centres will have the following as its objectives:

(i) Maintain public health programmes for the benefit of millions of displaced people.
(ii) Develop techniques for rapidly assessing people's nutritional status and for conducting surveys to identify populations in need of medical assistance.
(iii) Adapt traditional epidemiological techniques and public health programmes to suit the realities of disaster situations, refugee camps and scattered disaster affected communities.
(iv) Assess the death and injury potential of probable disasters and to develop strategies for preventing or mitigating the impact of future disasters. As a result, a considerable body of useful knowledge and experience can be accumulated.

Education, Training and Research

The volunteers have to be selected and trained to be active partners in health development programmes in the different countries. Effectiveness of primary health centres, health education and participatory learning in health education and training are vital to handling emergencies. The plan will provide for strengthening national research capabilities by promoting and coordinating research on regional priority problems related to social and economic development and by aiding research designed to facilitate rapid application of existing and emerging scientific knowledge. Research will be on behavioural, socio-economic and cultural aspects of health and disease. The development of human resources for health, the assessment and rapid application of existing and emerging health related strategies and information technologies to health programmes, health policy analysis and research, health promotion research and identification of the constraints in the implementation of health programmes will be included in the plan.

National Policy on Disaster Management

The plan will emphasise upon the member countries to develop a national disaster management policy based on the assessment of national priorities. The policy will include the assessment of natural disaster risk;

preparedness plans/emergency plans at the national and local levels and the access to warning systems. The national plan will tend to involve four possible approaches, viz., disaster relief, technological adjustments, comprehensive damage reduction and multiple hazard management (David, 2000).

The Regional Health and Disaster Management Centre will thus provide the member countries, the policy and the strategic direction, or in other words, the commitment and the mechanism necessary for agreeing on goals, priorities, plans, terms of mutual intervention, the sharing of resources, etc. necessary for encouraging all members to cooperate in bringing about improved health management during disasters, and in transferring the knowledge, skills and information that make it possible to workout what changes are needed and to carry them out (World Health Forum, 1998).

REGIONAL HEALTH AND DISASTER INFORMATION NETWORK

A disaster connotes randomness and a breakdown of information systems resulting in misery, deaths, depression and anxiety. The setting of a Regional Health Disaster Information Network can establish links for the disaster information mosaic.

The Internet may the availed for this. Using the Internet, the various countries can form a mega network by joining together their respective networks for health information pertaining to disasters. Setting-up of a Regional Health and Disaster Information Network will greatly help a country. Some of the advantages are:

- Provides a system of information transfer, especially, when the communication system on ground has failed. The neighbouring states can reach out for immediate relief to the affected state, if such information of requirement of the logistics is made available to them.
- Information about lifeline services like food and water supply, shelter, medication, etc. can be immediately conveyed to government and non-government organisations and the whereabouts of the victims can be transmitted through such network, besides through T.V., radio, and newspapers.
- The provision for and the access to correct information will not only lessen the morbidity and mortality level, but will also prevent economic losses.
- Help in setting up web sites which can be used by outsiders (relatives, NGOs, the Government, international organisations, etc.) to gain access to information regarding the disaster area.
- Provides linkages to the media/press.

- Creates networks of NGOs, which will increase efficiency of information transfer and reduce duplication of effort.

Disaster Information Server

The Regional Health Disaster Information Network will also set up a disaster information server, which will greatly help in providing data and information on disasters. The information will be accessible to all, that is, individuals, institutions, and countries. The information will be made available in different languages. It will contain knowledge based on the past and present experiences of various countries and will have national, regional and global advisors whom the institutions and individuals can contact.

Government, non-government, academic, commercial and such other organisations can play an important role in the creation of such an information system. Countries can set up home pages containing information on disaster and emergency medicine. All this will greatly help in providing rescue and administering emergency medicine in the very early phase of a disaster, thus reducing morbidity and mortality.

With the setting up of the information server, facilities such as e-mail, Internet-based educational and training programmes, telemedicine, epidemiological surveillance, disease telemonitoring, a health and disaster directory, and mailing lists can be provided.

CONCLUSION

The South Asian countries face severe problems arising out of natural disasters every year. These countries do not have adequate mechanisms—institutional or technological—to mitigate the adverse effects of such disasters. As such, the population remains extremely vulnerable to the adverse consequences of disasters. In the light of the prevailing conditions in these countries, this chapter identifies the common areas for technical cooperation among the South Asian countries in health and disaster management. With partnership among them in the areas of health and emergency management, it will become possible for these countries to collaborate in health and disaster management endeavours, coordinate in implementing common public health strategies and thereby strengthen the public health infrastructure. This partnership can be formalised by setting up a Regional Health and Disaster Management Centre and a Regional Health Disaster Information Network.

It is suggested that the member countries develop guidelines and modalities for setting up the centre and the network and may form a Task Force. The Task Force will include representatives of the national Governments, non-governmental agencies, media, academia, armed forces, and international agencies.

REFERENCES

Chintamani, Aparna, *Cyber Health, Health Action*, Oct. 1999.

David, Alexander, *Natural Disasters,* Dec. 2000.

Health Situation in South East Asian Region, World Health Report, 1996.

Internet site (http://www.pitt.edu/~GHD Net/Concept.html).

Internet site (http://www.who.dk/ch/Cor/annex-g.htm).

Kabra, Kamal Nain, *The Economic Times*, Nov. 9, 1999.

Mukhopadhyay, Alok, *The Economic Times*, Nov. 9, 1999.

Network Wizards, *Internet Domain Survey*, Jan. 1995, Internet Site (http://www.nw.com/).

Noji, Eric K., *The Public Health Consequences of Disasters*, 1997.

World Disasters Report, IRCRCS, 1996.

World Health Forum, WHO, Vol. 19, No. 3, 1998.

World Health Report, 1948–1988, 40th Anniversary Volume, 1992.

11

Implications of Macro Level Development Planning on Disaster Risk Reduction

—Kumar Prasad Poudyal

INTRODUCTION

Nepal is seismically active. It has recorded 4,000 occurrences of earthquake from 1985 to 1992 out of which 1,200 occurred in Kathmandu alone. Recent study has predicted that in Kathmandu, in the near future there can be an earthquake of similar intensity or more than that of the 1934 earthquake, capable of causing immense loss of life and property. Every part of the country is vulnerable to earthquake. The earthquakes of 1934, 1980 and 1988 inflicted severe damages in Nepal. Earthquakes cause severe damage to dams, roads, irrigation systems and other physical infrastructure besides taking heavy toll on human lives and property.

The geomorphology of Nepal is very fragile. Constant tectonic actions of different degree along with varying and intense weather conditions has a marked adverse effect on the stability of earth surface and river courses. Heavy precipitation, high degree of wetness and steepness of watersheds and river channels contribute to the magnitude of floods, which render Nepal one of the most severe flood hazard zones in the world. Besides these, Nepal faces several natural disasters such as landslides, windstorms, hailstorms, lightning, avalanches, Glacier Lake Outburst Floods (GLOF), debris flow, and drought which occur almost every year causing immense damage to life and property. Fires and epidemics are the other forms of disaster, which plague the country. There has been substantive loss of life and property in the year 2000 by various types of natural disasters. The situation of natural disasters in the country and the losses caused by them in terms of human life and physical infrastructures since 1983 calls for effective disaster mitigation.

In the year 2000 alone, various forms of disasters caused 394 deaths and damages to the tune of US $ 151 million worth of property. The total number of human deaths caused by various forms of disaster since 1983 to date is around twenty thousand. Natural disasters not only caused heavy loss of human life and property but also adversely affected development processes.

Thus, disasters represent a very serious issue in Nepal. Natural disasters, of course, cannot be stopped from happening but efforts can be made to reduce their impact. Effective measures are to be initiated in this respect. Disaster preparedness, response, mitigation, rehabilitation and reconstruction are the key areas of our concern.

IMPLICATIONS OF DEVELOPMENT PLANNING ON DISASTER

Assisting the helpless and helping the needy is the basic philosophy of the Nepalese people. Since early days, local communities have been providing disaster victims whatever they could. Guthi, Dharma Bhakari are examples of community mobilisation schemes designed by the community for combating disasters of the unforeseen future. As the community shifted towards modernity and the society began to develop in complexity, the traditional systems of the society started breaking up, to be replaced by new institutions in the society. The State began to realise the importance of the necessity for at least a minimum arrangement for future emergencies, as it is the primary responsibility of the State to protect the life and property of its people.

The responsibility of the State for disaster management until a few years ago was very limited, the scope being providing just the immediate relief and rescue at the time of disasters. The Ministry of Home Affairs and its network agencies held the primarily responsibility in this regard. Besides, a few of the philanthropic institutions extended help at the time of calamities.

Nepal started planned development process since the beginning of its first five year Plan in the 50s. However, disaster related issues could not be properly addressed until the 7th plan period.

Disaster related issues have come to be taken more seriously since the 80s. The Natural Disaster Relief Act (NDRA), 1982, and its subsequent ammendments can be taken as a departure in the State policy aimed at addressing disaster related issues in a systematic way. The Act defined the term *Natural Disaster* to include earthquakes, fires, storms, floods, landslides, heavy rains, drought, famine, epidemics and similar natural disasters. It also takes into account industrial disasters such as accidents caused by explosions, poisoning and other kinds of catastrophies.

The Act also defines *Natural Disaster Relief Work* as any relief work to be carried out in the area affected or likely to be affected by natural disasters in order to remove the grief and inconvenience caused to the people, to rehabilitate the victims of the natural disaster, to protect lives and public as well as private properties, to control or at least minimise the impact of the natural disaster and to make advance preparation thereof.

The basic features of the Act may be considered as the pronouncement of the Government reflecting its strong commitment in addressing disaster related issues in a coordinated manner amongst the related agencies, both government and non-government (NGOs). It was designed with the Ministry of Home Affairs as the nodal agency for carrying out all the responsibility of the governance related to it.

The Act entrusts The Central Natural Disaster Relief Committee (CNDRC), the highest organ of disaster management, chaired by the Home Minister as the nodal agency to formulate policies and plans related to disaster management in coordination with various other key actors. There is also a provision of a District Natural Disaster Relief Committee (DNDRC), which is mainly responsible for carrying out the post disaster activities at the district level. It helps the Government to formulate national policies besides implementing programme on natural disaster preparedness and mitigation, immediate rescue and relief works, strategies for data collection and dissemination and mobilising funds and resources. It has its network throughout for comprehensively combating natural disasters. There are 75 districts in the country, each headed by a Chief District Officer, who is responsible for the response to any natural disaster.

The Eighth Plan (1992–1997) was of great significance with regard to the issue of disasters. During this period, major policy announcement related to disaster management was made. In 1996, a Plan of Action for Disaster Management in Nepal was introduced which can be considered as a comprehensive document in this sphere. Four core areas of disaster that is, preparedness, response, mitigation and rehabilitation and reconstruction are addressed in the plan of action. The plan of action represents the matrix indicating the priority items of the activities involved, the time schedules and the executing agencies. Major thrust of the plan of action are on the following:

Plan on Disaster Preparedness. It touches upon measures related to Institutional arrangements, legal framework, geological, hydrological, and meteorological hazard assessment, environmental engineering, infrastructure specific and hazard specific preparedness measures, measures to strengthen firefighting capability, training, establishment of disaster management information as well as stockpiling the emergency supply.

Response. Evacuation, search and rescue, communication and transport, temporary settlement, health, nutrition and sanitation form the major concerns in this area.

Rehabilitation. Mainly addresses sustainable reconstruction and income generating activities.

Mitigation. It involves forming a mitigation cell in every key organisation, formulating a building code and a national land use and land cover plan.

The subsequent periodical plan of Nepal emphasised upon executing the above designed plan of action in the country as a continuous programme. Numerous policy initiations were announced to implement these plans of action.

Realising the urgency of an effective organisation, the Government constituted a new department to deal with disasters towards the end of the Ninth Plan period (1997–2002). Presently, the Department of Narcotics Control and Disaster Management under the Ministry of Home Affairs is the key agency for immediate response during the time of natural disasters. Thus, the department carries out disaster-related activities before, after and during disasters. Basically, the department undertakes the following tasks:

- Carrying out rescue and relief works in disaster affected areas
- Working towards reducing loss of life and property
- Mitigating the sufferings of the people
- Controlling and mitigating natural disasters
- Developing hazard maps
- Rehabilitating the disaster victims in coordination with the related agencies
- Arranging funds and resources to help disaster victims
- Undertaking disaster awareness campaign by conducting training and circulating publications
- Coordinating with the non-governmental agencies
- Collecting and analysing disaster-related data and disseminating information.

As Nepal faces serious problems due to water induced disasters, the Government has not only constituted a separate department for dealing with the matter but also entrusted some technical functions to other agencies, such as, early warning function to the Department of Hydrology and Meteorology, seismological study to the Department of Mines and Geology and so on. The Nepalese Royal Army and the Nepal Police force have standby arrangements for rescue and relief work, which is also carried out by various hospitals and other medical institutions. The Department of Housing and Physical Planning is responsible for enforcing the building code. Thus, presently, disaster related issues are seen as very serious issues, which concern almost every walk of life in the country.

Disaster related issues were almost untouched in the macro level development process for long. Disasters were not seen as a separate entity even during the ninth plan period. However, a few of the issues relating to disaster management surfaced in the 1980s. The enactment of the Disaster Act of 1982 definitely paved the way for disaster related issues to be considered seriously, with both, the center as well as the districts, being made legally obliged to address issues related to disasters in their spheres of activities. As discussed above, quite a wide range of activities related to disaster have been entrusted to various agencies. The Central, District and Regional Disaster Relief Committees have been constituted. Regular meetings are being organised at every level.

The adoption of a national plan of action for disaster management in 1996 has a very serious implication on macro level development planning on disaster. Prior to the declaration of this plan, there was no systematic observation on the necessity of the preparedness aspect of the disaster. Now in every relevant field of disaster, the preparedness part is being focused on and the Government is increasingly feeling the need for integrating disasters and effecting development in various related spheres.

Institutional arrangement bas been made in a very proper manner to address the issue appropriately. The relief committees as well as other agencies mentioned above are the outcome of the plan, which may bring about a very positive change in the coming days.

FINANCIAL ARRANGEMENTS

Even though it came as a realisation of addressing the issue of disaster in the macro development planning since long, resources have not been properly allocated for employing in this area. This issue has not been treated as a priority until now. For effecting immediate relief, there is the provision of Central Disaster Relief Fund at the center and the District Disaster Relief Fund at the district level. The central fund is controlled and operated by the Department of Narcotics Control and Disaster Management whereas the district fund is controlled and operated by the District Disaster Relief Committee. The Central Natural Disaster Relief Committee (CNDRC) has formulated certain norms and standards for the relief assistance of disaster victims. The District Committee provides relief assistance to the affected population as per these norms. Sectoral ministries are entrusted the responsibility of further reconstruction and rehabilitation programmes.

AREAS OF IMPROVEMENT

Given below are the areas where improvement is needed. The kind of improvement needed is also discussed.

1. Disaster Management activities are quite comprehensive and need the cooperation and participation of multiple actors. The Natural Disaster Relief Act, 1982, has been introduced since long and now the time has come to review it for

accommodating newly emerging issues related to disaster. It is felt that the definition of disaster itself is necessary to be reviewed and changed. There is immediate necessity of adopting the bylaws. The power and functions delegated to the district committee is very insufficient and need to be thoroughly revised.

2. The National Action Plan on disaster, which was adopted in 1996, needs to be overviewed and appraised. A number of action strategies already included in the Action Plan have not yet been adopted and this calls for a serious and sincere approach from all those involved to effect timely execution of the plans.
3. The institutional infrastructure for disaster management need to be strengthened. The newly born institutions such as the Department of Disaster Management and other specialised as well as allied institutions are to be streamlined, strengthened, and made capable of handling the issues they address in an effective manner.
4. The basic philosophy of disaster management is to make people aware of and be prepared for possible disasters thereby reducing the risk of hazard. Nepal lacks far behind in this respect and much more needs to be done to this effect.
5. Nepal is a developing country, which invests a huge amount of money in infrastructure related activities such as building roads, bridges and canals and for other construction works Nepal has not yet introduced the disaster vulnerability assessment which is much necessary before initiating such projects. Thus it needs to impose compulsory procedures for risk assessment as part of every major project. Records related to this may be required to be submitted along with periodic auditing reports.
6. Disasters affect all the people of an area, but the most seriously affected people are the disabled, children and women. Special care needs to be taken for protecting such vulnerable sections of the society. The Poverty Alleviation Programme should focus on disaster related activities also so that the economic needs of the people below the poverty line are taken care of. Like wise, the school curriculum should include the basics of disaster mitigation strategies so that the young generation acquires the required awareness. Similarly, the programmes initiated for women empowerment may also focus on the role of women in disaster risk reduction.
7. Search, rescue and evacuation are the other areas which require thorough overhauling. It is necessary to identify the proper agencies to be involved and to equip them adequately. Supply of necessary equipment as well as the arrangement for appropriate training is highly desired. In Nepal, rescue-training programmes are generally insufficient.

8. Proper communication and transport at the time of emergency is vital to emergency response. Nepal has been facing serious challenges in this respect. As Nepal is a mountainous country lacking proper transport and communication facilities, rescue operations are seldom carried out effectively, the main hurdle being lack of transport facility. Consequently, a number of people loose their lives for want of timely attention.
9. Arranging for temporary shelters to the victims as well as providing them timely medical care is another challenge faced by the country.

The effectiveness of early warning systems is an essential prerequisite in managing disasters in an effective manner. Though Nepal has taken several measures to address this issue, there is ample scope to develop it further. Since Nepal is a developing country, it needs the help and support of other international agencies and organisations for strengthening its warning system for prompt action effective results.

CONCLUSION

Nepal as a developing country is beset by challenges on many fronts. In accordance to its priorities with regard to disaster management, various schemes have been initiated at the macro and at the micro levels. However, whatever has been done is not sufficient and much more remains to be done. The endeavor should be to incorporate disaster mitigation components in the development planning initiatives so that the disasters do not affect development much and also development should not lead to disasters.

12 Disaster Risk Reduction: A Preparedness Approach

— **Uma Medury**

INTRODUCTION

Disasters, which have been on a rise, throughout the world, are the products of human ignorance or neglect of certain key developmental issues. The most important and decisive factor associated with any disaster is its disruptive impact that creates pervasive uncertainty, suffering and trauma. Disasters cause severe loss to lives and property, destroy infrastructure and productive capacity, interrupt economic activity and create irreversible changes in a country's natural resource base. Disasters retard the development process of the society by diverting scarce resources towards relief, reconstruction and rehabilitation activities. 90 per cent of the natural disasters and 95 per cent of the total disaster-related deaths world wide occur in developing countries while the fraction of Gross National Product (GNP) lost is estimated to be twenty times greater than in industrialised countries (Munasinghe, 1996). South Asia is among the world's most vulnerable regions to both natural and man-made disasters. A tough mesh of poverty, rampant and unplanned urbanisation, chronic malnutrition and night-marish population densities have trapped its people (Sapir, 2000). The impact of such disasters assumes severity especially due to ineffective prevention, preparedness, mitigation and response mechanisms, in dealing with such situations.

Of late, the approach of reviewing and assessing the effects of disasters, not just in technical and scientific terms, but in a holistic manner, encompassing humanitarian, social and economic concerns, is assuming significance. The aftermath of a disaster witnesses:

- Huge population and livestock losses.
- The traumatising of individuals and families to the extent that they cannot cope with the usual issues of personal and

community concern nor with the pressing need of repairing the damages caused by the disaster.

- Destruction of community facilities such as roads and bridges, public buildings, schools, hospitals, etc.
- Loss of income to individuals, commercial enterprises, local governments, because of destruction to income generating assets.

In view of such dangers, there is growing awareness and concerted efforts on the part of governments, globally, towards focusing more attention on disaster prevention and preparedness. Over a period of time, especially during the International Decade for Natural Disaster Risk Reduction (IDNDR) period of 1990–2000, disaster risk reduction has become the key focus of the efforts of several governments world wide. The main objective of their initiatives has been towards building hazard resilient communities and protecting people from the threat of disasters.

Risks are inherent with all disasters. Due to the inevitability of certain hazards, risk elimination, many a times, becomes an impossible task. Hence the broader goal is to reduce the risks disasters and hazards pose to the society. One needs to proceed from protection against hazards to the reduction of risks. An important step towards disaster risk reduction begins with the identification of risks and the principal hazards a society is likely to encounter, and assessing the vulnerability of areas; social, economic infrastructure, community and livestock.

Risk reduction is a culture that has to be shared by a variety of actors. The challenge of reducing the impact of all kinds of disasters is permanent or is bound to remain on a sustained basis and requires a multi-disciplinary and multi-sectoral approach. Hence it has to be incorporated in the development plans and strategies and requires an institutional framework at the local, national, regional and global levels. The initiatives of civil and community level organisations need to be taken cognisance of to strengthen risk management. Disaster risk reduction needs to be looked at as a convergence of prevention, preparedness and mitigation measures. Here, we examine the concept of preparedness and its principles and strategies from the point of view of a long-term objective of ensuring sustained support to the vulnerable people.

DISASTER PREPAREDNESS—CONCEPT AND SIGNIFICANCE

Disasters involve two key elements, namely the event and the vulnerable people. Disaster takes place, with the onset of an event affecting the vulnerable people in such a manner that their lives are threatened, harm is caused to the established economy, social structures, etc., thereby undermining the community's ability to survive. Preparedness lies in the ability to predict, and wherever possible, prevent disasters, lessen their impact as well as facilitate response and handling of their consequences at various levels. This involves efforts directed primarily towards reducing the vulnerability of households and communities in disaster prone areas and improving their abilities and capacities to cope with the effects of disasters.

Disaster preparedness needs to take cognisance of different types of vulnerabilities that a country or a community is exposed to. Some of the critical causes for vulnerability are:

- Lack of access to resources (material/economic vulnerability)
- Disintegration of social patterns (social vulnerability)
- Degradation of the environment and inability to protect it (ecological vulnerability)
- Lack of strong national and local institutional structures (organisational vulnerability)
- Lack of access to information and knowledge (educational vulnerability)
- Lack of public awareness (attitudinal and motivational vulnerability)
- Limited access to political power and representation (political vulnerability)
- Certain beliefs and customs (cultural vulnerability)
- Weak buildings or weak individuals (physical vulnerability) (Aysan, 1993).

The need for disaster preparedness is assessed by considering three basic questions:

(i) Do people know they live in a potential disaster area?
(ii) Do people know the risks?
(iii) Do people know effective ways to reduce such risks (Alley, 1993)?

This can form the basis for any disaster preparedness activity.

Disaster preparedness needs to pay attention to a number of variables including, economic, social, political, technological, and psychological. Any preparedness strategy aims at:

- Developing, strengthening and making accessible to people, early warning systems and making people alert and responsive to emergencies.
- Reducing the vulnerability of households, communities, in disaster prone areas and improving their ability to cope with the effects of disasters.
- Strengthening institutional mechanisms, capacities of the Government at various levels, non-government organisations (NGOs), communities, and other organisations in disaster preparedness and post-disaster response.

- Establishing networks and linkages between public and private organisations, NGOs, community-based organisations, community and other key stake holders to foster improved coordination of preparedness efforts.

DISASTER PREPAREDNESS—ESSENTIALS

Following are some of the most essential factors for disaster preparedness:

1. Disaster preparedness should constitute one of the primary activities of the Government as it is the most effective way of reducing the impact of both small and localised as well as large-scale disasters. Its role needs to be complemented by the efforts of other organisations.
2. Disaster preparedness is to be pursued or looked upon as an effective link between emergency response, rehabilitation and development programmes.
3. The need for effective disaster preparedness is to be felt by government organisations, non-government organisations, private sector institutions and the community.
4. Strengthening of the organisational structures and institutional mechanisms at the national, state and local levels for effective disaster preparedness.
5. Promoting multi-organisational cooperation, coordination, networking and partnerships.
6. Identification of persons, communities, households, etc. which are most susceptible to risk and to disasters through assessment and analysis of risks, vulnerabilities and capacities as a basis for prioritising the location and focus of disaster management programmes.
7. Generation of awareness through public education and by encouraging vulnerable communities to take preventive and mitigative actions. Ensuring that the information from prediction and early warning systems is assessed, comprehended and acted upon by the local communities.
8. Strengthening the ability of vulnerable communities to encounter the consequences of disasters through community-based disaster preparedness strategies that harness the existing practices, skills and coping mechanisms.

With the scope and magnitude of disasters also increasing with the knowledge relating to it, it is critical that the resources, capacities and institutions be harnessed and strengthened to prepare for and respond to disasters. Any preparedness programme or effort needs to be systematic, bringing various aspects within its fold. Important principles that are pertinent to planning a disaster preparedness programme include:

- Perceptions should be studied and opportunities created for people to modify their perceptions where necessary.
- Create strategies to rouse the curiosity of the individual and to encourage a general desire for change.
- Individuals and communities should be helped to compare existing ways with proposed innovations, relate innovations to the basic needs and overcome barriers to acceptance.
- Adopt educational methods that have a heavy emphasis on community involvement and participation.
- Learning by doing and develop participation in various activities related to the identification of disaster preparedness needs.
- Group approval influences adoption of new behaviour patterns. In traditional societies, most of the decisions regarding new practices are multi-personal decisions, the role of the family and other social groups being the determining factors.
- Behaviour is motivated. Motivation is the inner drive that propels human beings towards attaining a desired goal.
- Disaster preparedness behaviour is concerned with changes in knowledge, attitude and behaviour and the ultimate goal is sustained disaster preparedness behaviour.
- Since different agencies work simultaneously at the community level, it is necessary for them to come to an understanding in order to avoid the dissemination of conflicting advice.
- Psychological factors are not the only determinants in behaviour. They combine and interact with physical, social and other factors (Alley, op.cit.).

Preparedness and prevention strategies have become the need of the hour replacing the earlier relief syndrome. We are witnessing a shift from post-disaster assistance to pre-disaster preparedness, from readiness to mitigation, from dependency to self-reliance, from individual aid to community services and from relief to rehabilitation. Disaster preparedness involves a set of actions, which include:

- Preparation of a comprehensive disaster preparedness plan outlining the measures to be taken by the key stakeholders including government departments, before, during and after disaster strikes.
- Meteorological studies and dissemination of warnings well ahead of the event thereby enabling people to move to safer places. In India this has become more effective with the introduction of satellite based meteorological observations.
- Dissemination of information to the people through effective Information Education Communication (IEC) activities.
- Evacuation of people to safer places.
- On the onset of a disaster, measures to make available, emergency shelters, medical, food and first aid services, adequate stock of food and medicines and security arrangements.

- The post-disaster activities include damage assessment and immediate financial assistance, restoration of transport, power, water supply, tele-communication, etc.

Preparedness is not confined to or does not center on short-term measures which are taken during the pre-disaster phase, but extends to the period during the disaster and also, to the post-disaster period. Hence, it requires an improved understanding of the various complexities and risks involved.

An effective disaster preparedness programme must be:

- Bi-participatory in design
- Community-specific
- Based on an assessment of the information required
- Integrated with the existing disaster warning and response systems
- Established as an on-going process
- Include as a priority the most vulnerable people
- Include information on prevention, mitigation and long-term recovery (Alley, op.cit.).

Any disaster preparedness programme will be successful if it is done with the people rather than for the people. Globally, there is a shift in the strategy of development being attained through externally evolved solutions. Of late, the various ways in which people at risk manage and transform their lives in response to stresses and opportunities is a matter of global importance and general interest. Governments, no doubt hold the primary responsibility for protecting the people from risks and disasters. However, local communities and the civil society prone to hazards are often key initiators of important risk and disaster prevention actions. They must work through partnerships and networks to realise the vision of disaster resilience and to develop self-reliance in preparedness. Strengthened preparedness and response capabilities at the grass-root level results in better service delivery. Self-reliance is something that would appear contrary to the dependency culture. Preparedness will be the most effective if the community can be enabled to participate in decisions that affect their lives by cooperating and pooling their strengths and capacities.

DISASTER PREPAREDNESS: KEY STRATEGIES

There is a clear shift from superimposed efforts to attain development or tackle any crisis that is directed towards the people at risk. Now, there are sustained efforts to finding alternatives to crisis and stress. The realisation is gaining momentum that the global capacity to prepare for and respond to disasters needs to be strengthened. We now discuss under separate heads, some of the strategies that could be evolved in this direction.

Application of Information Technology in Disaster Preparedness

Advances in information technology now provide a plethora of resources for decision makers. This information needs to be tailored for specific needs and delivered in a timely manner. For example, the initiatives made in the state of Maharashtra in the aftermath of Latur earthquake in 1993, is quite remarkable. A comprehensive Disaster Management Information System (DMIS) implemented by the Maharashtra Remote Sensing Applications Centre (MRSAC), Nagpur, provides data related to various areas like climate, natural resources, socio-economic and demographic profile of the population, etc. which help greatly the district administration in enhancing the level of preparedness.

Planning for Building Disaster Resilient Communities

One of the key objectives of sustainable development is to enable the future generations to meet their own needs. When applied to managing disasters, it envisages reducing the vulnerability of people to risks through poverty reduction, promotion of economic activities, improved living conditions and thereby building communities resilient to disasters. Resilient communities may bend under the extreme stresses of natural hazards, but they do not break. They are consciously constructed to be strong and flexible rather than brittle and fragile (Godschalk, 1999). The United Nations Conference on Environment and Development at Rio in 1992 stressed the need for enhancement of capacity building for indigenous communities based on the adaptation and exchange of traditional experience, knowledge and resource-management practices in order to ensure their sustainable development (United Nations, 1992). It is increasingly being realised that the self-confidence and resilience of the local communities can be strengthened through harnessing of traditional knowledge, practices and values and integrating them with developmental activities.

The need of the hour is community empowerment which enables the people assume primary roles in setting goals, establishing programmes, developing priorities and managing recovery in times of crises. In the Central American Island of Honduras in 1998, a hurricane with sustained winds of 180 miles an hour resulted in severe flooding of many areas. Thousands of acres of land were swept away; a thousand miles of roads and more than a hundred bridges were rendered impassable. Relief and assistance came from all quarters. As the government began sorting out the offers of aid, drawing up project proposals and plans, the people, being accustomed to being self-reliant began putting back communities together on their own. It was expressed that, 'we want the government to lend us bulldozers to clear the streets. We will do the rest. We don't have money to rent machines like that, but we do have lots of strong arms and a willing spirit' (National Geographic, 1999).

The task of building disaster resilient communities is quite daunting. It needs commitment on the part of policy makers to plan and implement suitable programmes and to strengthen the capacities of communities to enable them to face the crisis. There is a need for suitable policy changes, system reform and appropriate practice development. The entire approach towards disasters has to be pro-active rather than reactive, effected through suitable policy and legislative framework with institutional and organisational support. Wherever possible, the best locally suitable preparedness practices need to be adopted. In the Saurashtra district of Gujarat, around 10,000 check dams have been built in response to the State Government's, Build Your Own Dam (BYOD) scheme. Under this project, the Government bears 60 per cent of the total cost of the dam, while the villagers shoulder the remaining costs borne as voluntary labour.

Jal Biradaris (Water Societies) are being formed in and around Udaipur laying special emphasis on utilising locally available water resources for the benefit of the local population through building up water shed structures and utilising the local resources for preventing the rain water from going waste.

In many states in India, watershed development activities are taking place. It essentially means storing and saving water in village ponds, building bunds, repairing old water harvesting structures and desilting the drains. Four years ago Madhya Pradesh initiated the largest watershed development project in the world. Gujarat is implementing the Sardar Patel Participatory Water Conservation Project. Earlier this year, Andhra Pradesh started the Neeru Meeru (water and you) Project, under which 5,260 watershed committees have already been formed. The biggest achievement of the watershed development programmes is that they empower the villagers for obviating the need for them to wait for the State and Central Governments to rescue them from drought that affects them year to year. Besides, such programmes represent the first step towards sustained economic growth (Business World, 1 January 2001).

The goal of building disaster resilient communities needs to be shared and realised by all key actors involved in the process. It should not be conceived of as the responsibility of the government alone, but as a collective effort.

Promotion of Sustainable Livelihood Strategies

Livelihood becomes a big question mark in the aftermath of disasters, especially for those who do not have a regular source of livelihood even in normal conditions. Sustaining livelihood assumes importance especially in the context of natural disasters.

Preparedness with reference to evolving any sustainable livelihood programme entails:

- Analysing the existing socio-economic conditions prevailing in the area
- Surveying the availability of infrastructural facilities
- Examining the occupational pattern
- Assessing the awareness levels of the community
- Reviewing the soil conditions, agricultural techniques and other environmental factors
- Gauging the mindset of the people—their culture, beliefs, myths, attitudes, and traditional wisdom.

A livelihood is sustainable when it can cope with and recover from stress and shocks, maintain its capability and assets and provide sustainable livelihood opportunities for the next generation (Sanderson, 2001). In any preparedness programme, generally, inadequate attention is paid towards developing sustained livelihood strategies, which provides the required self-sustenance to the community.

Any preparedness effort in this direction needs to focus on:

- Building basic, foundational and infrastructural support on which the community can rely in situations of disasters.
- Provision of governmental support at district, block and village levels in terms of physical and financial sources.
- Encouraging the formation of self-help groups (SHGs). For example, the efforts of an NGO PEDO (People's Education and Development Organisation) in Rajasthan, which seeks non-biased involvement of target community rather than doling out subsidies, are quite noteworthy. Micro credit was selected as a means for women's empowerment with the aim to reduce poverty in the area, discourage subsidy-dependent mentality of the local society, promote self-employment and development based activities, provide timely and need-based credit and encourage self-confidence of the poor. The entire saving and credit operation was left to the SHGs. They have reached out to 175 SHGs covering nearly 5,000 women, who were earlier considered 'non-productive and non-asset worthy'. Today they are regarded as agents of change and economic development (The Hindu, 2000).
- Creation of a chain of small-scale or cottage industries in different areas that could provide livelihood. These could be dispersed in different zones in the region, say, the non-disaster area zone, the less disaster intensity prone area and while a tertiary unit may be located in the disaster prone area.
- Retention and utilisation of distinct cultural patterns, traditional knowledge and wisdom of the community. For example, the Kutch craft of Gujarat, which uses colours, designs and stitches rooted in specific histories and life patterns, which needs to be preserved (The Hindu, 2002).

Sustainable livelihood strategies have to be developed in the larger context of disaster preparedness and mitigation. Merely making provision for employment generation, asset creation and resource mobilisation, cannot fully address the problem of livelihood. A long-drawn, well-analysed and thoroughly contemplated holistic approach towards sustainable livelihoods can help in formulating preparedness plans, which take into purview not just the economic aspect of livelihood but also the social and psychological aspects. A sustainable livelihood preparedness plan would include protection, conservation and better management of social and economic sub-systems in such a way that livelihood strategies can hold themselves intact even in the aftermath of disasters. These need to be formulated and executed in such a way that their longevity is ensured by their sustenance features.

Strengthening of Disaster Preparedness Capacities

The strengthening of the capacities of vulnerable people, especially at the local level requires a multi-pronged strategy through the involvement of various key actors, NGOs, community-based organisations, self-help groups, local government representatives and so on. The major areas in which capacities need to be built include:

- Basic knowledge pertaining to types and phases of disasters
- Vulnerability assessment, risk analysis and mapping
- Preparation of community resource maps, risk maps, disaster response plans
- Methods and techniques of community resource mobilisation
- Identifying and working with the vulnerable groups
- Deployment and utilisation of various types of resources during disasters
- Evolving ways for cooperation and collaboration in the preparedness and recovery process
- Effective leadership in the implementation of preparedness, response and rehabilitation programmes.

Any capacity-building programme needs to build up appropriate levels of knowledge, skills and attitudes amongst the concerned key players through appropriate Information Education Communication (IEC) activities.

Partnership for Disaster Preparedness

Partnerships involving public and private organisations, community-based organisations, NGOs and the community can be effective in promoting better coordination of disaster preparedness as well as response activities.

In the aftermath of the earthquake in Maharashtra in 1993, the Maharashtra Emergency Earthquake Rehabilitation Programme (MEERP) was started. Under this, the *Samvad Sahayaks* or village communication assistants and women's groups acted as an interface between the people and the administration. They initiated efforts to strengthen the capabilities of all those involved in the programme such as women's groups, representatives of gram panchayats, bankers, house owners and the community through information, education and training. The community-based reconstruction programme was introduced as a key to partnership between communities and government. The programme has the following components:

- Information dissemination
- Education and skills on earthquake reconstruction
- Motivating house owners to build earthquake resistant houses
- Forming collectives for construction management
- Settlement planning exercises and resource mapping
- Problem-solving of individual beneficiaries (SPARC-SSP, 1998)

The UNDP, in the aftermath of the earthquake in Gujarat, during May to August 2001, in the Patan and Surendra Nagar districts, in association with the All India Disaster Mitigation Institute (Ahmedabad), sought to improve the disaster preparedness through participatory mapping and action planning cycles, creating awareness on family disaster preparedness, cyclone preparedness, community-based disaster management, etc. (*refer* undpquakerehab.org).

The Delhi Government recently had taken the initiative to facilitate direct participation of citizens through Resident Welfare Associations (RWAs) in managing the Disaster Management Centres set up in the capital. About nine hundred citizen groups are said to have come forward to join this citizen-government partnership programme (The Hindu, 2002).

Such partnerships at appropriate levels of government provide a forum for the concerted efforts of the concerned groups to reduce disaster risks as well as associated costs. National and international partnership networks can facilitate rapid access to specialist expertise, which, in turn will help build and strengthen capacities for disaster risk reduction, access to information, communication and control of resources.

CONCLUSION

Disaster risk reduction is a multi-faceted activity that involves identification of risks, readiness to predict as well as wherever possible, prevent disasters, reduce their impact and respond to and contain their consequences. Hence any strategy towards attaining this needs to be placed as an integral part of the ongoing development planning and investment policy. Preparedness, a key step towards disaster risk reduction, is not confined to short-term measures that are resorted to only during the pre-disaster phase. It encompasses appropriate disaster legislation, operational planning, education and training, insurance and other long-term response measures and strategies. It is only through preparedness that people can be provided the needed leadership, readiness, support and assistance in pre-empting the consequences of disasters, mitigating the suffering they bring forth, organising effective response and launching appropriate recovery measures. A systematic campaign of disaster risk reduction needs to integrate the social, economic and psychological dimensions of mitigating the effects of disasters. The damage caused by disasters can be minimised by a change in the perception and behaviour of all sections of the society, striving towards placing safety in planning and development as a priority, and evolving towards more of a partnership rather than a top-down system in implementing disaster risk reduction measures. Disaster risk reduction shall be meaningful and effective only if it is accompanied by strong political will and implementation capability supported by adequate policy and system reform.

It is being realised that disasters, since they concern the individual, state, country and region, need to be responded to through a multi-faceted approach.

REFERENCES

Alley, E.E., *Combating the Vulnerability of Communities*, in P.A. Merriman and C.W.A. Browitt (Eds.), *Natural Disasters: Protecting Vulnerable Communities*, Thomas Telford, London, 1993.

Aysan, Y.F., *Vulnerability Assessment* (*Keynote Paper*), in P.A. Merriman and C.W.A. Browitt (Eds.), *Natural Disasters: Protecting Vulnerable Communities*, Thomas Telford, London, 1993.

Business World, 1 January 2001.

Godschalk, David R., et al., *Natural Hazard Mitigation Recasting Disaster Policy and Planning*, Island Press, Washington D.C., 1999.

Medury, Uma, *Coping with Disasters: A Community-based Approach,* in Pardeep Sahni, Alka Dhameja, Uma Medury (Eds.), *Disaster Mitigation Experiences and Reflections*, Prentice-Hall of India, New Delhi, 2001.

Menon, Vinod Chandra N., *Maharashtra Disaster Management Plan—An Overview* in (www.gisdevelopment.net).

Munasinghe Mohan, *Natural Disasters and Sustainable Development: Linkage and policy options*, World Bank, Washington, 1996.

National Geographic, *After the Deluge*, pp. 110–129, Vol. 196, No. 5, Nov. 1999.

Sanderson David, *Urban Household Livelihood Security,* in Pardeep Sahni, Alka Dhameja, Uma Medury (Eds.), *Disaster Mitigation Experiences and Reflections*, Prentice-Hall of India, New Delhi, 2001.

Sapir, Deborati Guha, *Disasters in South Asia* in Parasuraman and Unnikrishnan P.V. (Eds.), *India Disasters Report Towards a Policy Initiative*, Oxford University Press, New Delhi, 2000.

Society for Promotion of Area Resource Centres and Swayam Shikshan Prayog, *Rebuilding Communities: Participation of Women's Groups in the Earthquake Rehabilitation Programme*, 1998.

The Hindu, 27 January 2002.

The Hindu, 5 March 2000.

The Hindu, 5 May 2002.

United Nations, *Report of the United Nations Conference on Environment and Development*, Rio de Janeiro, 1992.

United Nations, *World Conference on Natural Disaster Reduction*, New York and Geneva, 1995.

Websites

(www.undpquakerehab.org/xcoordination.htm).

(www.oasorg/en/edmp).

(www.gisdevelopment.net).

13

Community Capacity Building for Risk Reduction in South Asia

— **Zubair Murshed**

INTRODUCTION

Not too long ago, disasters were viewed as one-off events and responded by governments and relief agencies without taking into account the social and economic causes and implications of these events. Disasters were considered as emergencies and handling them was the responsibility of the fire brigade, rescue workers, hospitals, Red Cross and military forces. However, over the past few years there has been an increased realisation amongst academia and the disaster response agencies that disasters and their impact are caused by a range of complex socio-economic and political processes underway in a society. From this perspective, the rise in disasters and their consequences is a result of a rise in people's vulnerability, induced by human-determined paths of development. Therefore the priority shifts to reducing people's vulnerability and managing risk that can lead to disasters.

Consequently, there has been a paradigm shift from a traditional relief and disaster preparedness focus where communities are considered 'victims' and 'beneficiaries' of assistance from outside experts, towards a more holistic and long-term approach incorporating vulnerability reduction and risk management concerns as part of the development planning process. This comprehensive approach recognises that the complex relationships and structures of society determine why certain groups of people are more vulnerable to disasters.

During this paradigm shift there has been growing realisation that disaster mitigation is most effective at the community level. The approach recognises that when disasters occur, the community people themselves are the first responders. It is always the immediate relatives, neighbours and other community people who come to help disaster victims and their families. The help from external agencies including the Government, NGOs, political parties and private sector comes later on.

The vulnerable communities and groups are to be the main actors in disaster management. They have the knowledge about their locality, history of the disasters in their place, and how vulnerability to disasters has changed over time. They have the right to participate in decisions that affect their lives directly. People's participation is basic because, safety, stability of livelihoods, well-being and disaster management is, in the last analysis, their own concern rather than that of 'experts' such as government departments, scientists, aid agencies, NGOs and specialists.

Therefore it is important that the capacities of the disaster prone communities are recognised and integrated with the externally designed risk reduction programmes and projects. Using top-down interventions alone for disaster mitigation is insufficient because such interventions often pay little attention to addressing community dynamics, perceptions and needs, may ignore the potential of local resources and capacities and may, in some cases, even increase people's vulnerability. Moreover, local communities are often either unaware of these formal disaster mitigation interventions or they find them inappropriate due to a lack of recognition of the community's vulnerabilities and capacities.

The paradigm shift during the past three decades motivated a range of organisations to design and implement community-based disaster management projects and programmes in the field. Mostly such initiatives have been designed by the local, national and international NGOs, so far. However, government agencies have woken up to the importance of community participation in disaster management. The upcoming programmes mounted by various government agencies in disaster prone regions are reflective of this change in the direction of government thinking.

The Bangladesh Red Cross, in collaboration with IFRC was amongst the pioneers to establish a highly successful community-based early warning system in the cyclone affected districts of the country. The initiatives of Intermediate Technology Development Group (ITDG), South Asia, in the drought prone parts of Sri Lanka for promoting rain water harvesting technology, the project on Community Awareness on Disaster Preparedness by the Indira Gandhi National Open University, New Delhi, India, in the five most disaster prone States of India, the work on relief auditing, public awareness and community training conducted by the Disaster Mitigation Institute jointly with flood and drought prone communities in Gujarat and Rajasthan in India, the programmes of CARE Bangladesh and India, the activities of Thardeep Rural Development Programme (TRDP) for the drought prone communities of Tharparker, Pakistan and the projects of the Nepal Society for Earthquake Technology (NSET) and ITDG for flood and earthquake prone people are but few examples in this regard. Various exemplary cases could be observed in neighboring Southeast Asia. The work of Citizen's Disaster Response Center (CDRC) and Centre for Disaster Preparedness (CDP) in Philippines requires particular mention. PACT, Cambodia and IFRC, Vietnam are also cases for consideration.

Despite such innovative and exemplary initiatives by various institutions, there still is much to be done considering the vast geographical territories, population, and infrastructure available in South Asian countries. To this effect, the ADPC and Duryog Nivaran jointly designed the course on Community-Based Disaster Management (CBDM) in 1997 to promote sharing of systemised knowledge, information and experiences amongst the participants and the resource persons. So far, nine international courses have been conducted. Under the livelihood Options for Disaster Risk Reduction Project, the ADPC and ITDG have taken initiative to institutionalise the course at the national level in India and Sri Lanka in order to reach the wider audience.

The presentation in this chapter draws heavily from the curriculum of ADPC's CBDM course and the work of other agencies mentioned above. It emphasises the importance of sharing strategies for community capacity building among the government officials and others who are concerned with initiatives for disaster risk reduction.

Features of Community-based Disaster Management Approach

Some key features of the community-based approach to disaster management and risk reduction are as below:

Vulnerability causes disasters. It looks at disasters as a question of people's vulnerability. It defines a disaster-event as *the manifestation of an interaction between an extreme physical or natural phenomena and a vulnerable human group.* A disaster occurs when a hazard strikes a vulnerable community or group whose inherent capacity to withstand or cope with its adverse effects and impacts is not enough. The community-based approach intends to address the root causes of people's vulnerabilities and to transform the structures that generate inequity and underdevelopment.

Indigenous knowledge and local capacities. It recognises people's existing capacities and aims to strengthen these. Despite people's vulnerability, they still have capacities and are not completely helpless in times of disaster. It is the people themselves who initiate the necessary steps to survive even before aid givers arrive at a disaster scene. They have adapted coping strategies based on previous experiences in dealing with disasters. Although disasters may deprive people physically of food, shelter, crops, tools, they always will have some resources left. These may be some recoverable goods or, as is often the case, the skills and attitudes they possess. It can also be the presence of a strong community organisation or an elaborate family support system, which allows them ample leeway to cope with the disaster event. The modern way of living with its emphasis on scientific and technological fixes could not appreciate the importance of indigenous knowledge and skills. It is imperative to build on these capacities and cumulatively increase them, as they are the point of departure for developmental disaster response.

People's empowerment. It accepts that disaster management is primarily the concern of the victims. Therefore their participation and empowering them are essential. People must be involved and should be encouraged to participate fully in all aspects of the process to bring about development, including disaster management. People oriented disaster management means developing peoples' potential and capacities so that could cope with disasters on their own.

Context specificity. Instead of applying a standard set of responses for all communities, the community-based approach opts for specific actions depending on the situation and the community in question. It largely depends on the kind of hazards that the community has to cope up with and what preparedness and mitigation measures they need to undertake. These can be responses with immediate benefits such as an effective warning system, an evacuation plan, diversification of crops or long-term mitigation efforts like tree planting, community alliance building and advocacy for resettlement and land rights. This process can take several years and some of the root causes might not even be eliminated in a lifetime. It may require the effort of several generations.

No quick fixes. The community-based disaster response framework considers capacity building as a long-term process, and this is not always compatible with the project-based funding principles of most donors. Donors release funds according to a particular disaster stage for administrative or political convenience. The community's responses do not follow these disaster stages; vulnerable communities continuously develop coping mechanisms and survival strategies before, during or after disasters strike. They address vulnerabilities and not the disaster per se. This sometimes makes it difficult to provide financial support for the community's needs since extending the required support does not match the donor's framework or policies.

Puts premium on organising communities. Organisational and social support networks are the most crucial coping mechanisms people rely on in times of crisis and emergency. They make for not only sharing food, farm animals, farm tools or other resources, but also for coordinating decision-making and managing community-wide activities, organising evacuations and monitoring emergency response. Supporting communities in forming disaster response organisations strengthens the organisational capacities of communities at risk to enable them to take action towards reducing their vulnerabilities. This enables them to effectively participate in the activities during, after and before disasters. The tasks of community organisation include preparing and sharing the Counter Disaster Plan with all community members, monitoring disaster threats, networking and coordinating with local government officials, issuing warnings, managing community-wide evacuation, search and rescue operations besides planning and conducting relief delivery operations with aid agencies and implementing mitigation and rehabilitation activities.

STRATEGIES FOR COMMUNITY CAPACITY BUILDING

The process of building up the capacity of disaster prone communities is long-term and multi-dimensional in nature and requires innovative approaches and motivated individuals. It could be frustrating at the worst and exhilarating if the results are positive. Few core areas such as planning (participatory), community organisation, community training, public awareness, community risk assessment and community early warning system could be described as people-centered. Otherwise there are no standard fixes for building the capacity of a certain community upon, the process being context specific. The things that worked in one context may not be relevant to another. Solutions cannot be replicated, approach can be. Commitment and patience are the key equipments for someone who intends to explore community-based disaster management. The sharing of the different strategies discussed herein can serve only to give a few ideas to start with. It does not and cannot cover all the challenges, difficulties and problems, which could arise in the field.

Community Risk Assessment

The process of community capacity building for disaster risk reduction starts with community-based risk assessment. Assessment of prevalent hazards and the vulnerabilities and capacities of the community is done with the active involvement of local people. By applying participatory techniques, people are engaged in a process of the analysis of the situation and the risks they face. The process of community risk assessment aims empowerment of the people by developing their analytical, decision-making and problem-solving skills and by encouraging local level action. The prime objective of the application of Participatory Risk Assessment (PRA) tools in community risk assessment is not data-collection, but empowerment of people. Once people learn to analyse the situation and identify the causes of their problems, they could also find solutions.

Disaster Management Orientation

Organisation of disaster management orientation sessions involving hazard prone communities can help to build contact. The orientation would increase people's awareness on disasters and on how disaster responses can be delivered in a way that is people-centered and development-oriented. The topics could include community disaster situation, clarification of basic concepts like hazard, disaster, risk, vulnerability, capacity, etc., besides those emphasising the importance of community participation in disaster management and disaster response strategies and activities.

Disaster Preparedness Training

Disaster preparedness training can also be used as an entry point to organise community disaster response organisations in high-risk areas. The training could include subjects like review of disaster management concepts and community-based approach, assessment of vulnerability and capacity assessment of the community and identification of preparedness and mitigation measures. The disaster preparedness training could initiate two major processes in the community; the formation of a community disaster response organisation and the formulation of a Counter Disaster Plan.

Community Disaster Response Organisation (CO)

Disaster Preparedness Training makes people aware of the benefits of organised action. Organising a community into a Disaster Response Organisation requires time that exceeds the scope of one disaster preparedness training. So, follow up support is necessary like with the other inputs for organisational development and leadership training. The range of the organisational expression of a Community Disaster Response Organisation varies from being a disaster response committee to being a Community Organisation, which may later take on development issues and concerns. The aim of a CO is to enable communities to become better prepared for impending disasters and to become resilient in the long-term. In this way, communities will be able to undertake actions to address their vulnerabilities to disasters. These actions are laid down in the community's Counter Disaster Plan. The following are the main activities of a functional CO:

- Design and share Counter Disaster Plan with all community members
- Monitor disaster threats, conduct drills, and draw lessons to improve the counter disaster plan
- Networking and coordination with the local government, political and religious parties, NGOs and other groups
- Mobilise resources for disaster management from outside the community
- Issue warning and manage evacuation
- Involve community members in search and rescue
- Conduct damage, needs and capacity assessment (DNCA) and report to disaster agencies for assistance
- Coordinate, plan, and implement relief delivery operations with aid agencies
- Advocacy/lobbying regarding disaster response related issues
- Facilitate the identification, selection, and implementation of disaster response activities with the community.

Counter Disaster Planning

A Counter Disaster Plan or a community-level contingency plan helps to consolidate the community's efforts to prepare for coming hazards. The plan provides guidelines for operation and clarifies roles and responsibilities before, during and after disasters happen. Community Organisation forms the plan on the basis of results of the Hazard, Vulnerability and Capacity Assessment. This plan contains the following elements:

- Identification of pre, during and post disaster community requirements to address vulnerabilities
- Identification of available resources and capacities the community can build on or has to mobilise from outside (boats, vehicles, communication equipment, evacuation site, volunteers, etc.)
- The organisational structure of the community organisation
- Roles and responsibilities of the CO leaders and committees
- Policies, decision-making mechanisms and operational guidelines
- Warning system
- Evacuation and evacuation center management plan
- Mitigation measures like reinforcement of houses, improving drainage, additional maintenance on footbridges, or crop diversification can be included in the counter disaster plan.

The Counter Disaster Plan also contains the following particularities or annexes:

- Timetable to implement the plan or to conduct drills to test the efficiency and effectiveness of the plan
- Master-list of community members (names, family composition, age, gender)
- List of volunteer teams
- Hazard, Vulnerability, Capacity Assessment summary and hazard map
- Directory of key-people, NGOs, local officials, religious groups than can be contacted
- Organisational structure and functions and responsibilities of all committee.

Design a Community Specific Warning System

Warning is a positive action-oriented stimulus to alert people about an impending hazardous event or circumstances in their location, which may threaten their safety and security, and which requires an adaptive response. Very often, official warning signals from the national level do not reach communities or if they do, they are rather late or they are not trusted or understood by the local people. A community-based warning system will contribute to people's safety. The hazard assessment provides information that the community uses to design a warning system. The community-based warning system helps to avoid panic and to enable villagers to prepare for evacuation in an orderly manner. The vulnerability assessment identifies the most vulnerable groups in the community. Warning should be given first to these groups. Further, the medium to relay the warning message should be appropriate to these groups. The capacity assessment identifies the resources and skills that can be used in warning the community: e.g. existing organisational structure, means to relay warning (drums, gongs, firecrackers, runners, megaphone, etc.), available communication equipment, coordination with local disaster coordinating councils, etc. Cyclone Preparedness Programme of the Bangladesh Red Cross Society provides the best example of a successful community-based early warning system, which is now functional for more than 30 years.

Design an Evacuation Plan and Conduct Drills

Evacuation is an organised movement of people from an area at risk to a safe place. Not all hazards require evacuation to protect life and properties. Evacuation is needed in case of floods, typhoons, fire, volcanic eruption, landslide, earthquake, or armed conflict. Evacuation does not happen all at once but is designed in stages following the alert levels of the warning system. An effective evacuation may have the following stages: (1) Warning about impending hazard, meaning that people should prepare for evacuation; (2) Order to move to assembly/pick-up points; (3) Actual evacuation from area at risk to safe location; (4) Stay at evacuation centre; (5) Return to former or relocate to a new place. Steps involved in designing of an effective community-based evacuation plan are as below:

1. Identify a safe place for the evacuation centre
2. Identify shortest and safest route
3. Identify and prepare alternative routes
4. Identify pick-up points or assembly points for people
5. Place 'road signs' along evacuation routes
6. Prepare master list of evacuees an check at each pick-up point if group is complete
7. Prepare evacuation schedules and groupings in case transportation will be used
8. Set provisions and plans evacuation of animals and other properties of evacuees
9. Organise an Evacuation Management Plan and form committees among community members
10. Identify and prepare requirements during evacuation (transport, gasoline, food, water, medicine, road signs, communication system, etc.)

Emergency Response Training

Different strategies can be applied to increase the preparedness capacity of communities at risk. The strategy depends on the kind of hazard threatening the community, the accessibility of the community, and on the existing organisational capacity of CO. Training on Emergency Response is particularly essential for communities that encounter recurring hazards, are exposed to multiple hazards, become inaccessible to outsiders during the emergency period, or do not receive much outside assistance but have an existing CO. The Emergency Response Training can be a separate training or be integrated in the Disaster Preparedness Training. The training could cover two major topics: (1) *Damage, Needs, Capacity Assessment*: what is the purpose of conducting DNCA? what kind of data have to be collected, methods for data gathering, guidelines for reporting information to aid agencies? and (2) *Planning and conduct of relief delivery operation in the community or evacuation centre*: prepare master-list, identify relief items, design distribution and accountability (tickets/coupon) system, plan schedule, location and physical arrangement for relief operation, mobilise volunteers to secure orderly distribution, make a programme for the relief operation, tasking.

The Emergency Response Training will enable community members to respond systematically to the aftermath of an emergency situation and to identify the nature of assistance (what and when) they need from outside. From the community's perspective, the preparedness training helps them in timely evacuation to a safer place. The emergency response training helps them to assess the impact of the disaster in their community, to identify the kind of aid they need, and where to get it. This could speed-up the delivery of emergency relief tremendously. Also, the appropriateness of relief is guaranteed, as well, because needs assessment is more accurate, regularly updated, and faster than when outsiders have to come. This is especially true for areas that have become inaccessible due to the disaster.

Mitigation

Training of masons, farmers, village leaders and other professionals on mitigation. The training of different local professionals could play a very crucial role in mitigation of disaster risks. The training of masons in flood and earthquake prone areas on building disaster resistant houses, for example, could help reduce the loss of lives and property enormously with minor cost implications. In the same way, the training of masons and household members on construction of water harvesting tanks for improved water conservation and water purification could reduce the periods of water shortage. Imparting awareness to farmers on drought or flood resistant crops is another instance in this line. The work of ITDG South Asia in Sri Lanka and that of Thardeep Rural Development Programme (TRDP) in Tharparker, Pakistan, are examples in this regard.

Diversification of Crops According to Different Planting Season

Many disaster prone communities depend on only one type of crop, which is harvested only once a year. Due to the occurrence of hazards this crop is damaged and as a result, they regularly face acute food shortages and people are forced to depend on seasonal out-migration. It is therefore important to identify the potential to harvest other crops in different months of the year. It may involve the assessment of suitability of land for cultivation of other crops, availability of water and orientation and training of farmers to harvest new crops. This can help mitigate the effects of failure of the main crop on farming families due to prevalent hazards; e.g. drought or flood. If the main crop fails, the farmers still can harvest products during the two other seasons. As a result, the number of months people face food shortage could be reduced. Interesting work is being done in Philippines and Vietnam by the agriculture ministries and extensions services.

Diversification of Livelihood Sources

Diversification of livelihood sources can mitigate the effect of disruption of the people's main livelihood. Some livelihoods are extremely vulnerable to particular hazards. To reduce people's vulnerability, support could be provided to these communities with alternative livelihoods that match the people's capacities and that are not sensitive to the existing hazard. This can help them earn an income during the time when their main source of production was down due to the occurrence of hazard. One should be conscious of gender roles and responsibilities when selecting an alternative livelihood. Whether the settlement was urban or rural also play an important role in availability and provision of alternative sources of livelihoods.

Propagation of Disaster Resistant Crops

Disasters disrupt farming and damage certain crops, which may result in the form of food shortage. In old times in the face of food shortages due to disasters people depended on wild crops like certain beans, leaves, and root crops. These crops can still be found in many regions although some of them have become scarce and are gradually disappearing. Therefore it is essential to propagate these, especially among the younger generation, who is hardly aware about the existence of such crops. It is important that the knowledge and skills of old people are tapped regarding these plants for their promotion of mitigation measures. The re-introduction of the already forgotten indigenous crops and the campaign to cultivate them could awaken and remind local residents of the importance of such crops. They are easily grown and do not require a lot of inputs. Their use can be propagated by planting community nurseries/demo-farms and through educational campaigns. Disaster resistant crops and other indigenous crops can serve as a staple food source in times of disasters.

Seed Banks and Nurseries

Establishment of seed banks and nurseries at the community level can ensure a stable supply of seedlings, seeds, cuttings and other plant materials. Seed stocks can be used in times of emergency for rehabilitation of damaged croplands. Most seed banks focus on traditional rice and corn seeds, which are more resistant to pests and less sensitive to changing climatic conditions. These varieties are slowly disappearing because hybrid varieties are promoted in the market. This highlights the importance of community-based seed banks. Nurseries propagate fruit trees, forest trees, forage trees, bamboo, and other plants useful for people in times of crisis. Cultivating utility trees around homes and land, maintaining forest reserves for food, animal feed and cash, improving watersheds, and protecting water sources are long-term interests of vulnerable people. Seed banks and nurseries could strengthen people's existing livelihoods and increase the presence of fallback resources in the community. The CO's agriculture committee can manage the community's seed bank and nursery. They should be provided training on seed bank and nursery management, record keeping. They can also formulate policies for the approval of the farmers regarding repayment of seeds, operation and maintenance of the nursery, and the sustainability of the seed bank and the nursery.

Post Harvest Facilities

Marginalised communities cannot always enjoy the benefits of a harvest. After paying the landlord back, they face problems of inaccessible and expensive grain mills, storage problems (attacked by rats and insects, or rotting due to damp conditions), or they have to sell seasonal perishable crops at a low price. It is essential to help communities to address post-harvest problems to reduce losses, to reduce expenses for milling, and to maximise the availability of vegetables, fruits and root crops by processing and preserving them for times that they are less abundant. In this way people's coping strategies are strengthened. This can be done by helping construct appropriate storage facilities to avoid rats from entering and eating the rice grains, the main problem in times of drought. Seeds can further be protected from insects and damp conditions by adding charcoal and pinewood splints. Provision of cheap and easily accessible milling facilities could be a strategy. This can improve the availability of food in seasons of disaster.

Encourage Proper Land use Management and Sustainable Agriculture Practices

The fast deterioration of environment due to outside pressures (logging, mining, multinational plantations, and encroaching settlers) increases the vulnerability of people to various hazards and many times leads towards introduction of new hazards. Therefore, conservation of environment through tree plantation, rehabilitation of watersheds, mitigation of soil erosion, and replenishment of forest reserves is important. It can be done through organising tree-planting campaigns to raise awareness on environmental issues among a broader public. It is important that people have land use policies which should indicate where remaining forest should not be touched, where 'slash and burn' can still be practised, where permanent farms can be established applying contour lines, and where water sources can be tapped for irrigated farming. In times of crisis, forest products can be used to a limited extent. In this way, local people could try to control and manage their direct surroundings for future generations.

Community Health Workers

The presence of capable Community Health Workers who can provide first aid, and who are knowledgeable on mother and childcare, preventive health, nutrition and sanitation, can make a big difference in the community in the wake of a disaster. It could reduce casualties and the risk of epidemics by providing medical first aid and taking other necessary actions for public health. Among community members, people could be selected who have skills related to health and nutrition, or who are motivated to learn these, and who have sufficient time and commitment to serve as Community Health Workers. They should be trained on Basic Health and Nutrition, Advanced Health Training (diagnosing, cures, minor surgery), First Aid & Home Remedies, Acupressure, Disease Prevention related seminars, Pharmacology (on correct use and storage of western and herbal medicine).

Village Pharmacy and Medicinal/Herbal Garden

A village pharmacy aims to make medicines and first aid immediately available to communities, which have no access to government basic health services. They could produce herbal medicines but also purchase the most common and basic (western) drugs. The pharmacy should also have a first aid kit for emergency purposes and can only be accessed by the trained Community Health Workers and the Health Committee members. The Health Committee together with members of the Executive Committee should formulate policies to manage the pharmacy.

Improve Mobility During Disaster Situation

Disasters can greatly hamper the mobility of community people, affecting their livelihood. Mitigation efforts to improve mobility can range from simple solutions to more expensive and technically more complicated structures. In urban poor areas where people experience floods for weeks or months, elevated foot paths improve mobility; children can keep on going to school, and vendors can keep on selling their home-made products. In remote mountainous areas, communities become inaccessible during cyclones and after earthquakes and landslides, footbridges and safer trails can enhance mobility of people. Also, livelihoods are more secure; fields are located at both sides of rivers; farm inputs and products can be transported easily and continuously.

Provision of Critical Facilities and Infrastructure to Disaster Prone Communities

The disaster prone rural and urban communities in South Asia many times lack the critical infrastructure. When infrastructure does not exist, the poor invest in their own makeshift community infrastructure, which is destroyed in disasters. This makes the poor more vulnerable and the impact of disastrous events greater on them. Adequate physical infrastructure is a key to realising opportunities to increase income and improve standards of living. When infrastructure is at risk, it means that jobs are at risk and in turn, revenue collection. Influencing infrastructure investment policies in South Asia could help reduce vulnerability of hundreds of disaster prone communities.

EMERGENCY RESPONSE

Search and Rescue

Organised communities with a functional disaster response committee can benefit from the presence of a community-based search and rescue team. It is essential that the search and rescue team formulates its counter disaster plan and practice rescue maneuvers. They will also need to have proper equipments; e.g. life jackets, flashlights, batteries, ropes, megaphones, first aid kits, and rescue boats. They should be trained in undertaking proper search and rescue operations.

Relief Delivery

The establishment of a local level relief goods warehouse and an organisational arrangement to collect, store, maintain and distribute relief could help mitigate the impact of a hazard on local people. Standard relief items could be stored in advance. In addition to other food and non-food relief items, the storage of plastic sheets and other materials for establishing emergency shelters is very important. Those people whose houses were destroyed due to earthquake, flood, landslide or civil conflict will immediately need to reside in emergency shelters. The outside help may take few days or weeks in some cases. Relief items purchased locally can be more appropriate culturally and would be cheaper due to no or low transportation costs involved. Donations from individuals, schools, religious institutions, companies, and support groups could also be collected. Non-useful donations could be sold and cash earned this way could be added to relief funds.

Improving the Quality of Public Relief

Costs of relief in South Asia compete with sustainable development allocations. For example, money spent on plastic sheets to cover damaged roofs means less money is allocated to build new homes for the homeless. Repeated relief allocation and use have not brought about region wide improvement in the quality and standards of relief in the voluntary sectors. Poor relief means, poor recovery and rehabilitation. Non-development oriented relief means the continued vulnerability of the poor. Therefore it is imperative that victims and recipients conduct audits of relief delivery and they are engaged in the design and distribution of relief after disasters. Such initiatives could help make relief more appropriate and meaningful.

Evacuation Centre Management

Some disasters cause long-term displacement. Evacuees gather from different places and often have to share a common evacuation center. This can lead to chaotic and unhealthy conditions if no management system is put in place. People might become dependent, depressed and even violent if they have nothing to do but sit and wait for support. The participation of evacuees in camp management is important, since it is people's right to be involved in decisions that affect their lives and shape their environment. People's involvement in evacuation centre management is a way to rebuild people's confidence and capacities after what they have experienced during the disaster. Life in an evacuation centre is very different from the 'normal' life people lead before they enter the camp. While being engaged in the day-to-day management of an evacuation centre, people can rebuild a community and get the energy to look ahead again. They will require an evacuation centre management committee. Other committees like; the health committee, security committee, networking/public information committee, and education/training committee will support it.

Mobilisation of the Less Vulnerable Sectors

Linking the vulnerable communities with the less-vulnerable sectors can also be very useful strategy to reduce people's vulnerability. The less vulnerable groups get the opportunity to participate in the development endeavors of the vulnerable sectors. They could be students, teachers, doctors, nurses, dentists, religious organisations, scientists and technical experts, drivers, companies, and media personalities. The involvement of the less-vulnerable sectors in disaster response is not limited to financial and technical assistance, but covers a wide range of support:

e.g. material resource generation, financial resource generation, human power and moral support.

CONTINUING RESPONSES

The previous list of community capacity building strategies has been categorised in community risk assessment, preparedness, mitigation and emergency relief for practical reasons; but, except for emergency response, all other strategies can be implemented regardless of the actual disaster event. These strategies do not focus on the disaster event itself, but the vulnerabilities of the affected communities. In this way, disaster response is developmental in nature, since the responses strengthen people's existing coping strategies and their capacities. However, to sustain the initial gains of the long list of interventions, community organisations should be continuously supported through activities that are not always considered disaster responses, but are essential in the process of capability building and for addressing root causes of vulnerabilities. These responses take place continuously because addressing root causes requires long-term attention.

Continuous CO Capability Building

In order to sustain the implementation of above-mentioned risk mitigation strategies, it is must that the organisational and management skills of the Community Organisation are developed as well. Its members should be trained and couched in leadership skills (facilitation of meetings, conflict management, speaking in public, assessment and planning skills, etc.), and skills related to project management (monitoring system, financial management, recording, documentation, etc.). Community organisations that are strong enough to undertake lobbying and advocacy campaigns should be trained in negotiation skills and knowledge on human rights and other laws related to the issue.

Public Information

Continued public information is important from two angles. First, there is a need for continuous awareness raising programmes for the disaster prone communities, so that they are aware of the prevalent risks and they prepare to cope with them. Second, target of the public information campaign should be the less vulnerable group so that their support could be mobilised when needed to help the disaster-affected communities.

Networking, Advocacy and Alliance Building

To influence the policies and transform the structures that make certain groups of people vulnerable to disasters, building alliances with like minded groups and undertaking advocacy and networking activities will have to be a permanent activity of those agencies who are interested in reducing people's vulnerability and building community capacity to enable them to cope with disasters.

CONCLUSION

The experiences of different agencies in south and south east Asia and other parts of the world demonstrates that involvement of disaster victims in planning, implementation and monitoring of risk reduction programmes could greatly help in reducing vulnerabilities of people and building their capacities, thus reducing the disaster risks in the long-run. Their involvement helps them analyse their situation and design and implement activities appropriate to the local socio-economic, environmental and cultural context, thus creating a sense of ownership amongst the disaster victims and building there confidence in their own capabilities to cope with disastrous situations. The programmes of CARE Bangladesh, Cyclone Preparedness Programme of Bangladesh Red Cross, the Flood mitigation programme of the PACT Cambodia, the DMI's initiatives on auditing of public relief and advocacy, the Thardeep Rural Development Programme's counter disaster projects in Pakistan and the Citizens' Disaster Response Network's experience in the Philippines are but few examples in this regard.

The important role of communities in disaster management is strongly supported by the United Nations International Strategy for Disaster Reduction (ISDR) whose vision is to 'enable all communities to become resilient to the effects of natural, technological and environmental hazards'. This shift from a top-down relief and response approach to a more inter-sectoral risk management approach has begun to influence the way disaster management programmes are now being planned and implemented. Many high-level policymakers from the government sector and international agencies are recognising the importance of Community-Based Disaster Management (CBDM).

The British Government's Department For International Development (DFID) developed a livelihood framework, which view people as operating in a context of vulnerability. The Disaster Preparedness-European Community Humanitarian Office (DIPECHO) developed an Action Plan for South East Asia in 1999, which identified the need to provide an institutional arrangement in targeted countries for training of national, provincial, and local governmental and non-governmental institutions to enable them to incorporate community-based disaster risk management in their programmes. ADPC with support from DIPECHO aims to meet this need in south east Asia.

It is becoming clear that the nature of the vulnerability of communities is complex and varied. Hence, there are no straightforward solutions for risk reduction. It will take concerted efforts at different levels and across different sectors to improve our understanding of the linkages and to devise effective mechanisms for disaster risk reduction. The communities themselves need to first be aware of the importance of disaster mitigation. It is then necessary to go beyond awareness and impart skills, which can translate this awareness into concrete practice. Finally, community-based disaster mitigation depends on a favorable political environment that promotes and supports this participation process.

14

Disaster Risk Reduction through Capacity Building of the Community and Panchayati Raj Institutions

—Pardeep Sahni
—Alka Dhameja

INTRODUCTION

Disasters play havoc with the lives of people. They cause excessive losses to humanity and infrastructure. Due to disasters, normal life is often thrown out of gear and the existing patterns of regulatory and development administration suffer heavily. The economic, social and psychological dimensions of the wrath of disasters adversely affect the environment around. As the frequency of natural disasters in South Asia is on an increase, the nations in this region are compelled to utilise their scarce resources for the purpose of relief and reconstruction. The ferocious rage of the recurrent catastrophes has hampered the developmental process and economic growth of these countries.

There are elements at risk with regard to each disaster. Risk is not an inherent property of a hazard alone. The weak structures are more at risk. It depends on the fury of the disaster as well as the vulnerability of the affected region. Thus, the local communities are required to be prepared to face the aftermath of the disasters effectively. The first step in this direction is to undertake vulnerability analysis, which brings to light the elements at risk such as the population, buildings and infrastructure. The most vulnerable members of the community are the expectant and lactating women, single women, children, and old, disabled, handicapped, sick and ailing people. Their needs have to be kept in view while making the analysis. Likewise, the physical vulnerability elements have to be recognised by the community for the purpose of initiating specific measures to reduce the extent of losses in their regions. The community should also identify the potential threats in order to cope with the intensity of future disasters.

The key objectives of Disaster Risk Reduction as highlighted in the Natural Disaster Reduction Initiative by the National Science and Technology Council Committee on Environment and Natural Resources constituted in U.S.A. in 1996 include:

- Identification of natural hazards and assessment of the associated risks and costs.
- Application of new technologies to local and regional warning and dissemination systems.
- Improvement in predicting and understanding natural hazards.
- Transfer of technology for natural disaster reduction to public sector managers and private sector businesses and institutions.
- Development of awareness on natural disasters through training, education and research.
- Strengthening of disaster mitigation capabilities of industry; and local and state governments.
- Enhancement of post-disaster response and recovery capabilities.
- Increase in the export of natural disaster reduction technologies.

THE ORISSA CYCLONES AND THEIR IMPACT

Two big cyclones hit the coast of Orissa at Gopalpur and Paradeep on the 18th and the 29th of October 1999. Out of 30 districts in the state, the 14 coastal districts were heavily affected. The study, on which this chapter is based, has been conducted in two districts of Orissa—Jagatsinghpur and Kendrapara, with special reference to one worst hit block from each district namely, Ersama in Jagatsinghpur and Mahakalpada in Kendrapara. The data have been collected from both primary and secondary sources.

Orissa has experienced floods, droughts and cyclones over the past many years. But, the super cyclone on 29th October 1999 was totally different. Traumatised people of the 14 coastal districts swear that even their forefathers must not have witnessed such a cyclone. It is believed to be the most severe catastrophe in the last two centuries. The assessment of damages is summarised in the Tables 14.1 and 14.2.

This chapter is devoted to the topic of disaster risk reduction, particularly managing the situations perpetrated by cyclones. The focus is on capacity building of the community and Panchayati Raj Institutions (PRIs) for disaster risk reduction. Special reference to the capacity building of the community is being made because in any disaster situation, the community is the worst affected. Unless the people are sensitised, made aware and prepared to respond effectively to disasters, risk reduction becomes difficult. Thorough analysis of the needs, vulnerability, risk, and resources of a local area is a must to come out

TABLE 14.1 Areas affected

Geographical area affected	*Number affected by the disaster*		
	Super cyclone of October 29–30, 1999	Cyclone of October 17–18, 1999	Total
Districts	12	2	14
Blocks	97	29	126
Gram Panchayats	1,846	198	2,044
Villages	14,000	3,407	17,407
Population (lakhs)	125.69	23.1	148.79
Agricultural land (lakh hectares)	17.33	1.21	18.54

Source: 'Voluntary Sector in Relief and Rehabilitation', Orissa Disaster Mitigation Mission, Bhubaneswar.

TABLE 14.2 Extent of damage

Indicators	*Type of damage*	*Estimate of damage*
Humans	Death	30,000
Animals	Death	300,000
Individual houses	Washed away	50,000
	Fully collapsed	1,000,000
	Partially collapsed	500,000
Families	Affected (Million)	1.5
All Crops	Loss (Rs. Million)	12,000
Paddy	Crop loss (Million MT)	2.0
Trees in orchards, backyards, etc.	Severely damaged (Million)	100
Coastal shelter belt	Damaged (km^2)	200
Mangroves	Damaged (km^2)	50
Entire tree cover	Loss (Rs. Million)	150,000
Total loss	In Rs. Million (Minimum)	300,000

Source: 'Voluntary Sector in Relief and Rehabilitation', op. cit.

with a short-term as well as long-term disaster risk reduction plan. This task can best be done by the structural institutions existing in the form of PRIs at the village, block, and district levels. Thus, there is an urgent need for capacity building of the PRIs for disaster risk reduction.

CAPACITY BUILDING OF THE COMMUNITY

In disaster management and also in development activities, decisions have to be taken for determining the work plan, preparedness, availability of skilled manpower, and arrangements needed for supervision and monitoring. The only difference in these two is that in case of disasters, the decisions and strategies should be ready before the occurrence of the catastrophe and that the time and scope of everyone's participation is very little due to the emergency situation. The exact time of arrival or magnitude of the impact of a disaster is never known, but technical development has made it possible to forecast it in most of the cases. Therefore, disaster management entirely depends on the preparedness and the warning system, emergency relief operations, rehabilitation and reconstruction.

Community Preparedness

Preparedness is crucial in disaster management and mainly depends on systematic planning. Planning should be targeted to save lives and mitigate the damage in any disaster. Disaster response programmes are the plans prepared for proper utilisation of time and resources. The programmes have to be strictly observed due to availability of limited resources and shortage of time. Preparedness has to be accomplished prior to the occurrence of a calamity.

Areas that need to be looked into for the purpose of community preparedness are:

- Assessment of the devastation caused by a disaster on the basis of experience
- Establishment of the warning and communication systems
- Arrangements for search, rescue and evacuation
- Supply of emergency drugs and food
- Temporary shelters and its management
- Planning for rehabilitation and reconstruction.

The following are important basic considerations with regard to community preparedness:

- Assessment of possible devastation in case of a disaster
- Type of aid and cooperation needed
- Assessment of local help
- Nature of external aid required
- Type of assistance and cooperation that can be offered by other organisations.

The characteristics of community preparedness should be in the following lines:

- The plan envisaged should be definite and target-oriented

- Activities should be vividly descriptive and continuous
- Responsibilities and duties must be specifically described
- The plan must be envisaged in harmony with the ideals, objectives, goals and aspirations of the community.

In preparedness planning, the local people should be organised to enable the following

- Formation of committees at the village level and distribution of responsibilities.
- Involvement of the local committees in the warning system.
- Participation of the local committees in emergency relief aid, supply and delivery.
- Participation of disaster victims in the management of temporary shelters.
- Building up community awareness on hazards.

The most affected in a disaster situation is the community. In order to draw up a viable Community-Based Development Plan (CBDP) strategy, it is necessary to highlight such components, which need to be addressed for better outcome of the result of appropriate preparedness on the part of the community.

Coordination with the Community

Before going into the issue of pertinence of the community, it is essential to discuss what a community exactly means. A community is a group of people sharing common ideas, resources and environment. It has leaders who may be elected, appointed or unanimously selected members of political, religious or formal associations. The community has common goals and aspirations. There are majorities and minorities, insiders and outsiders, and groups within groups, which can be positive or negative in a community. A community has complementary skills and often enters into strategic alliances.

It is therefore pertinents to explain the need for a *Coordination Council* as well as what such a council is expected to do in ensuring the participation of the community in disaster management. It is normally observed that the disaster preparedness programme should be acceptable to the community. This requires the government agencies; volunteer groups or NGOs from outside and the local NGOs to coordinate amongst them as well as with the community. Such an action removes mutual misconceptions and fears and averts undesirable competition between them. There is a greater need for coordination regarding mitigation actions, research, resource mobilisation and utilisation and for networking among the various actors involved in disaster management at the local level and also in case of external agencies operative in the field.

The NGOs should form a Disaster Coordination Council to ensure a meaningful coordination at the state, district and village levels. At the grassroots level, NGOs and Community-Based Organisations (CBOs) are far more acceptable and effective than the Government. In times of a disaster, they can exercise first aid and search and rescue operations more promptly and efficiently than Government agencies. On the other hand, the Government has more resources, equipment, and transport required to implement these activities.

Thus, there is an urgent need for an effective coordination strategy between the manpower and organising capability of the NGOs, and the resources and initiatives of the Government. Since many Government institutes are involved, there is a threat of getting stuck with innumerable coordination meetings. There should be only a single coordination council at each level for all, so as to avoid too many meetings and conflicting decisions at different fora. It would also save manpower and wastage of time. Components of coordination that can act as controlling factors are:

Work plan. The method of coordination must be made clear, subject to the approval of all concerned.

Role. Everybody's roles, responsibilities, management and facilities should be clearly identified and written down to determine the method of operation of all the groups.

Priority list. The priority list should be made final with the consent of all concerned.

Information collection and dissemination. Garnering of information should be done methodically, scientifically and ethically. There is a need for coordination between all the agencies involved in information gathering and dissemination.

Coordination needs to exist at various levels for efficiency and greater impact. A coordination plan amongst the various agencies involved should be evolved as a part of disaster preparedness measures. Local and area based coordination are significantly important at the overall planning stage.

Community Awareness

The members of the community, PRIs, NGOs and government officials feel that there will be more effective community participation if the community is aware about the vulnerabilities and risks involved in various types of disasters affecting them.

Awareness is the most important aspect for community participation in disaster preparedness. An alert community will take active part in disaster reduction or mitigation programmes and will provide inputs in terms of local knowledge and available resources. Through community awareness programmes, the community becomes more informed, alert, self-reliant and capable of participating in all activities of disaster management as well as in the programmes of the government agents and NGOs. Some important means for creating community awareness are:

- Media and press
- Short films and folk songs
- Posters, cartoons, charts and photographs
- Training camps
- Short street plays in fairs, religious functions and other celebrations and on occasions of public gatherings
- Lectures and debates at schools and colleges
- Special lectures by community leaders or well-known persons in the area
- Group discussions.

Community Participation

Meaningful involvement and participation of the community is concerned with achieving the power to influence individuals' livelihood. A meaningful participation empowers people so that they can take their own decisions—political, administrative and financial; and also helps government authorities and NGOs in implementing programmes related to various facets of disaster management. There is definitely a need to have a proper match between the goals and methods of these agencies and the needs and capacities of the beneficiaries. This is possible only through community participation. The advantages of community participation are as follows:

- Systematic identification of problems
- Innovative ideas/solutions
- Motivated participation
- Sense of belonging
- Less expensive decision-making
- Better utilisation of local resources
- Faster communication
- Participatory decisions at local level
- Effective and speedy monitoring
- Cost effectiveness
- Less dependence on the Government
- Involvement of all classes in the local community.

Community participation stimulates and encourages people of all social standards to be aware of their expertise, power, usefulness, responsibilities, and resources so as to ensure their spontaneous contribution. This is necessary to mitigate destruction and damage to individuals, families and society, evolve an alternative to Government activities, achieve self-reliance, observe a programme and keep it going systematically, identify the problems of the beneficiaries and determine possible solutions, and to make proper use of indigenous resources.

Although spontaneous participation is ideal, the NGOs are the best external agencies to motivate and encourage the community to initially participate in the process. Once the people take the initiative, they begin to participate spontaneously.

Community participation is necessary for the following reasons:

- Most of the actions required for disaster management or preparedness are at the individual or community levels.
- The State has limited resources and during disasters, these are just not sufficient, hence the need for complete participation of the community is indispensable.
- The process of participation of the people motivates them to attain self-sufficiency and to reduce their dependence on others.
- It facilitates regular review of the progress of the activity and helps in guiding the programme along a well-charted direction.
- It helps the implementing agency to interact and exchange views with the community, identify their problems and offer necessary assistance.

The idea of community participation is not of any recent origin. Traditionally, members of the community used to identify their problems and also used to utilise their local resources to solve these problems through collective efforts. However, the idea that people can be made to participate in development planning through proper educational methods is a modern one.

The barriers to community participation could be *social* (social structure, cultural traditions, beliefs, values, and attitudes of the people) as well as *administrative* (bureaucratic authorities, rules, regulations and complexities). The principles to be followed to ensure community participation are:

- The development workers should be provided with appropriate training so that they can work as motivators of the community.
- The motivated village community should identify their formal and non-formal leaders.
- The formal and non-formal leaders should be supported to identify and prioritise the problem areas and to look for solutions.
- Local initiatives should be undertaken and local resources must be identified and mobilised as far as possible.

Community participation can be ensured in the following phases:

- Formation of Disaster Management Committee at the village level
- Pre-disaster planning
- Establishment of local warning dissemination system
- Provision of training

- Identification of local resources (physical and human) and their use
- Formation of the volunteer groups
- Construction of shelters, their maintenance and management
- Search, rescue or salvage and first aid facilities
- Liaison and linkages with local administration and other related organisations
- Construction of houses and repair and installation of tube-wells and latrines
- Promotion of employment generation schemes as well as increase in savings and local loan sanction.

Community participation can be measured through the following ways in which it manifests:

- Resource mobilisation/utilisation
- Awareness
- Values and beliefs promoting cooperation
- Type of socio-economic structure
- Economic status of the people
- Locus of decision-making
- Benefits and distribution
- Legal factors
- Policy-related factors
- Political pressures.

Participation of Women in Capacity Building Strategies for Disaster Management

Women and children are the first and foremost victims of natural disasters. The responsibility of women in the household is far greater than that of the men. They are responsible for looking after children, and rearing of livestock and poultry. The women also do kitchen gardening, cultivation of crops and preservation of seeds.

Even during disaster periods, women have to perform their normal domestic functions like procuring drinking water and fire wood, cooking, looking after the health of every member of the family, drying seeds and food crops, preventing household belongings from possible losses due to disasters and many other difficult tasks. Some women may be pregnant and some of them may deliver babies during this period, which may endanger their health and life. In some local communities, women may not appear in public, but they have a major role in important social and economic spheres.

Women generally have had fewer opportunities to play any role in disaster mitigation. There are many obstacles preventing women from becoming positive role players, but a lot of signal benefits could be gained through greater involvement of women in disaster management.

Some unique characteristics of our communities make it imperative to lay special importance on the participation of women in formulating plans for disaster preparedness. There is a considerable difference between the socio-economic activities performed by men and women in disaster situations. To incorporate all information in the formulation of any disaster management strategy, equal importance should be given to both.

The nature of work that men and women perform, the gender relationship, respective skills of men and women and their social and economic roles vary from community to community. These may also be interdependent, and at the same time, little may be known about their domain. Hence, planning cannot be dependent entirely on the information supplied by men, nor should the information be collected about one from the other. For successful implementation, the formulation of local plans has to lay special emphasis on the role of women in making a significant contribution and ensure their participation in disaster mitigation.

COMMUNITY-BASED DISASTER PREPAREDNESS: A HOLISTIC APPROACH

Community awareness on disaster preparedness assumes predominance because disaster prevention, preparedness, mitigation, and relief are largely dependent on the capability of the communities to withstand disasters. Disaster preparedness needs to be the way of life in such communities if they have to survive the aftermath of different catastrophes. There is an urgent need to build the capacity and capability of the local communities by empowering them with coping capacities and increasing their self-confidence through recognition and enhancing their knowledge, practices and values so that all this falls in line with the developmental activities. As disaster preparedness is of prime importance in reducing the load on disaster relief, it should be considered as an integral aspect of development policy and planning.

There is, thus, a need to promote and strengthen human and institutional capacity building for coping with natural disasters by way of disaster preparedness as preventive measures are most effective when there is participation at all levels. Involvement of community in all facets of disaster management becomes imperative and therefore, there is a need for a holistic approach for community-based disaster preparedness. This can be achieved by strengthening the capabilities of communities through the 3 major stages of disaster management.

Pre-disaster Stage

This stage involves the following major activities:

- Understanding disasters and their effects
- Preparation of preparedness plan
- Promotion of mitigation measures

- Risk analysis and vulnerability analysis
- Resource assessment and mobilisation
- Implementation of insurance schemes.

The community representatives of PRIs, NGOs, and Government organisations have stated that the elements at risk in case of a cyclone are:

- Thatched houses
- Mud houses
- Weak houses
- Light weight houses
- Loose and poorly attached building elements (chips and boards, etc.)
- Telegraph, television and electricity poles
- Sign boards, fences, trees, etc.
- Fishing boats
- Women, especially pregnant, lactating, and single
- Old people
- Ailing and handicapped persons
- Children
- Fishermen

The vulnerability of these elements may thus be an important consideration in the plans for the pre-disaster stage of managing cyclone inflicted disasters.

During the Disaster

Following are the major considerations for the period during the disaster:

- Search, rescue and evacuation
- Shelter for victims
- First aid
- Distribution of food, water, medicine and fodder
- Clearance of debris
- Movement of injured to hospitals
- Disposal of dead humans and animals
- Sympathetic attitude towards victims
- Assisting rescue teams
- Security of property
- Information dissemination and checking of rumours
- Immediate damage assessment
- Filing of claims.

Post-disaster Stage

The post-disaster stage considerations are:

- Damage assessment

- Economic rehabilitation
- Social rehabilitation
- Protection of women and children
- Disasters and development.

CAPACITY BUILDING MEASURES

There are various capacity building strategies that could be put to effective use for reducing the risk of disasters. Let us examine these.

Preparation of Community Action Plans

Human being is a social animal by nature and necessity. During danger, he attempts to protect and save his own life and property and also that of his friends and relatives. Collection of individuals in groups ultimately constitutes the community, which has certain common goals, ideas, ideals, aspirations and resources.

There is no denying the fact that the community, which is affected by disaster, needs outside help and intervention, may be that of the Government, NGOs or other communities for relief, rescue, response, and rehabilitation. But along with this, it becomes imperative for the community to take its own remedial steps to face the challenges posed before it by a disaster. The community needs to be prepared for contributing to the response and recovery activities being taken up by other stakeholders.

The community has to be active and ready, not passive. It must work in accordance with an age-old saying that 'God helps those who help themselves'. Its involvement and participation in the activities initiated by the Government and the NGOs would help these agencies to perform better. It would also lead to better development orientation, which certainly will go a long way in disaster mitigation and preparedness. The community must take adequate initiative to draw up its own action plan. Certainly, it may require the help of NGOs, Gram Panchayats and some local-based officials to complete the job. But an action plan drawn by the community will definitely be more realistic and in accordance with the local needs of the area.

The action plan must be a comprehensive document including preventive measures to be taken up by the community for reducing the occurrence of disasters to as much extent as possible. The plan should refer to mitigation comprising structural and non-structural elements. It needs to entail a preparedness scheme enlisting up-to-date counter-disaster measures, provisions for warning dissemination by making use of local mechanisms like beating of drums, passing of information through mouth, etc. It should have provisions for emergency action, such as evacuation of people and livestock to safer places, ensuring ways of communication to inform the block and district level officials after the disaster strikes; and conducting consistent and thorough awareness campaigns.

The action plan must have a provision for educating people through all possible formal and non-formal ways and also for training programmes for the Disaster Task Force members, Gram Panchayat members, NGOs and local officials. The government officers, PRIs and NGOs operating in the area concerned can facilitate such preparedness measures. It must include a list of elements at risk in the village including the number of human beings, livestock, thatched roof houses, mud roof houses, and weak structures.

The action plan must make a comprehensive vulnerability analysis of the area to determine the number of vulnerable people like children, women, especially expectant and lactating women, single women, old people, handicapped and disabled people, etc. It must also prepare a list of the number of livestock like cows, bulls, buffaloes, goats, sheep, pigs and dogs. The action plan should clearly spell out the ways for undertaking the activities assigned to all the members of the community which comprise the Disaster Task Force. These activities include all the activities mentioned under the 'During the Disaster Stage'. Other people in the community can help the DTF members as well.

The action plan should also refer to mechanisms, which need to be developed by the community for comprehensive damage assessment, economic rehabilitation, social rehabilitation, and rehabilitation of women and children. It should be prescriptive in order to correlate disasters and development and should also highlight the development requirements to be undertaken in the area so that the Government and NGOs can work in accordance with the community needs. Such developmental programmes must have components of disaster mitigation incorporated in them. The action plan of the community must also spell out the mechanisms for monitoring and evaluating various facets of disaster management undertaken by the Government, PRIs and the community.

Strengthening Educational Activities for Capacity Building

The community needs to be made more aware about various facets of disasters. This is possible through a well-knit mechanism with specific roles assigned to various agencies for providing formal as well as non-formal education to the community. The formal educational curriculum, right from the primary level onwards, should have discussion on various aspects of disaster management. The government bodies like the Department of Education, Board of Secondary Education, universities, etc. must reframe their syllabi and include disaster management education in their curricula. Besides, there have to be exclusive as well as joint training programmes for the government officials, the NGOs, and the PRIs.

For the members of the community who did not get the chance to be educated, there is a need for non-formal education through videos, documentaries, group discussions, and public meetings. The institutions like Gram Panchayat, village schools, Primary Health Centre, Anganwadis, etc., along with NGOs and other social as well as religious groups need to play an active role in this regard.

Formation of Active Community Disaster Task Force (DTF)

Each community must have its own Disaster Task Force comprising of 10–12 members. These members should be chosen on the following criteria:

- The members must be young and healthy persons
- There must be 2–3 women representatives in the DTF
- There must be a member of the Gram Panchayat in the DTF
- The members must be credible and integral so that they have a wider acceptability in the community
- The members should not be going to distant places for work during the day time
- The members should have strong feelings and commitment to community service and volunteerism.

The Task Force members should be capacitated and trained through the initiatives of local NGOs, Gram Panchayat and local officials for activities like dissemination of warning, evacuation of people and livestock, management of shelter, storage of food, fuel, medicine and fodder, search for people stranded in and around the villages, and for the evacuation of people trapped in their houses after the disaster.

Provision of first-aid to the injured, arrangement for community kitchen for distribution of food and water to the people assembled at safer places, clearance of debris so that the rescue team and relief material can reach the local area speedily, assistance to the response teams in the disposal of dead human bodies and dead animals, ensuring the safety of public and private property, collection and dissemination of proper information to the concerned officials/media, and also exercising check on the spread of rumours, undertaking rapid damage assessment and helping the community in filing its claims for compensation are some of the other activities that the DTF should take up on priority basis.

Besides, the DTF must be active in creating as well as sustaining community awareness on disaster preparedness, establishing liaison with Gram Panchayat, local officials and local operating NGOs, developing area based indicators for monitoring the self-help capacity of the community as well as its coping mechanisms for disaster management practices and for coordinating with as well as monitoring the activities of NGOs and local officials so that the agencies concerned get proper feedback about their relief and response operations in the area.

The DTF members are to be sensitised towards the following in the different phases of a disaster.

In the pre-disaster phase, the DTF may carry out activities for (a) preparing a Village/Community Action Plan, (b) Receiving of warning and its timely dissemination throughout the village, (c) Checking on cyclone shelters and other safer places before the disaster, (d) Keeping a stock of food, water, utensils, medicine, milk powder, candles, match boxes, etc., (e) Ensuring that sanitation facilities are usable and in place, (f) Arranging a community kitchen in order to avoid too many people cooking in the shelter, (g) Maintaining information about fishermen, (h) Identifying safe routes, (i) Arranging transportation, (j) Preparing a rescue kit, (k) Mobilising stock like water pouches, food grains, dry rations, medicines, etc. from the Government, (l) Ensuring stock of medicines like paracetamol tablets (500 mg) for fever and pain, acetylsalicylic acid tablets (300 mg) for relief from pain (not on an empty stomach); cotrimoxazole tablets (20 mg to 160 mg) for pneumonia; tetracycline capsules (250–500 mg) for cholera, metronidazole tablets (200–400 mg) for amoebic dysentery, chloraquine tablets (250 mg) for malaria, antacid tablets for gastritis, promethazine tablets (10 mg) to stop vomiting, diazepam tablets (5 mg) for acutely depressed patients, ORS Packets for diarrhea, cotton, gauze, and antiseptic solution for injuries, splints for any fracture, plaster for injuries, Whitefield Ointment for fungal infection, benzyl benzoate lotion for scabies, gentamycin eye/ear drops for any eye or ear infection, methergine tablets or injections (0.2 mg/ml) in case of bleeding after delivery, and chlorine tablets for purification of water, etc., and (m) Stocking temporary building materials like bamboo, rope, tarpaulin, and asbestos sheets, etc.

During a disaster, the DTF should involve itself in discharging on a war-footing, initiatives for (a) Cross checking of the warning received on radio with the government officials, (b) Moving stocks of food, water, utensils, medicine, etc. to the shelters and safer buildings, (c) Organising space to house the evacuated, (d) Maintaining strict hygiene practices in the shelter, (e) Stopping people from moving out of shelters during the cyclone, (f) Picking up fishermen still at the coasts, (g) Helping the vulnerable groups in collecting their belongings before evacuation, (h) Evacuating livestock well in advance, (i) Looking after the medical needs of the people, (j) Providing first aid to the victims, (k) Transporting seriously injured or sick to hospitals, and (l) Disposing dead bodies of human beings and animals.

In the post-disaster phase, the DTF should apply itself to (a) Disseminating information regarding ways to counteract diseases through remedies, (b) Informing people about the places they can go for help, (c) Replenishing stocks of food, fuel, wood, and medicines, (d) Disinfecting the shelter through out the stay and before leaving, (e) Clearing the roads immediately of fallen trees and debris and attempting to restore communication and transportation, (f) Providing counselling to the victims, (g) Isolating cases with infectious diseases and preventing them from spreading, (h) Spraying disinfectants in the village to prevent the spread of epidemic, (i) Checking for water contamination, (j) Making damage assessment, (k) Helping people in filing claims for compensation, (l) Helping and monitoring the rehabilitation activities undertaken by the Government and NGOs, (m) Recording the lessons learnt from the immediate cyclone and preparing accordingly for future, and (n) Preparing development schemes for the village in keeping with its requirements in order to mitigate disasters.

Creation of Alternative Livelihood Options

During the course of a study conducted by the authors on 'Livelihood Options for Cyclone Risk Reduction with Special Reference to The Orissa Super-Cyclone', the Communities suggested the following measures for building more secure livelihoods after the cyclone:

- Disaster victims' needs must be considered
- Rehabilitation packages should be put in place by the Government
- *Targeted Feeding Plan* for older persons, widows and families without any earning members needs to be implemented
- Special shelters need to be built for older persons displaced in the cyclone
- The dignity for older persons needs to be kept in view
- Methodology to ensure equity needs to be operationalised
- Method of prioritisation of needs must be determined through objective analysis of vulnerability
- Rehabilitation must be adaptable to the local needs and environment of the locals, especially the weaker sections of the affected region.

During the course of this study, the communities suggested the following measures for building more secure livelihoods after the cyclone:

- Measures to improve agricultural production
- More focus on animal husbandry
- Provision of adequate seeds and fertilisers
- Draught animals to be provided to farmers
- Arrangements to be made for providing tractors or power tillers on a hire basis
- Dairy cattle to be procured
- Arrangements for restoring insemination centres for local breeds of cattle have to be made
- Alternative cropping pattern in areas affected by salinity needs to be implemented
- Reclamation and treatment of soils

- Ecological balance needs to be maintained through large-scale plantation
- Wage employment under various Rural Development Schemes
- Financial assistance to artisans for restoration of their vocations.

The important sources of livelihood during rehabilitation identified by the community include:

- Supply of seeds
- Supply of fertilisers
- Horticulture plantation
- Livestock rearing
- Food for Work Programme
- Assistance for purchase of tools
- Assistance for repair or construction of work-sheds.

A *Livelihood and Employment Restoration Programme* launched by Oxfam in collaboration with the European Union aimed at:

- Restoring the traditional livelihoods of the affected populations through the provision of food/cash for work.
- Facilitating work for community-based rehabilitation and restoration activities, such as reclamation of agricultural land.
- Providing a model for Food-for-Work Programme that is capable of reaching the most vulnerable and marginalised so that other larger-scale programmes can integrate vulnerability and equity perspectives.

The support provided by a number of NGOs working in Orissa during rehabilitation and the support extended by the Government of Orissa has led to the strengthening of the following livelihood options in two major sectors—the farm and non-farm sectors:

Activities in the *farm sector*:

- Cultivation of paddy, wheat, ragi and groundnut
- Growth of vegetables and potatoes
- Kitchen garden nurseries
- Plantation of mangoes, mangroves and pipal.

Activities in the *non-farm sector* for engaging more people in the following occupations:

- Fishing
- Fish drying
- Pig rearing
- Coir production
- Rearing of cows, buffaloes, bulls
- Bamboo production
- Lime making
- Poultry

- Goat rearing
- Boat building
- Making fishing nets
- Shrimp farming
- Basket making
- Mat weaving
- Broom making
- Masonry
- Carpentry
- Plumbing
- Electrician work
- Blacksmith work
- Toy making
- Pottery
- Appliqué work
- Petty trading.

CAPACITY BUILDING OF PANCHAYATI RAJ INSTITUTIONS

Low-income levels, inadequate to ensure a quality of life compatible with physical well being characterise rural India, which encompasses three-fourths of India's population. The frontal attack on rural poverty should be spearheaded through the active role of Panchayats in dealing with the development and disaster issues to reach out to the most disadvantaged sections of society, provide them with avenues of employment, be it self-employment or wage-employment, and improve the infrastructure related to their life support systems. Democratic decentralisation of power has been one of the avowed objectives of the Government through effective implementation of the principles of Panchayati Raj. The elections to the 3-tier bodies of the Panchayati Raj administration have led to a continuous devolution of the powers and functions of the Panchayat bodies making the concept of decentralisation truly meaningful. The Panchayati Raj Institutions (PRIs) can effectively deal with all matters pertaining to rural development.

The health of Indian democracy depends on the sound working of the PRIs. They are responsible for the economic prosperity, social progress and political development of the country. The objectives of Panchayati Raj Institutions are decentralisation, development and social change. These Institutions act as the catchment areas for the discovery of new leaders. It is natural to expect the new elites to justify themselves as the end-generating catalysts of a new transformation in terms of the objectives of Panchayati Raj.

The Panchayati Raj Institutions in India have been entrusted with major responsibilities of bringing about development of the rural areas through participatory approach. This ensures the greatest chance of sustainability as it builds on local capacities, ensures compatibility with development plans and addresses people's needs effectively. The representatives elected at the local level are responsible for social justice, economic development and implementation of schemes at the grassroots level. The training of these elected representatives along with the government officials at the village level on various aspects of disaster management enhances their capabilities to understand the impact of the disasters on the development efforts at the local level.

Strengthening of PRIs

Panchayati Raj Institutions in India operate at the district, block and village levels by the nomenclature of Zilla Parishad, Panchayat Samiti and Gram Panchayat respectively. The PRIs have got a Constitutional status through the 73rd Constitutional Amendment and have been involved to play a role of prominence in micro-level planning, which gets integrated in the state plans and finally in the Union plan. The PRIs, more so, the Gram Panchayat is the body, which operates at the village level and is required to take care of the village needs of regulatory and development nature. The involvement of the community in all facets of disaster management can be effectively ensured through the Gram Panchayat and two other upper institutions. There is a growing need for qualitative strengthening of PRIs, especially the Gram Panchayat and also building their capacity and capability by various measures adopted and put in practice by the Government and the NGOs.

The members of the PRIs are to be adequately sensitised, trained and oriented for better understanding of disasters, their consequences, need for preparedness, etc. Tailor-made training programmes that are to be conducted at the block level must provide due focus on techniques of hazard mapping, risk assessment, vulnerability analysis, resource analysis and mobilisation.

The members of PRIs at the village and the block levels need to be sensitised for building a basic database for each village. It should include an analysis of the area considering its history of disasters, proximity to sea shore/river bank, weak points of embankments near the village, population (especially referring to the women, children, elderly, sick, handicapped and disabled persons), number of concrete houses, buildings and shelters in and around the village, cattle population of the village, high land places in village, road accessibility of the village, etc.

There is a need to mobilise adequate funds earmarked for the PRIs, especially the Gram Panchayats and the Panchayat Samitis, for undertaking activities pertaining to generation and sustenance of community awareness on disaster preparedness, repairing and maintenance of village public buildings, storage of food stuff, water, fodder, etc. Funds need to be made available for the collection of rescue items like boats, rafts, rope, ladder, candles, match boxes, kerosene, hurricane lamps, buckets, hooks, tyres, anchors, first aid box, cutting saw blades, hammers, nails, etc.

The Gram Panchayats need to be oriented for and also be imparted skills for motivating community and building its morale in order to be of help during a disaster situation. The PRIs work has to be facilitated not only by local administration and the NGOs, but also on their own. The PRIs have to adhere to the techniques of owning their responsibility and coordinating with all stakeholders. The development planning undertaken by the PRIs must have disaster mitigation components in-built in every development project being conceived by them.

In accordance with their mandate, as per the Eleventh Schedule of the Constitution, the PRIs should have Developmental Schemes in the areas of (a) Land improvement, land consolidation, soil conservation; (b) Minor irrigation, water management, watershed development; (c) Animal husbandry, dairying and poultry; (d) Fisheries; (e) Social forestry and farm forestry; (f) Minor forest produce; (g) Small-scale industries including food processing industries; (h) Khadi, village and cottage industries; (i) Rural housing; (j) Drinking water; (k) Fuel and fodder; (l) Roads, culverts, bridges, and other ways of communication; (m) Rural electrification; (n) Non-conventional sources of energy; (o) Poverty alleviation programmes; (p) Education, including primary and secondary schools; (q) Technical training and vocational education; (r) Adult and non-formal education; (s) Health and sanitation, including hospitals, primary health centres and dispensaries; (t) Family welfare, social welfare, welfare of the weaker sections; (u) Women and child development; (v) Public distribution system; and (w) Maintenance of community assets.

Capacity building of the PRIs and strengthening of the financial position in keeping with their fixed and specified role and responsibility to work for the overall development of the village with focus on disaster mitigation through these schemes will go a long way in preparing and managing disasters effectively.

Panchayat level committees comprising representatives from different political parties/social organisations in the area can be formed by the Gram Panchayats to help the Government and NGOs in their performance in response actions, relief distribution, and rehabilitation programmes. The Gram Panchayat must give due help and cooperate with the Disaster Task Force in the village for its better functioning. Its major role should be to prepare the community and solicit its support and cooperation for every task.

The broad objectives of the PRIs' capacity building exercise should include:

- Strengthening the capabilities of Government officials at the village level and the elected representatives in the areas of disaster prevention, preparedness, mitigation, and response.

- Developing the capacities of various local officials and elected representatives in disaster preparedness, relief and rehabilitation.
- Stimulating government officials to enlist the support of local institutions, elected representatives and NGOs for community awareness as well as assisting the officials and local institu-tions to enlist the required support from other relevant quarters.
- Reinforcing the skills of government officials and elected representatives for appropriate hazard assessment, vulnerability analysis, resource analysis, and local capacity assessment.
- Garnering basic data from Gram Panchayats for analysis to make Micro and Macro level action plans.
- Preparing disaster management and development plans at the district, block and village levels.

Increasing Knowledge Base (Inputs) of the PRIs

There is a growing need to increase the knowledge base of the elected members of PRIs about various facets of disaster management. It will help them supplement their traditional knowledge with scientific techniques and methods. Efforts need to be made to sharpen their existing skills as well as to inculcate new skills in them for effective output and outcome. Since the number of PRI members is too large, it will be possible and cost-effective to reach out to them through the distance-learning mode. Self Instructional Material (SIM) needs to be developed in keeping with the prevalent needs. The SIM has to be more application-oriented and less theoretical in nature. It is suggested that the following SIM may contain the following modules each one specially developed with focus on practical implementation of theoretical concepts:

Introduction to Disasters

Reference in this Module shall mainly be on understanding disasters like seismological disasters, hydrological disasters, wind-related disasters, geo-physical disasters, chemical disasters, nuclear disasters, environmental disasters, and fire accidents.

Focus on Risk Assessment and Vulnerability Analysis

This Module will have a special focus on understanding hazard, risk, and vulnerability. Emphasis shall be on socio-economic vulnerability factors like gender, poverty, livestock, population displacement, commercial agriculture, erosion and flooding, shanty urban settlements, etc. Besides, the reference shall be on strategic development for vulnerability reduction, fundamentals of risk assessment and risk analysis techniques.

Module on Disaster Preparedness

This module will deal with the importance of prevention, preparedness and mitigation in disaster management. Special focus shall be on the nature and important components of disaster preparedness, characteristics of preparedness plans and the mechanism for formulation and implementation of plans. Disaster mitigation strategies, major trends in mitigation policy, ethical guidelines in disaster mitigation and managerial skills for mitigation shall be adequately dealt with.

Module on Disaster Response and Disaster Medicine

The components of this Module shall include essentials of disaster response, relief measures, management of relief commodities and services, roles and responsibilities of various agencies, immediate financial relief, managing human response, and recovery. Focus in the Module on Disaster Medicine shall be on the meaning, significance and components of disaster medicine. It will comprehensively deal with health preparedness and planning, and community health management in disasters, socio-psychological interventions in disasters, epidemiology in disasters, and health response and recovery.

Module on Rehabilitation and Reconstruction

The knowledge component of PRI members shall be enhanced through this Module, which shall have due focus on development and disasters with emphasis on economic rehabilitation, social rehabilitation, psychology of reconstruction and rehabilitation and measures for long-term recovery.

Preparation of Audio-Video Programmes

Depending on the nature of the component to be dealt with, special audio-video programmes need to be made for enabling the PRI members to grasp and visualise for themselves as to how the response and recovery measures are put to effective use. Such programmes are to be developed in local languages preferably using local settings and conditions.

Conducting Training Workshops

Need-based training programmes with more focus on practical exercises are to be conducted at the district level involving one or two representatives from each block. These will be in the form of comprehensive Trainers' Programmes, wherein the participants will be sufficiently trained to equip all the members of their Panchayat Samitis and inculcate requisite skills needed to face the disasters. The Panchayat Samiti members will in turn do so for the Gram Panchayat members.

The major outputs of such an exercise would be:

- Orientation, sensitisation and development of the PRI members for disaster preparedness and integrated development.
- Development of human resources in the government offices and elected institutions at the local level for effective and efficient disaster management.
- Imparting skills to the government functionaries/elected leaders to enlist support of the NGOs and communities to cope with disasters in a planned manner.
- Initiating the process of preparation of disaster management plans at the village level.
- Preparation of District, Block and Village Level Disaster Management and Development Plans.

CONCLUSION

Capacity building of the community and PRIs essentially draws sustenance from viable livelihood options available in the area. Creation of livelihood options could go a long way in disaster risk reduction. Sustainable livelihood approaches start from developmental standpoint and place the topic of livelihood at the centre of the discussion. Livelihood options include the capabilities, assets, and activities required for a means of living. A livelihood is sustainable if it can guard itself from stresses and shocks and also maintain or enhance its capabilities and assets without undermining the natural resource base. Operating within the vulnerability context, under the considerable influence of transforming structures and processes, poor people choose and implement livelihood strategies using their livelihood assets.

Attempt needs to be made towards creating more sustainable livelihoods with due emphasis on food security and sustainable use of the natural resource base. Increased well-being with regard to non-material goods such as self-esteem, health, cultural heritage as well as resistance to shocks and seasonality are also quite essential. This could be achieved through systematic community-based preparedness programmes, which give due emphasis to community participation, gender sensitivity, education schemes, and awareness campaigns. Capacity building exercises to educate, train, sensitise, orient and equip the PRIs, especially the Gram Panchayats and Panchayat Samitis will go a long way in generating the environment so essential for disaster risk reduction and for development of alternative livelihood strategies.

15

Assessing Abilities for Disaster Risk Reduction: Discovering Possibilities and Exploring Options

— **Hina Shah**

INTRODUCTION

In recent years, many rain water harvesting/collection skills have been forgotten, with the increase in piped water supply. However, population growth and economic development have forced people to settle in areas that have scarce water resources. In certain parts of the dry zone such as the coastal belts with salt-water intrusion into the ground water and some uphill localities in the wet/dry zone, rain water harvesting remains an important and sometimes the sole source of water.

Anjar and Halwad are blocks of Kutch and Surendranagar districts of Gujarat in India. Though the average rainfall received by the state of Gujarat is 500 to 600 mm per year, the average rainfalls in these blocks are only between 200–250 mm. Agriculture is difficult, and only certain types of crops can be grown, which are not dependent on ground water resources as such resources are not available. Dug wells are there, of course, but they deplete so fast that the water is available only for a limited period of 4–5 months in a year.

The Anjar block faces severe drought conditions for 3 years out of 10, and moderate/mild drought for 4 years out of 10 on an average. Hence, water shortage is a way of life, and the lives of women are very tough due to their water collection responsibilities, which can take up 4–8 hours of a day.

While people in such areas of Gujarat were struggling to survive the drought, on 26th January 2001, an earthquake struck, which turned out to be one of the most dreadful and nightmarish in the history of Gujarat.

The earthquake had superseding effect on human life, environment, infrastructure, individual and community assets, movable property, communication facilities and the psyche of the people. The loss was severe. Life came to a sudden standstill.

Understanding Vulnerability and Capacity (Rehabilitation Needs Assessment)

The International Centre for Entrepreneurship and Career Development (ICECD) team initiated a quick survey of the worst effected interior villages in Anjar and Surendranagar districts, to glean the extent of damages—social, economic and psychological—through community meetings and interactions with the affected people.

The survey identified the following needs of the areas and the people there:

1. The study indicated that there were serious gaps in addressing the issue of disaster mitigation and the planning process for it.
2. There was massive damage to houses (reduced to rubble) and in many cases, this was coupled with loss of life within the family. The issue of providing the victims temporary shelter till a permanent rehabilitation work plan for housing was finalised and implemented was a pressing one.
3. The most severely affected issue was the livelihood of the people, especially the vulnerable groups (farmers, artisans, landless labourers, women, etc). They had not only lost their assets and confidence, but had absolutely no direction about what to do and how.
4. The resources and facilities for day-to-day life as well as for agriculture were also greatly affected. The need was, therefore, to restore community facilities and water resources and to provide adequate support to reestablish agricultural farming, livestock and allied activities, which were, and are, the major sources of livelihood in these villages.
5. The basic need was to systematically identify village-wise, activity-wise and specific group-wise, the needs for socio-economic rehabilitation of the people and develop short-term as well as long-term rehabilitation plans.

ICECD—ACTING AGAINST DISASTER IN KOYBA AND LAKHAPAR

Even when recovering from such a tragedy seemed impossible, it was actually to be made possible, by encouraging the human spirit and survival instinct. This was the scenario when ICECD—The International

Centre for Entrepreneurship and Career Development—stepped in. At this stage, ICECD took up the challenge of:

- Ensuring economic empowerment for those who lost their means for survival.
- Providing peoples with opportunity and access to drinking water, livelihood options, and self-employment.
- Enabling people to recover from the shock and assisting them in lead a normal life with dignity.

Project Implementation Strategies

This ITDG-ICECD project was started in April 2001 and detailed activity plans for rain water harvesting (RWH) and livelihood programmes in two villages, namely Koyba and Dhawana in Halwad block of Surendranagar district in Gujarat, India. Based on the initial plans, promotional work was done in these two villages. During the promotional work it was understood that because of certain caste and cultural problems, work would be difficult in Dhawana. Hence Lakhapar village in Anjar taluka of Kutch district, Gujarat was selected with ITDG concurrence to replace Dhawana. Technical experts and expert trainers took up the task of project implementation.

Poor rainfall over the previous couple of years along with serious drought conditions and damages to existing water sources had made water availability the biggest concern. Hence this intervention was widely welcomed by the villagers. On the basis of budgetary provisions, it was decided to make 10 rain water harvesting tanks each, in both the villages. These were to be kind of demonstration units, which would create among the people the impact of and the awareness of such structures. Along with this, the livelihood options were to be improved by providing training and reviving the existing livelihood aspects. New opportunities were also tapped for this purpose. The approach and strategies for the project implementation included the following:

Approach

Participatory approach. The specific needs assessment was conducted for appropriate development intervention in close association with the community. People in both the villages actively participated and contributed their time and labour for the planning and construction of RWH structures. Their participation was even more pronounced in the livelihood projects.

Providing demonstration effect. The planned development intervention for creating rain water harvesting structures and training the people in utilising livelihood options were initiated to provide a 'Demonstration Effect', through people's participation, awareness creation and actual implementation of the project activities. The capacity building process of implementation of these projects provided them confidence and a new direction to work towards socio-economic empowerment.

Sustainable development. Through capability building and training men and women, people developed the ability to plan their economic activities in such a way that it would provide the maximum income. Further, right from the start, there was contribution by the beneficiaries in building the rain water structures. This was indicative of the help people would receive in building such structures on their own in the times to come.

Gender balance. In view of the critical role of women in managing such disaster situations the project aimed to achieve a gender balance. Gender sensitivity was integrated at all the stages of the project i.e. planning, implementing and follow up and 50% women participated in both the projects.

Strategies

The following strategies were incorporated in the implementation plans:

- Identifying specific needs of the people
- Creating awareness for water harvesting and livelihood alternatives
- Making people psychologically ready and technically capable to combat disasters
- Training and capacity building of men and women in the villages
- Ensuring community participation
- Demonstrating impact of sustainable water and enterprise development Interventions.

Implementation of Rain Water Harvesting Tanks

Based on the ITDG Technology and adopting it to local conditions, rain water harvesting activities were carried out through the construction of underground storage tanks. Tanks having a capacity of 8000 litres were constructed with the help of local masons as well as identified beneficiaries. Ten such tanks have been constructed in each of the two villages, Koyba and Lakhapar. Following steps were initiated in the process of construction of RWH structures:

Activities Undertaken

(i) Local awareness was created by organising Gramsabha (village community meeting).

(ii) Two-day awareness workshops were organised, to provide knowledge to the people about the various benefits of these structures in their lives.

(iii) Identification and selection of household and beneficiaries were carefully done keeping in view the following criteria:
- Economic back ground—they should be from poor or low income families
- Backward community—priority to such people
- Family size—not exceeding 7 members
- Acceptance and participation—responsibility of maintenance.

(iv) Construction of rain water harvesting tanks of 8000 litres capacity in Koyba and Lakhapar villages.

(v) Handing these over to the community/individuals for maintenance and use.

Participation and Contribution

During the actual construction of rain water harvesting tanks, people contributed in terms of time, labour and materials. The contributions came as follows:

- Actual digging work and related labour
- Helping local masons during construction work
- Contributing materials in form of sand and grit
- Providing appropriate place near the house.

ICECD technical experts with the help of experienced masons demonstrated the construction of 2 tanks initially. During this period the community members, beneficiaries and local masons were involved and were trained on the job. Thereafter, local masons themselves, with the full participation of the benefiting families, carried out the construction of the remaining eight tanks.

Handing over to Community

A local committee representing men and women from different strata was formed before initiating the work. This committee and the beneficiaries involved were oriented and made aware of the responsibility in maintaining the tanks. Ultimately the responsibility was handed over to individual household owners, to be supervised by the committee. Once the tanks were built, a short training on *how to effectively keep the stored water clean and how to use the water* was organised for these 10 specific families.

Implementation of Livelihood Training

Promotion

As a result of the earthquake, people had either lost their assets or had lost all livelihood options. After carrying out a needs assessment study, some potential livelihood options were identified. These were:

- Petty trades like grocery shop, etc.
- Servicing venture like repairing of household electrical gadgets
- Craft related articles like mirror work, cloth printing and sewing
- Agricultural related activities like supply of seeds and fertilisers.

Information on various livelihood options also included dairy and animal husbandry related activities. These were explained through promotion pamphlets and awareness workshops for men and women. Further, close interaction and counseling were provided to the interested people through support institutions.

Identification and Selection of Potential Participants

Specific criteria were set for the identification and selection of potential participants. These can be summarised as under:

- Participants were selected from the age group of 18–45 years and from poor and low-income strata of the community.
- Gender balance was observed involving both men and women while the selections were made. Selection was based on the entrepreneurial attitude and behavioral competency of the people.

Enterprise Development Training

A livelihood-training programme for enterprise development was conducted to impart the knowledge and skills the participants would need for developing their business capabilities and initiating and managing new economic activities.

The 10 days training programme included the following inputs:

- Identifying and assessing livelihood options
- Assessing feasibility
- Role, function and responsibility of an entrepreneur
- How to conduct a market survey
- Project planning, formulation and preparation
- Technical requirements of machinery, equipments, materials, etc.
- Motivation and behavioral inputs
- Small enterprise management including marketing, financial management, etc.
- Information and linkages on funds, market and technology.

The training used participatory methods and simulation games including fieldwork, which facilitated learning for illiterate people.

Support for Setting up Livelihood Activities

Once the training was over, the participants were provided guidance on procuring material, machines, and setting up of business ventures. Both individual counseling and group counseling were conducted to resolve operational problems faced by trained people in setting up the livelihood activities.

Experienced trainers of the ICECD, with the help of local experts from Banks, District Industries Center, Agricultural Department and Animal Husbandry Department provided need-based support in conducting the training and guiding people to set up their own enterprises.

Achievements

The major achievement of the exercise undertaken was that the interventions in Lakhapar, Anjar Taluka, Kutch District effected improvement which may be summarised as:

Rain water harvesting tanks. Ten rain water harvesting tanks/structures were built and are now being used by the community. Most of these participants belong to the backward community especially the scheduled castes. The benefits directly reached 10 families (65 members). Indirectly, awareness has been created amongst at least 500 people in the village.

Livelihood training. All most everyone in Lakhapar is involved in agriculture activities. 55 participants were selected and trained for livelihood activities.

- Training related to agriculture was imparted to 20 small and marginal farmers.
- Training for Micro Enterprise Development was imparted to 35 women.
- Micro enterprise has already started with 15 women.

Interventions in Koyba, Halwad taluka, Surendranagar District had the following results:

Rain water hrvesting tanks. 10 rain water harvesting structures have been built and are now used by the community. The beneficiaries belong to the backward community.

- The benefits directly reached 10 families (55 members).
- Awareness was created amongst atleast 300 people in the village.

Livelihood training

- Training related to agriculture was imparted to 30 farmers.
- Training for Micro Enterprise Development was imparted to 23 people.
- Micro enterprise has already started with 7 participants.

OBSERVATIONS AND LEARNING

Initiating and implementing this unique pilot project on rain water harvesting and livelihood options in two earthquake affected villages involved a lot of challenges and learning. Some useful observations are summarised as follows:

1. Rehabilitation work in earthquake-affected areas initially taken up by all agencies tended to be excessively welfare-oriented. This created too much expectation amongst the community and the feeling prevailed that everything will be given freely and will be readily available to them. In view of this, the promotion and awareness creation for developmental interventions in RWH and livelihood was carried out very carefully, devoting ample time and focused efforts, since it called for the people to realise the need for their contribution and partnership for sustainable development.
2. Participation and involvement of local community was the key factor. However, because of their emotional, psychological and economic conditions, resistance to participation was quite high initially. Further, the nature of their participation also needed to be properly planned. Hence, a participatory approach involving creation of wide based awareness and identifying the right people was taken up.
3. Rain water harvesting is not a concept new to the Indian culture, but its adoption to local condition needs systematic development interventions. The community did not achieve full benefits in Lakhapar, as the rain pattern in Kutch changed dramatically that year. It was experienced that in the absence of sufficient rains in the late phase of the monsoon (Sept.–Oct. 2001), after using much of the water from the constructed tanks stored in the initial part of the monsoon especially for the construction of temporary houses, the community could not store sufficient water subsequently. However, in Koyba, all tanks were fully filled up, to be used in course of the year.
4. Livelihood training and agriculture related activities provided the right boost to people's motivation in Lakhapar. This intervention has changed the very life and thinking of people. Women could harness their potential and for the first time ventured into business. This provided lot of confidence to them. However linkages for finance with banks are not smooth and many had to start their Micro Enterprise on a small scale with their own resources. ICECD has provided micro credit for starting some enterprises from its micro credit fund.
5. The 'Demonstration Effect' of the project was very positive. More and more people of the community having realised the need, and having seen the benefits derived by targeted groups, are now are approaching ICECD to help them. New linkages will have to be brought in for continuing such activities at village level for larger community outreach.

Areas of Concern

The drought was the problem until last year, while the earthquake represents the prime disaster of early 2001. Hence the focus of the support was to rehabilitate people through a variety of interventions.

In the context of the work undertaken, some important issues emerged during the project implementation for future consideration. We will discuss some of these issues.

Disaster Management

- Though triggered of by natural events such as a drought or an earthquake, disasters within disasters are generally man-made. The impact of a natural disaster is aggravated by unwise human actions, which should be carefully prevented. Often, natural disasters are the result of unwise and selfish human actions and behaviour.
- The social development of people and their communities in disaster prone areas has been grossly ignored. The appalling human conditions in such areas can best be fought through the weapons of self-reliance, literacy, preventive health care and equitable pricing structures for rural food producers thereby helping people to respond to their own basic needs, within their environment.
- Allocation of work should specify who will do what and clarify the organisation as categories under the public sector, NGOs, army or civil administration. Otherwise natural disasters lead to chaos, anarchy and destruction due to confounded confusion.
- Most of the scientific efforts and money devoted to natural disasters are spent on studying the climatological and geological aspects of disasters—over which humans have very little control—rather than on studying the wide range of human actions over which human beings do have some control and which can be analysed and planned so as to develop precise response mechanisms to pre-empt the harmful effects of disasters.
- Since disaster is a state subject, the State Government is responsible for rescue, relief and rehabilitation but national resources are available and plans to use these should be formed in advance.
- There should be plans for every district, in which response mechanism should be clearly spelt out; what, how and when the relief is to be given, should be specified.
- Researches have concentrated on the physical aspects of disasters rather than the social and political aspects; governments have been lured into putting their trust in physical prevention and mitigation measures. The social side is usually attended to only

long after the event. This attitude needs to be changed by creating people's organisation ready to respond.

- Disaster management process and methods raise questions for the various groups interested in disaster relief measures and rehabilitation. These measures can succeed only if the people empowered to deal with the disasters are sensitive, honest and have integrity and vision for better future.
- Most problems in the rural areas are seen as being under separate categories. Disaster prevention and mitigation should be treated primarily as an important aspect of all round development.
- The very essence of disaster mitigation is long-term planning, while the very essence of the rural poverty is the struggle for day-to-day existence. These aspects clash after the disaster when people are worried about day to day affairs while planning programmes talk of long-term measures.
- Effective disaster mitigation and prevention depends on long-term planning for development towards a more sustainable and less vulnerable society.
- Some agencies are working for new types of community-based disaster prevention programmes. Improved self-reliance in respect of food and fuel, access to land, employment, better health and nutrition, family planning and education are important components of disaster prevention. Disaster prevention and the elimination of poverty are closely linked, as are poverty and environmental degradation.
- Journalists should seriously consider whether it is enough to merely count the dead and describe the damage or whether the media is morally obliged to look behind the scenes of destruction to expose ways in which people, politics and social systems have been responsible for mishaps within any given disaster so that such instances do not recur.

Rehabilitation, Displacement and Development

Disaster often forces the otherwise stable, sedentary population to move away from their original and established places of home and work. It uproots people from their original land and creates psychological stress leading to many dysfunctional consequences. Very often, population does not share social and economic benefits of developmental projects equally.

- All developmental efforts imply planned change to improve existing conditions, whereas rehabilitation is designed to restore the status quo. This makes it imperative to understand the term rehabilitation in a much wider perspective.
- Development and rehabilitation efforts depend on the target groups (men, women, youth, children) their perception and

awareness of the situation, such as the fears and apprehensions of the affected people about the possible problems to be faced by them, acceptability of the proposals by the host population, fulfillment of government promises and understanding about various developmental projects.

It would be interesting to understand the impact of disaster at different levels.

Psycho-sociological consequences of displacement. It is universally acknowledged that displacement is stressful and sometimes leads to traumatic conditions. In spite of this, the psycho-sociological issues involved in the displacement of people after the earthquake have often been neglected.

- Rehabilitation programmes often lack specific components of the aspects of mental health of people. There is no evidence of discussion of mental health problems and their implications for assessing cost and benefit of development projects.
- Human beings are constantly seeking gratification of their physiological, security and social needs in unique manners through their established sources of livelihood, social relations, social status, kinship ties, etc. Any change or blockage, though temporarily, in satisfying these needs results in varying degrees of tension and stress in different people.
- Fear of change of source of livelihood leads to 'occupational disruption'. This subsequently leads to low to high degree of 'occupational redundancy'. This loss, even if substituted by new ways of managing earning of livelihood, often leads to a state of mental stress. Loss of agricultural land by the landowners, though made up by 'economic compensation' and also by the offer of new occupational opportunities under various 'rehabilitation programmes' by the Government, could not fulfill the psychological loss such as, that of the social status characteristic of 'feudalistic culture' or the village constitution which provided the victims, psychological satisfaction.
- Displacement after disaster also forces individuals to change the pattern of social relationships. Individuals have to undergo an entire process of resocialisation and adjustment in a new social milieu. This leads to loss of existing social relationships. For this reason, in some instances, inspite of building them homes, the affected population has not occupied the new housing facilities.

Management of Displacement and Rehabilitation Process

Minimising the adverse effects of disasters forms the primary focus in achieving the efficacy of various objectives of rehabilitation projects. The success of these projects depends upon, how effectively the disaster is managed, and how the affected population perceives the various rehabilitation programmes as appropriate and adequate sources for the gratification of needs. This makes it imperative to plan, design and implement strategies as 'preparedness measures' and 'rehabilitation programmes', to cope with specific aspects of a disaster at appropriate stages of the rehabilitation process.

Strategies for minimising psycho-sociological consequences at various stages of the rehabilitation process are suggested.

Promotional stage. Preparedness measures and coping strategies need to focus on all the factors constituting a disaster so as to counter its overall impact on the population. The rehabilitation process must be pursued with a developmental objective rather than merely as a coping mechanism for the situation. All preparedness measures and coping strategies at each stage of the process must reflect the achievement of the developmental objective. Therefore, appropriate interventions at Government level, NGO level, affected population level and community level, become very important.

Implementation (manifest) stage. Careful and timely handling of cognitive and affective aspects through appropriate preparedness measures would greatly reduce the psychological tension and stress generating dysfunctional behaviour at the manifest stage. In many cases, the Government has taken an unduly long time in deciding the compensation package. Evidences suggest that people have resorted to mass agitation for compensations assured under certain development projects. Such a situation can be effectively handled through negotiations between the Government and affected population. All infrastructural facilities such as land, houses, schools, hospitals, etc., should be provided to the people with their consultation.

Rehabilitation stage. Rehabilitation programmes necessarily need to be designed according to the expectations of the affected population. Rehabilitation programmes should make use of available skills and talents and also focus upon providing new skills and competencies to the affected population, to equip them to face the outer world. Efforts should be made to develop commitment to self-support and sustenance, to eliminate dependency syndrome at the onset itself through psycho-therapeutic/health programmes implemented in the form of family counseling and stress reducing exercises, with community programmes and other socio-occupational programmes forming an essential complement to the rehabilitation package.

Rehabilitation aftermath. Effective management of earlier stages will greatly reduce further demands by the people from the government.

It is believed that the perspectives presented herein are functional and would be able to provide guidance in overcoming many of the deficiencies of the current management approach to problems associated with rehabilitation.

16

Disaster Risk Reduction through Disaster-Resistant Construction Techniques

— Alka Dhameja

INTRODUCTION

The combination of extreme environmental changes and rapid social, economic and institutional transition has triggered a spate of disasters in the last few decades. The pernicious effects of disasters cannot be fully washed away; their negative consequences, however, could be reduced through effective disaster risk reduction methods. One such method is to practise disaster-resistant construction process. Despite several efforts at the government, non-government and community levels, this type of construction is not being followed in disaster-prone areas. Compromises on building design, quality of construction as well as blatant flouting of building codes have resulted in heavy losses of life and property during disasters. Many houses and buildings crumble as if made of a pack of cards in the event of an earthquake or a cyclone. Inundated houses are a common sight during floods. Lack of knowledge, awareness and training on construction techniques has largely compounded the problem.

There are various pertinent structural and non-structural measures to combat disaster risks. As per the World Disasters Report, the structural measures are meant to keep disasters away from people, whereas, non-structural measures aim at keeping people away from disasters. Disaster-resistant construction is a potent structural disaster mitigation strategy for reducing disaster risks. Analysis of disaster-resistant construction process for specific disasters could go a long way in reducing the risks of catastrophes. The factors that need to be kept in mind at the time of planning for disaster-resistant housing for earthquakes, cyclones and floods are: (a) Location and Siting (b) Layout and Design, and (c) Construction Techniques.

EARTHQUAKES

Seismic movements cause relative differential movements in all vertical and horizontal structures of a house making it sway in different directions. Therefore, for safety against collapse during earthquakes, proper reinforcing of concrete and masonry with steel bars is considered crucial. The idea is to increase the ductility, deformability and damageability of houses so that they perform better during earthquakes without collapsing. The various codes on earthquake-resistant design, which lay emphasis on the structural integrity of all elements of the buildings, diaphram action in roofs and floor, provision of R.C.C. bands at plinth, sill, lintel and roof levels, and reinforcement around openings and critical sections need to be strictly adhered to.

Location and Siting

Buildings constructed on sites with open and even topography are usually less damageable in an earthquake than those on strip-shaped hill ridges, separated by high hills, and steep slopes. Steep sites may have problems of landslides and rock falls and should either be avoided or effectively improved, if adopted. Terracing and constructing retaining walls may improve buildings on a steeply sloping site. A site liable to liquefaction or subsidence may be improved by stabilisation by sand piling, etc. before going in for construction work. Very loose sand or clayey type of soil has a tendency to undergo compaction and should be therefore avoided.

Layout and Design

Buildings, which are structurally strong to withstand earthquakes sometimes, fail due to inadequate foundation design. Tilting, cracking and failure of superstructure may result from soil liquefaction and differential settlement of footings. Certain types of foundations are more susceptible to damage than others. For example, isolated footings of columns are likely to be subjected to differential settlement particularly where the supporting ground consists of different or soft types of soil. Mixed types of foundations within the same building may also lead to damage due to differential settlement.

Analysis of building frames for the lateral earthquake loads could be carried out by a number of methods.

- (i) Approximate methods based on statical equilibrium making the frame statically determinate by a number of assumptions.
- (ii) More accurate methods using plane frame approximation but considering stiffness of the beams and columns.
- (iii) Computer analyses using two-dimensional or three-dimensional idealisations.

Buildings as well as their structural frames are subjected to torsion, especially when the centre of gravity of masses above any storey and the centre of stiffness of the elements of the storey do not coincide and there is, thus, an eccentricity between the actuating and the resisting forces. This happens even in symmetrical frames, so the torsional shears and resulting moments in the elements should be analysed.

Usually, the horizontal seismic forces as determined earlier are sufficient for designing buildings. In some cases, however, vertical seismic forces also become important and should be considered in design, either alone or in combination with the horizontal seismic forces. The structure should be designed to have adequate strength against earthquake effects along both the horizontal axes. This can be achieved either by using columns of square sections or by orienting the major axes of the columns or shear walls along the two axes of the building. The rigidity distribution in a building along the vertical direction should also be regular. Another important feature is simplicity and symmetry in the overall shape of a building. A building shaped like a box, say, rectangular both in plan and elevation is inherently stronger than one that is L-shaped or U-shaped.

Modern buildings in the alluvial soil areas in Europe and Japan use a new concept of shifting the centre of gravity in the earthquake-resistant design. The plan design includes placement of a steel weight on the top of the building. In view of an earthquake situation, a provision is made for tremor-resistant effect. In actuality, what happens is that when the building starts swaying, sensors in the basement are activated. The computer sends a message to the hydraulic power device on the terrace, which in turn, causes the steel weight to move in such a way so that the motion caused by the earthquake is offset. Thick pillars of concrete and steel are buried deep below the foundation. As a result, buildings stay erect when hit by an earthquake.

Construction Techniques

Earthquake-resistant construction techniques could aid in lessening the destructive aftermath of earthquakes. Needless to say, structural engineers must approve all construction projects and the laying of foundation must depend on the nature of soil in the area. The endurance of earthen houses during earthquakes is generally poor. The performance of brick buildings during an earthquake depends on the type of roof, mortar used and overall quality of construction. The performance of stone masonry buildings is the worst during earthquakes. The wooden dowels generally improve the performance of constructions during earthquakes. Some of the points that need to be kept in mind at the time of designing general earthquake-resistant constructions are:

(i) RC columns should support the long walls in the building. All the junctions should be secured with the help of steel so as to impart extra strength to the walls. Load bearing walls can be strengthened with steel reinforcements along with continuous horizontal ties at the sill, roof and plinth levels. Columns or shear walls should run preferably continuously from foundation to the roof. In general, large window and door openings in the walls of a building tend to weaken the walls. Therefore, fewer and smaller the openings, lesser the damage the building will suffer during an earthquake. The wall thickness should not exceed 450 mm with mud mortar and 350 mm if cement mortar is used. The stones on the inner and outer 'wythes' should be interlocked with each other.

(ii) Use steel frame door shutters instead of natural timber. Brittle materials crack under load; the addition of steel reinforcements can add ductility to brittle materials. Masonry and concrete, for example, can be made ductile by proper use of reinforcing steel. Doors, especially those of bathrooms should open on the outside. Suspended ceiling should preferably be avoided. Where necessary, it should be light, adequately framed; and securely connected. Use tie beams to connect individual column foundations or combined column footings or provide rafts or piles as needed for the loads. Provide X-bracings in planes of rafters to withstand tremors. Also, provide adequate iron straps on wood truss joints. 'Through' stones or bonding elements of full-length equal to wall thickness should be used at an adequate distance horizontally.

(iii) Projecting parts like large cornices, parapets, and all types of ornamentation should be avoided as far as possible; otherwise, they should be properly reinforced and firmly tied to the main structure. The mortar should preferably be cement-sand (1:6), otherwise, lime-sand (1:3) or clay-mud of good quality. Ceiling plaster should preferably be avoided; otherwise, it should be as thin as possible and applied with care to ensure good adhesion.

(iv) Since the earthquake force is a function of mass, the building should be as light as possible, consistent with structural safety and functional requirements. Roofs and upper storeys of buildings, in particular, should be designed light. Main switch in the building is a must, so as to cut off electrical supply to the whole building at the time of an earthquake. Also, install circuit breakers as most fires are caused by short-circuits.

(v) A new concept of isolating the structure from the motions of the ground is now emerging. This is known as base-isolation. For simple buildings, this can be achieved by reducing the coefficient of friction between the structure and the foundation or by placing a flexible link between them. One long-standing method has been to make a house on short posts, which rests on large stones. So when the earthquake strikes, the posts are pin-connected to the top and the bottom of the building and are thus able to sway to and fro. This causes a substantial reduction of the lateral forces and thus, isolates the building from the high frequency movements on the ground. However, in case of massive earthquake, this technology is not very effective.

As far as earthen houses are concerned, the soil used in construction should be clayey. It should have good cohesion and dry strength, but should not show fissures on drying. The height of an earthen house should be restricted to one storey or one storey plus attic. The height of the walls should not be greater than eight times the thickness. Longer walls should be strengthened by intermediate pilasters/ buttresses. Hand-formed walls should preferably be made tapering. The footing should preferably be built by using stone or fired brick laid in cement or lime mortar. Alternatively, it may be made in lean cement concrete with plums. The wall above foundation up to plinth level should preferably be constructed using stones or burnt bricks laid in cement or lime mortar.

In the case of wooden houses, attention must be paid to proper seasoning and chemical treatment of timber before use, fire proofing details and maintenance, using tight joints having galvanised iron straps with nails/screws/bolts, starting the timber construction above masonry plinth/pedestals above the high flood level and bracing the wood-frame in vertical as well as horizontal planes to prevent its distortion under the lateral seismic forces. The foundation may be strip type or isolated pedestal type and the superstructure is to be fixed to it. The plan of the building should be surrounded and divided by bearing wall line. All bearing wall lines of the upper storey should rest on the bearing wall lines of the lower storey. Horizontal struts and diagonal braces should be used to stiffen the frame against lateral loads due to earthquake and wind. The wall covering may consist of matting made from bamboo, ekra reeds and timber boarding.

For brick and stone buildings, in addition to the general earthquake-resistant features, the total width of openings in a wall should not be more than 50 per cent of wall length in one and two storey houses; and not more than 40 per cent in three storey buildings. Width of piers between openings or from openings to wall corners needs to be not less than half the height of the openings. Use of steel or wooden dowels going into the walls, meeting at corners or T-junctions need to be provided for effective bonding between them. The diagonal bracing may be framed into the verticals or nailed to the surface. It will be beneficial to use holdfasts screwed to the wooden members for providing positive connection between the frame and the brick for preventing out of plane falling of the brickwork during earthquake shaking.

CYCLONES

Different types of buildings behave differently under the impact of cyclones. Fully-engineered structures are damaged as a result of overturning of bridges, collapse of high steel towers, blowing off of doors, windows and roofing material. Severe damage to marginally engineered structures is caused due to lack of proper connection between walls and roof and improper strength of the structure, blowing off of railings and parapet walls and twisting of transmission towers. Non-engineered structures simply collapse in the aftermath of cyclones. Weak connections among various elements of the building leads to failure of structures.

Location and Siting

As far as cyclones are concerned, the building sites, which offer shielding from high winds such as main landward side of a hill range, are preferable. The building site should be much above the probable maximum tide level or the ground should be raised to that level. Piles could be used if site consideration requires so. The sites, which lead to wind concentration, should be avoided. The water retreating to the sea applies substantial scouring; therefore, pitching of slopes will be desirable. Sites, which lead to wind concentrations such as protruding ledges on hill slopes, should be avoided.

Layout and Design

The foundation should be taken up to a firm natural soil level, so that the resistance under the footing may not be lost due to flooding. Piles should be used if a site consideration requires so, from the bearing capacity or scouring point of view. All dynamically sensitive structures in cyclone-prone areas such as chimney stacks, specially shaped water tanks, transmission line towers, etc., should be designed following the dynamic design procedures given in various IS codes.

A considerable blocking of wind can be achieved by designing windbreaks on the windward sides of the buildings while planning for cyclone-resistant construction. In case of normal prevailing strong winds, proper orientation of buildings ensure minimum exposure to the wind. It is desirable to place the smallest face of the building across the prevailing high wind direction. In the layout plan of settlements, if successive rows of buildings are spaced apart at less than seven times their heights, the wind movement will be reduced. More plantation and afforestation around cyclone vulnerable areas gives greater depth of protection and must be therefore taken care of at the layout stage.

A symmetrical building with a compact plan-form is more stable. In case of designing for a group of buildings, cluster arrangement can be followed in preference to row arrangement. However, in certain cases, both may give rise to adverse wind pressure due to tunnel effect. Extensive studies need to be conducted to look into this aspect. For individual buildings, a circular or polygonal plan shape is preferred over rectangular or square plans, but from the viewpoint of functional efficiency, often a rectangular plan is commonly designed.

Load effects shall be determined considering all critical combinations of dead load, live load and wind load. In the design of elements, stress reversal under wind suction should be given due consideration. Members of flanges, which are usually in tension under dead and live loads, may be subjected to compression under dead load and wind, requiring consideration of buckling resistance in their design. Corresponding to the building dimensions (length, height, width) the shape in plan and elevation, the roof type and the slopes as well as projections beyond the walls, determine the coefficients for loads on all walls, roofs and projections. It is desirable to calculate the wind loads on the various elements normal to their surface. Decision on the lines of resistance, which will indicate the bracing requirements in the planes of the roof slopes, at the eave level in horizontal plane, and in the plane of walls needs to be taken at the layout stage.

Construction Techniques

The following points are important with regard to constructions that may be exposed to cyclones:

(i) The footing should preferably be built by using stone or fired brick laid with lime mortar. Alternatively, it may be made in lean cement concrete with plums. Where a building is constructed on stilts, it is necessary that stilts be properly braced in both the principal directions. This will provide stability to the complete building under lateral loads. Knee bracing will be preferable to full diagonal bracing so as not to obstruct the passage of floating debris during storm surge. In case of hilly regions, where construction is made after cutting terraces on the hill slopes, it is essential that for the stability of slopes, a minimum edge distance of the foundation from any terrace be kept and that the foundation should rest on the natural firm strata. Further, proper drainage of the area should be ensured for surface water to flow unobstructed.

(ii) The roof projections should be kept to a minimum, the larger projections must be tied down adequately. For the purpose of reducing wind forces on the roof, a hipped or pyramidal roof is preferable to a gable type roof. Flat roofs should be avoided. In areas of high wind or those located in regions of high cyclonic activity, light weight and low pitch roofs should be avoided and if necessary, must be strongly held down to purlins and rafters.

Pitched roofs with slopes will not only reduce suction in roofs but would also facilitate quick drainage of storm water. Openings just below the roof level should be avoided except that opposite walls should have two small vents without shutters so as to prevent suffocation in case the room gets filled with water and people climb up onto lofts or pegs. The openings should have strong closing/locking arrangement and glass/wooden panels should be securely fixed. In case of roof tiles, the overlap joint along the edges should be provided in cement mortar. A cement mortar screed, reinforced with galvanised chicken mesh, may be laid over the entire tiled roof. Precautions should be taken to prevent the rain water from soaking the wall so that the mortar is not softened due to wetness.

(iii) One of the most damaging effects of strong winds or cyclones is the extensive breakage of glass panes caused by high local wind pressure or impact of flying objects in the air. A broken glass pane of windward side opening increases internal pressure abnormally and may lead to a chain of events including a roof failure. It is therefore necessary to provide well-designed glass panels. In cyclonic regions, where the exposure to high wind and gustiness is sustained, it is recommended that in designing glass panels, any relief, by way of increase in permissible stresses on account of the consideration of wind load, be not allowed. Recourse may be taken to reduce the panel size to smaller dimensions. Also pasting thin plastic film or paper strips can strengthen glass panels. This also helps in holding together the debris of glass and prevent them from flying in case of breakage.

(iv) The proper connection of roof framing to the vertical load resisting elements i.e. wall or post, is equally important for the overall stability of the roof. Care is particularly needed while connecting roof trusses to R.C. columns or masonry walls in cyclonic regions, by providing properly designed anchor bolts and base plates. Typical connection of wooden framing to wooden post is through cyclone bolts or metal straps. In order to reduce the induced flutter/vibration of the roof during cyclones, it is recommended that all members of the truss and the bracings be connected at the ends by at least two rivets/bolts or welds. Further, the cross bracing members should be welded/connected at the crossings to reduce vibrations. In case of thatched roof, it should be properly tied down to wooden framing underneath by using organic or nylon ropes in a diagonal pattern. For connecting the wooden members, use of non-corrodible fixtures should be made. If non-metallic elements are used, these may need frequent replacement.

(v) The thin reinforced concrete slabs, may also be subjected to uplift under wind speeds. Larger slabs require holding down by anchors at the edges, and reinforcement on top face. As a guide, there should be extra dead load (like insulation, weathering course, etc.) on such roofs to increase the effective weight. A through and through tie of bamboo or timber flat should be provided along the edges of sheeted roofs, in addition to intermediate ties for long roofs. After a cyclone warning is received, a rope net, properly anchored to the ground should preferably hold down all the lighter roofs. It will be useful to provide damp-proof course at plinth level to stop the rise of pore water into the superstructure. Through stones of full-length equal to wall thickness or bonding elements should be used in every 600 mm lift horizontally. Use of bonding elements of adequate length should also be made at corners and junctions of walls to break the vertical joints and provide bonding between perpendicular walls.

(vi) The wood used in cyclone-resistant houses should be well treated with preservative, so that it is durable against weathering and insect action. Use of a waterproof plaster on the outside face of walls will enhance the life of the building and maintain its strength at the time of cyclone or high wind as well. Resistance to corrosion is a definite requirement in cyclone prone sea coastal areas. Painting of steel structures by corrosion-resistant paints must be adopted.

(vii) Height of the stone masonry walls (random rubble or half-dressed) should be restricted to two storeys in lime-sand mortar and one storey when clay mud mortar is used. Wherever necessary, the walls should be reinforced with vertical steel reinforcement or buttressed. The height of the earthen building should be restricted to one storey only in cyclone area and to two storeys in other zones. The wall thickness should preferably be about 350 mm, and the stones on the inner and outer wythes should be interlocked with each other. The wall above the foundation up to plinth level should preferably be constructed using stone or burnt bricks laid in cement or lime mortar.

(viii) Door and window openings in walls reduce their lateral load resistance and hence should preferably be small and more centrally located. The total width of all openings should not exceed one-third of total length of a wall. Openings in any storey shall preferably have their top at the same level so that a continuous band could be provided over them including the lintels throughout the building. Openings in load bearing walls should be within an adequate distance from the inner corner for the purpose of providing lateral support to cross walls, with the storey height up to the eave level. The locking arrangement of shutters should be sturdy and the door or window frames be securely fixed in the walls using hold fasts so as to resist the local wind pressures.

FLOODS

Due to the increasing population pressure and economic compulsions, a large number of people shift to flood prone areas and engage in building activities, thus obstructing the natural function of flood plains in carrying away the excess water during the flood situation. The extent of damage to houses depends upon the particular type of flood they are exposed to. It is a well-known fact that the flood hazard reduces as the distance from river or water channel increases on account of increase in the elevation of land surface. The regular encroachment of flood plains has led the concerned authorities to exercise some control on buildings in such areas.

During coastal floods, many houses subjected to a large mass of water striking them with force collapse totally. This kind of damage is typical of coastal waves followed by cyclone. The sea waves may reach several meters above sea level and strike the structures with great force. Flash floods also bring upon sudden pressure on the structure due to the high velocity of water, often causing complete destruction. They are also capable of devastating land by eroding the upper strata of soil or sand leaving behind a rugged surface, which is unsuitable for construction. Even buildings constructed with R.C.C. may get damaged. 'Kutcha' structures made up of thatch, mud, etc., are wiped out completely. In locations where the houses are built on loose soil, complete collapse of the structure is likely in the event of inundation.

Riverine floods are characterised by low velocity and long duration. Prolonged inundation is usual with riverine floods. This may result in softening of the soil and as a consequence, uneven sagging of roof, resulting in the damage of roof covering and supporting systems, cracks in the walls, sinking of floors and scouring of foundations owing to the slow and steady flow of water. The standing water causes deterioration in the finishing of houses, like in the painting, distempering, plastering, etc. All the timberwork too gets soaked leading to all-round decay.

Layout and Design

Layout and design for flood-resistant construction lays emphasis on flood-prone area planning. The rural areas traditionally have a large amount of vegetation and afforestation resulting in reduced runoff in small watersheds. Thus, there are chances of flash floods due to heavy rainfalls in such small catchments. Changes in the land-use pattern have a great impact on the hydrology of the region and in increasing the flood potential of the catchment areas. The urban areas have a lot of paved surfaces as roads, residential areas, etc., and these give rise to flash floods situation in these areas. Damage to individual buildings and structures may be prevented to some extent by incorporating in their design, the ability to withstand the nature of inundation and high water velocity.

Layout of the settlement should be such that it does not block the free flow of water. Ground drainage and escape lane should be incorporated in the settlement layout. Growing of vegetation and afforestation in the catchments areas and along the riverbanks should be promoted. A thick plantation around the individual building will help. The buildings should be designed in such a way that the collapse of one building does not affect the adjacent ones.

Location and Siting

Taking into account the variability of the flood hazard in the river, the zones are demarcated into prohibited, restricted and warning zones. The construction of houses should be restricted to the zones involving least risk. Development of any kind in the prohibited zone should be totally disallowed to avoid damage to property and to avert flood situation upstream threatening the settlements there. A limited building activity and planned agricultural activities can be taken up in the restricted zone. In this area, minimum ground floor level, flood proofing arrangements, etc. should be provided in the houses. People wishing to settle down in the warning zone must be informed of and advised on the risks involved. Buildings should be constructed on the best bearing soil and on the highest ground available locally.

Construction Techniques

To withstand the negative impact of floods, flood-resistant construction techniques must be strictly followed in flood-prone areas.

(i) Houses should not be constructed in low lying areas, wetlands, lagoon mouths, edges of island lakes. Flood plains, narrow defile or gorges, downstream banks and flow ways should be avoided.

(ii) The houses, if built on raised mounds, should be thoroughly compacted and made of locally available soil. When houses are built on silt, they must have a rigid framework construction or should be adequately braced to reduce the risk of damage of overturn. Houses constructed with mud must be given waterproofing treatment.

(iii) The roof levels should be sufficiently above the danger level in order to provide emergency protection. The choice of roof type should be such that it gives protection for a minimum period of one day to the people who take shelter while waiting to be evacuated to safety.

(iv) Many of the traditional methods are still valid and should be adopted because they are cost-effective and environment-friendly. Provision needs to be made for channelisation ponds and floodwalls.

(v) Repairing and strengthening of the damaged and undamaged buildings need to be taken up regularly. This could vary from superficial repair to structural strengthening.

DISASTER-RESISTANT CONSTRUCTION, REHABILITATION AND DEVELOPMENT

Disaster-resistant construction must go hand in hand with the rehabilitation and development activities in the affected areas. The rehabilitation package should incorporate housing and infrastructure redevelopment, social rehabilitation programme and economic rehabilitation schemes. An effective disaster risk reduction plan must consist of preparedness schemes for health services, resurrection of educational activities and rehabilitation of disadvantaged sections especially, the women, children and elderly. Disaster risks could be effectively reduced if the development of infrastructure, agricultural rehabilitation, growth of alternative employment opportunities, livelihood options and viable disaster-resistant construction methods are given due emphasis in the development schemes.

Disaster-resistant construction does not merely revolve around the technical aspects of construction. Houses or buildings, even when constructed as per the bye-laws, codes and scientific techniques remain uninhabitable unless they take into account the traditional building methods, the socio-cultural patterns of living and the psychology of the people in the disaster-prone areas. Disaster-resistant construction process must therefore be participatory. It should involve the Government, NGOs, self-help groups, International Aid Agencies and the community at large in the planning, design and construction stages.

Latur is a case in point, where the aftermath of the earthquake in 1993 witnessed large-scale development and reconstruction work. However, due to lack of involvement of the community, most of the houses remained uninhabited for a long time. The entire approach toward implementing reconstruction and rehabilitation programmes in disaster affected areas cannot meet the basic objectives, unless it involves the target group in the planning, execution and evaluation of development schemes. For instance, the systematic implementation of the Maharashtra Emergency Earthquake Rehabilitation Programme (MEERP) in 1994 could not really involve the community and the NGOs. In fact, it only legitimised policies that were essentially paternalistic, top-heavy and manipulative.

Some efforts in the past have focused on the need for going in for contextual housing. Recognising the urgent need to create awareness and to train the local artisans in the cost-effective earthquake-resistant technologies, HUDCO has launched a comprehensive training programme in Jabalpur District in Madhya Pradesh in 1997. Habitat Polytech has organised several rounds of training for local masons under master trainers to teach them methods of safe house construction using innovative ways. In Tembhi, Latur District, HUDCO has adopted a novel approach to building houses through in-site rehabilitation. This has been done in order to preserve the emotional linkages of the villagers to the village. Improving on the traditional village square, existing roads, etc., it has retained the original character of the village while making it safer.

Disaster-resistant construction, though a very pertinent disaster risk reduction method, becomes redundant if seen in isolation of the broader economic, social and cultural requirements of the target groups. All risk reduction strategies draw sustenance from development-disasters interface. Unless the aftermath of disasters is turned into a development opportunity with due focus on the requirements of the target groups, availability of infrastructure and mobilisation of the resources and agencies involved, disaster risk reduction through disaster-resistant construction will only remain a rhetorical exercise.

REFERENCES

Andharia, Janaki, *Social and Institutional Aspects of Disaster Management and Risk Reduction*, Reflections on MEERP—The Latur Earthquake, 1994.

Disasters and Development, HUDCO's Initiatives, 1998.

Earthquake-resistant Buildings, Dos and Don'ts, The Week, April 1, 2001.

Guidelines on Improving Cyclone Resistance of Housing, Building Materials and Technology Promotion Council (BMTPC), Ministry of Urban Development and Poverty Alleviation, Department of Urban Development and Poverty Alleviation, Government of India, New Delhi, 1999–2000.

Guidelines on Improving Earthquake Resistance of Housing, Building Materials and Technology Promotion Council (BMTPC), Ministry of Urban Development and Poverty Alleviation, Department of Urban Development and Poverty Alleviation. Government of India, New Delhi, 1999–2000.

World Disasters Report, International Federation of Red Cross and Red Crescent Societies, 1999.

17

Community Participation in Disaster Mitigation and Risk Reduction (with Special Reference to Cyclone)

— **Hrusikesh Harichandan**

INTRODUCTION

Disasters are natural processes as old as human existence. The occurrence of disasters like cyclones, floods, earthquakes, volcanic eruptions, tornados, etc., cannot be prevented. Scientific development and improved technology can detect the formation of cyclones, the probability of earthquakes, etc. Some disasters like tornados still cannot be forecasted in developing countries. It is a fact that most disasters happen between the Tropic of Cancer and the Tropic of Capricorn where most developing countries lie. So disasters are unavoidable by South Asian countries. Human disasters such as competition for nuclear power in the developing countries, militant activities, etc. are equally alarming.

Disasters are of national and international concern only when human beings are affected and social life disturbed. Disasters are often believed to be fated. Their impact is widespread due to the population explosion, over-exploitation of natural resources, industrialisation and so forth. The issue whether disasters are the result of development is debatable. Programmes and schemes have been launched, without much focus on the risk aspects of the development planning.

In the aftermath of a disaster, there are attempts from every corner (Governments, NGOs, International communities) for long-term mitigation plans. But eventually the dominance of other priorities suppresses government action. Disaster management has become an *event* rather than a process of development. Linkages in mainstream programmes can reduce cost on relief heads by improving local capacity and preparedness measures. For example, school buildings constructed in cyclone prone areas can be used as rest sheds during cyclones, provided RCC structures are built. But in many schools, asbestos roofs which are of no use during cyclones and other disasters are used. The 73rd Constitutional amendment has given enough scope to people's representatives to develop local plans and need-based programmes. Yet there is still a long way to go to achieve results. Similarly, centralised planning needs to be analysed from the perspective of disaster reduction. The narrow approach to managing risk has become a priority without consideration of a long-term strategy to handle issues. The traditional approach of dealing with disasters in isolation is still used by governments, NGOs and international donors. This approach of disaster mitigation is adhoc, relief-centred and dependent. So there is a need to understand the responsibilities of various stockholders in reducing the impacts of disasters and to make communities aware of counter disaster management plans.

COMMUNITY RISK PERCEPTION

Risk perception is extremely situational and differs from individual to individual at the community level.

Characteristics

(i) Perception varies from person to person
(ii) People take risks because of non-availability of choice
(iii) People make decisions only at the eleventh hour of disaster
(iv) People leave carrying the valuables
(v) Community people even government never thinks of a disaster, which may occur in the next 5 years and media focus on the issues, are rare
(vi) Some communities are more vulnerable than other.

Disaster and Community Understanding

The definition of a disaster from community perspective is:

Disaster = Hazard × Vulnerability

which implies,

Vulnerability × lesser Capacity = More disaster

Vulnerability × better Capacity = Less disaster impact

(The factors influence victims at community level in cyclone disasters.)

Vulnerable Areas

The following factors engender vulnerable areas and human inhabitation of such areas:

(i) No/inadequate disaster warning and poor response from community
(ii) Ruined natural protection measures
(iii) No information base on disasters
(iv) Production/seasonality not maintained
(v) Self help capacity not improved
(vi) Regulation not in force
(vii) Inadequate community contingency plan
(viii) Lack of political willingness.

Important Areas of Concern and Community Responsibility

Habitation in Vulnerable Areas

In rural villages or urban slums, the dwellers have problems of housing, drinking water, health, hygiene, etc. Similarly in coastal areas fishermen folk and farmers, live in vulnerable areas due to the fertile landmass, assured fishing and highly profit oriented prawn culture. Inspite of many problems people live in the vulnerable areas because they make a living there. No motivation and force can restrain them to leave that area. Thus restricting habitation in low-lying areas or disaster prone areas is the primary responsibility of the community. For this, the following are essential:

(i) Demarcate the areas affected in the last cyclone disasters that can remind people to raise the plinth of their own huts as a precaution. Making vulnerable lands habitable by additional filling should be checked by all means.
(ii) Legal course of action to be taken on the encroachment by concerned department.
(iii) Public institutions, market and other such infrastructures constructed by individuals or by government should be opposed by public opinion.

Warning Measures and Community

Science and technology has helped mankind in forecasting the potential disasters and weather forecast. In order to generate better action from the community and to restrict miscommunication, it is essential to:

(i) Monitor the warning through constant watch of the media.
(ii) Form and make responsible community-based information groups on warning such as the Bangladesh Cyclone Preparedness Programme and Bangladesh Red Crescents Society (BDRCS).
(iii) Keep watch and ward to CWDS in the community.
(iv) Be regularly informed about recent warnings from local police stations.
(v) Effect appropriate dissemination mechanisms.

Natural Protection Measures

Forest cutting, destruction of mangroves and other wind brakes have been a threat to the ecosystem in most of the developing countries in the coast. The *global warming* adding new dimensions to disaster events such as changing seasons with too less or heavy rain, big cyclones and heat waves (Orissa heat wave taking 2,000 lives in the month of May and June in 1998). The unpredicted behaviour of rain causing high flood due to hydro-electrical projects, irrigation dams in one hand and in other hand heavy silt deposit in the riverbed raising the water level has increased the effect of floods in the countryside. The insecurity increasing day by day is due to lack of integration in development initiatives. The community has greater role in reducing further depletion. The following must be encouraged:

(i) Protect mangroves in the coast line and the forest.

(ii) Protect river mouths from silt deposit to allow better drainage of flood water.

(iii) Take joint forest management programmes in the region and increase plantation in household and at the community level.

(iv) Negotiate with the industrial houses on environment protection packages.

Reduction Measures

Reduction in impact of disaster on the victims is the end result of preparedness. In disaster reduction the most vulnerable sections like women, pregnant women, children, aged, handicaps are most vulnerable. Improvement of capacity is essential among the community members to mitigate the disaster impact. Relief and external assistance reach to the sites, after disaster, for those who might have survived by fighting few hours with the fury of the nature. So individuals need to be skilful, courageous and alert in the changing concept of disaster management. Under capacity building programmes, organisational as well as training programmes are essential. These can be achieved by:

(i) Preparing Task Force and assigning them special responsibility.

(ii) Taking expert service in training volunteers on first aid, search and rescue.

(iii) Taking support of district or local administration in keeping food stock and other basic amenities before the seasons (flood or in cyclone).

(iv) Preparing a clear plan in reaching to the most vulnerable.

(v) Discussing with the Government as regards their counter disaster mitigation plan.

(vi) Proper preparation for advance action before the season not before the disaster.

Political Willingness

Security of life, food and shelter to its citizen is the primary duty and responsibility of the Government. Relief materials, money and manpower can not bring solutions unless permanent efforts are made for the disasters. It can be only achieved if there is political willingness. A strong advocacy in focusing the social dimension of the problem is necessary for global thinking. Political willingness should be in consideration of the following:

(i) The local experiences and the action of the community should be brought to the national level to have a wider view.
(ii) Reform in education syllabus emphasising disasters and local experiences are to be initiated.
(iii) There is no cooperation between the action programmes and the researchers. It is high time to have action research on disaster themes.

Model-stake holders model of disaster mitigation has the following main characteristics:

- Equal partnership (community, NGOs, INGOs and Government)
- Clear role clarity
- Bottom up approach

Spider model roles have the following characteristics:

- Responsibility interlined
- Directive in nature
- Top down planning

Public awareness programmes could achieve the following:

- Public knowledge about preparedness measures
- Improve public response to preparedness and warning
- Achieve support for disaster plan action
- People are made aware on the possible dangers and local solution
- Dissemination of information
- Re-deciding the cropping pattern on the flood prone zones
- Facilitate stock of medicines at community level (minimum)
- Availability of transport facility
- Relief administration and distribution to the most vulnerable.

Awareness generation programmes are to bear the following characteristics:

- The programme desires to have impact by informing the nature of hazards of the disaster, practical knowledge on the preparedness measures, improving human attitude to warning.
- Information is power but in disaster management one must be clear about the audience and focus on the most vulnerable.

The message should also focus on the potential effects of the disaster.
- Audio and visual aids can impact for a long time than only trainings and formal meetings. Theme posters pamphlets can be displayed in common places.
- To select appropriate channels of communication that can be locally accepted.

For propogating such programmes, less costly and effective media are, folk dances, culture activities, Padayatras, Street theatres, Public speeches, public addresses, cycle rallies, public competitions, dry exercises, meetings, etc. The electronic and print media can also be employed. Television advertisements and caption, radio talks and bulletins, photography, pamphlets, posters, leaflets, etc. can be effected through these media.

Capacity Building

Improving the human capacity to manage stress, panic and fearfulness can have impact in reducing the vulnerability. In the community, the less capable are women, children, the aged and the handicapped. They are vulnerable because of less physical strength and lack of information. Besides, women in the villages are deprived of organisational avenues. In the community-based disaster management programmes it is desired to form women groups and organise discussions, training and exercises and competitions to improve their capacity in the following areas:

- Disaster preparedness training and role in the pre, during and post disaster stages
- Search and rescue operations
- First aid and home nursing.

Community need assessment involves understanding and improving knowledge on Community Based Disaster Management methods such as:

- PRA techniques
- Risk assessment
- Questionnaire
- Meetings with occupational groups, women and other vulnerable sections.

Survey and damage assessment is needed in the aftermath of disaster to re-stabilise the community providing basic amenities and rehabilitative measures to bring normalcy. The focus of the damage assessment is on:

- Human loss
- Livestock loss
- Property loss
- Infrastructure loss.

Human behaviour and response involves management of human behaviour in the following four stages:

- Pre-impact
- Warning
- Impact
- Recovery.

Community is the worst affected in any disaster. A disaster derails the normal life pattern of the victims and their families. The social and economic consequences of the disasters are of greater magnitude and it takes a very long period of time for the concerned community to be on normal path. For disaster risk reduction, it is imminent for the community to adopt culture of prevention, preparedness, response, and recovery. It could be achieved mainly through the ensured participation and thorough involvement of the community in all such measures initiated by the organised stakeholders. There is much felt need for creation and sustenance of community awareness on disaster preparedness.

18

Armed Forces in Disaster Risk Reduction

— **Rajive Kohli**

INTRODUCTION

Disasters and emergencies are fundamental reflections of normal life. They are consequences of the way societies structure themselves, economically and socially; the way societies and states interact; and the way relationships between decision-makers are sustained. It is essential to make a distinction between hazards and disasters, and to recognise that the extent to which a hazard turns out to be a disaster is essentially a measure of society's vulnerability. The hazard types have their peculiarities. While many occur infrequently or impact a very small population, a few hazards occur often but in areas that are accustomed to dealing with them and therefore seldom become disasters. However, from the perspective of a disaster victim, it is not particularly useful to distinguish between minor and major disasters. Where natural hazards become disasters, they become potential focal points for intervention by the military forces. The Armed Forces are dedicated and professional organisations with a tradition of being involved in the socio-developmental roles of nation building. Their physical presence in the remote, inaccessible and backward areas and in most of the hot-spots gives them a unique opportunity to be involved with the problems of the local communities. The soldiers coming from different areas of the country constitute a model of national integration which acts as a catalyst for the progress of an area by its sheer presence. A soldier is known to be hardy, ingenious, skilled, disciplined, and resourceful so as to be relied upon to handle any type of adverse situations. Apart from the main responsibility of defending the borders of the country, the Armed Forces render assistance to civil authorities when called upon to do so. After natural calamities, they help in relief work, the maintenance of law and order and restoring essential services. Whenever the impact of a disaster goes beyond the control of local authorities, invariably the defence forces are called upon to be pressed into service.

However, the involvement of the armed forces in civilian functions should be reserved as the last resort for the government. Military roles imply actions where the civilian functions ceases and hence too much reliance on the armed forces weaken the civilian democratic set-up. The involvement of the armed forces is generally uneconomical in the wider sense as it diverts it from its primary role of defending the country from external aggression. The range of responsibilities and tasks likely to arise in dealing with disasters underlines the need for a carefully organised and systematic approach to disaster management. Disasters can have wide-ranging effects on a country, its government and people. Therefore the primary responsibility for coping with disasters must rest with the government. Government organisations and resources necessarily have to bear the brunt of countering disasters. Such government organisations must be prepared to accept a substantial additional workload to be discharged under immense pressure. They also have to cooperate with a variety of other organisations and may also have to have additional requirements not necessary for their normal activities. Disasters require immediate response and the armed forces of a country play a vital role in imparting disaster relief and humanitarian assistance. Being a major organisation involved in disaster management, their contribution towards disaster risk reduction cannot be undermined.

Disaster risk reduction serves to strengthen democracy, create prosperity and realise human potential by recognising the need to develop, implement and sustain shared, comprehensive disaster management strategies and programmes to reduce the vulnerability of our populations and economies to natural and man-made disasters. In this regard, there the need to expand the community of stakeholders engaged in the formulation of early warning systems, the management of risk and integrated, sustainable development strategies.

As a matter of policy, a disaster management system is needed which would aim at disaster risk reduction. This system should not control other departments or tell them how to do their jobs, but should ensure that at all times the resources and operations of the various agencies are coordinated to produce the best effort. Hence, the major requirements for disaster risk reduction is to have an understanding of the significance of disaster, especially at key levels; a clear assessment of the disaster threat; and an effective disaster management policy. This system should have all the requisite agencies including the armed Forces to be an integral part of it.

Within the resources of most nations, the military assets represent unique technological and logistical capabilities that can be mobilised on short notice in a self-contained, self-sufficient and highly mobile fashion, to support life-saving relief efforts. Nearly all nations have at one time or the other, used their military assets and capabilities for national disaster relief. A number of nations have also done so in international relief operations lead by the UN, regional organisations, or bilaterally, having recognised that natural or man-made disasters can pose as lethal a threat as that of war, to the stability of any society. Disasters do not recognise national borders and call for international cooperation to prepare for and to contain the challenges and horrors associated with them.

Generally, a layman may understand most of the uniformed services and organisations to be part of the Armed Forces, though the expression 'Armed Forces' includes the Army, Navy and Air Force of the country, but not the police. The term 'troops' or 'military' generally refers to the Army, but in exceptional cases where the Navy or Air Force are called upon to render aid, these terms will also refer to them. Disaster management is primarily the function of the civil authority as already mentioned and as far as possible it should be done by the civil authority with the aid of resources available to them or can be obtained by them. Although using the troops too frequently and too long for managing disasters adversely affect their operational training and tend to undermine their morale and efficiency, the Armed Forces have played a vital role in combating many a disaster, providing prompt relief to the affected people even in the most inaccessible and remote areas of the country. The organisational strength of the Armed Forces, their high sense of discipline, the enormous manpower at their disposal and their excellent human resource management makes them an extremely useful tool in providing disaster relief. There are occasions when the scale and intensity of disaster goes plainly beyond the capabilities of the civilian administration and the defence forces are necessarily used to provide assistance. They are most useful in providing timely assistance in delivering food, water and medicines to the affected areas, erecting temporary bridges, roads and for other emergency, life saving services.

The Armed Forces are trained for the primary task of defending the country against external aggression. The constitution and the legal framework provide for the Armed Forces to render assistance during disasters/calamities whenever the situation is beyond the capability of the existing civil administration to handle. Disasters, by definition, are events, which are beyond the coping capacity of the local people. Hence, during a disaster it might be imperative that the Armed Forces are to be involved. Although, at present no specific disaster legislation exist within the country, the assistance of the Armed Forces is provided to the civil government for various internal duties, both, in times of peace and in times of war. This is enunciated under the subject 'Aid to Civil Power' both, in orders issued by the Central Government (Ministry of Defence) and also by the respective Services—the Army, Navy and Air Force. Instructions for the guidance of the concerned civil and military officers have been laid down by the Ministry of Defence so as to enable organised and clearly defined support from the Armed Forces and also to provide necessary sanctions (financial and otherwise) for the deployment of defence resources. The civil authorities may seek the assistance of the Armed Forces for:

- Maintenance of law and order
- Maintenance of essential services
- Assistance during natural and other calamities such as earthquakes, floods, famine and fires
- Other types of assistance, including assistance in development projects.

The term 'Aid to Civil Power' applied in a wider sence pertains to all the above types of assistance. In the restricted sense, it implies only that, for the maintenance of law and order, the executive magistrate of the highest rank who is present, may call upon the Armed Forces, and every officer of the Armed Forces is required to obey the requisition of the magistrate in such manner as he thinks fit. Normally, in case of a disaster, the State Government through the concerned Central Ministry, may approach the Ministry of Defence for assistance, which would issue necessary direction. Anticipatory action may be taken in good faith by units of the Armed Forces, when requested by the local administration, with *ex post facto* sanction being provided by the authorities. These forces must, therefore, be requisitioned under a central act and not under a state act.

Maintenance of Essential Services, whereby assistance by Armed Forces is provided only to run an essential service, when it has been declared to be an essential service by the Central Government, is under the provision of the Armed Forces (Emergency Duties) Act 1947. During an emergency, all the resources available with the administration may be inadequate, and the multifarious resources that can be mobilised by the defence services are called upon to bring prompt succour to the grief stricken population. Therefore, it is mandatory for every administrator, manager, social worker, and the general citizen to understand the role of the defence services in managing disasters. Disaster legislation or plans should represent a system of command and control for the utilisation, integration and coordination of the defence forces in aiding the government and the people in their efforts to protect the population, prevent damage, provide prompt and appropriate assistance, and achieve recovery during disasters.

The three wings of the Armed Forces have been actively involved in containing various disasters despite their primary commitment to the defence of the country. The Army, being the largest in number, is invariably involved, providing its expertise and a range of resources. Its manpower is used for all kinds of support and control needed to restore infrastructure damaged by a disaster. The Air Force is obviously in the lead in providing the helicopters and other aircraft needed during a disaster for transporting people, arial dropping of essentials, locating distress areas and for aerial assessment of the situation. The Navy is called upon for providing divers, boats in the coastal areas and help at sea. All the forces have trained and organised manpower for judicious employment during the emergency in quick response. Additionally, all military units carry out voluntary service in the vicinity of their locations as a welfare development activity and a good will mission. This could includes local disaster preparedness and capacity enhancement also.

Calling the defence forces for rendering aid primarily requires the civilian administration taking a deliberate decision to involve the defence forces in a situation which it deems to be serious enough to involve them. Under such trying circumstances all the agencies involved are to contribute the best of their efforts in providing succour to the affected population. In view of the fact that the Armed Forces are mainly intended to defend the country against external aggression, it is necessary that they are called upon only when it becomes unavoidable and imperative for the civil authorities to enlist its assistance.

The important issues to be kept in mind while aiding the civil authority during disaster situations are discussed below.

Cooperation. All agencies involved in disaster management operations must provide full cooperation to one another. There is the need to pool all available resources like medical, transportation or communication facilities at a particular place. Also, understanding the capabilities of all involved goes a long way in deriving the maximum benefit. All actions from preparation to operations should be undertaken jointly with constant exchange of information. Liaison, coordination and dissemination of plans should be ensured even to the lowest level of administration as well as the military. In all situations, clear enumeration of the specific tasks is the first essential step in the interest of well-coordinated action. What is to be accomplished by the military and what, by the civil authorities should be clearly established so that there is no scope for any misunderstanding between the military and the civil authorities regarding their respective tasks. The Armed Forces are to play a supportive role in relief operations and should be provided with proper liaison and coordination with all agencies involved. They can play an important role even in the mitigation, rehabilitation and prevention stages of disasters.

Planning and preparation. Planning is carried out at various levels i.e. at the national, state and local levels. The Ministry of Defence and the service headquarters are involved at the highest level. The command headquarters of the services interact with the state governments through periodic civil-military conferences. At the local level, the sub-area headquarters and units coordinate with the local administration, the police and others. Planning for disaster management has to be done incorporating all the actors. It should provide a course of action approved by all the agencies involved to pool the resources and to take the necessary actions as incorporated in the plan. Full familiarity with the task necessitated is possible by timely preparation. This involves joint planning, reconnaissance and rehearsals. The State Government, which has the primary responsibility of disaster management, should involve the Armed Forces in the planning process. The plan needs to be periodically updated considering the changing operational responsibilities of the Armed Forces. The success of any plan for disaster management involving the assistance of the Armed Forces depends on the cooperation between the concerned civil and military authorities through all the stages; from initial preparation of the scheme to the final application. To ensure this aspect, it is necessary that all plans be issued with the signatures of both these authorities. Periodic informal conferences between corresponding civil and military authorities are of importance. Liaison during aid to civil authorities is essential and could best be effected by placing the headquarters of civil and military authorities together or by the exchange of liaison officers. Typical organisation of State Government with corresponding military authority are given below:

Subject	*Civil Authority*	*Military Authority*
General Policy	Governor, Lieutenant Governor or Chief Commissioner, Minister in charge, Secretary of Department concerned	Army Commander, Area Commander
General Policy in the Division	Divisional Commissioner	Sub Area Commander
All matters relating to aid	District Magistrate and Collector	Station Commander

Humanitarism. The needs of the affected population should always be the foremost priority in any disaster-related action. Even while being fully committed to defending the borders and to internal security duties, the Indian Armed Forces have always responded readily to any disaster that puts human life at stake. Special efforts have been made to inculcate and foster humanitarian ethos amongst the Armed Forces with regard to the possibility of their frequent involvement in sensitive and delicate issues.

Economy. Extensive training is what that accounts for the multi-operational ruggedness, skill and resourcefulness of the Armed Forces. Consequently, as an aid to disaster management, the Armed Forces are more expensive than the other resources available in the country for the same purpose. Whenever the services of the Armed Forces are sought, the cost for it is borne by the concerned state government. Considering the cost, minimum essential military resources should be utilised in disaster relief operations and their participation terminated at the earliest possible. Also, they must only be used when other agencies are unavailable or unable to meet the contingency. For menial and labour intensive tasks, local labour must be utilised as it is cheaper and also provides at least temporary income to the locals.

Military operations to support disaster assistance are based upon *principles for military support* which are consistent with general military doctrine and confirmed acceptable by the civil authorities. Some of the principles underlying for military support to civil authorities are outlined hereunder:

- Armed Forces assistance will be requisitioned only when the situation can not be handled by civil administration.
- When needed, the Armed Forces will provide immediate response.
- While responsive to the direction of civil authorities, the military chain of command will remain in place and in force.
- Aid will be requisitioned by civil authorities on a stated task (mission) basis.
- Liaison and coordination will be effected throughout the period of the assistance mission.
- Advance planning and training will be conducted.
- Military and civilian resources will be integrated as necessary to effectively accomplish tasks.
- The Armed Forces will be released from the support mission as soon as the civil administration can take control of the situation.

PROCEDURES

Though the Armed Forces are always prepared to come to the aid of the disaster stricken population, it would do well in avoiding complications if the administration follows the rules and regulations laid down in seeking their involvement. Assistance is provided by the Armed Forces with the approval of the Central Government, as the use of Armed Forces is a function of the Central Government. Whenever troops are called in for such tasks, the State Government, or the civil authority through the State Government, should report the fact to the Central Government (Ministry of Defence and Ministry of Home Affairs) without delay. For grave situations warranting immediate requisitions like maintenance of law and order, the magistrate of the highest rank may requisition directly to the nearest military authority in his jurisdiction. Sufficient information should be given while requisitioning troops so that the military commander can work out the resources required. The sanction of the Central Government must be obtained within 10 days of employing the troops.

In a strike situation, State Governments may seek for military assistance for the maintenance of essential services. They must furnish the following information while requesting military assistance:

- Whether the strike has been declared illegal, furnishing the number and date of notification in this regard.

- Whether all civil resources have been explored before military assistance is sought to run essential services.

Although assistance by the Armed Forces during natural calamities is generally provided on sanction by the Central Government, in case of immediate necessity when reference to Central Government is not practicable, the local military authority may, at its discretion, comply with a request from the civil authority and report to the higher military authorities. The requisition should be made in writing to avoid subsequent complications. Any assistance sought of the Armed Forces in such circumstances would obviously be of an unforeseen nature and consequently no previous plans can be made. However, if such exigencies are addressed in a disaster management plan that has been approved by all the participating agencies, the incorporation of the armed forces becomes easier.

The military units remain under the command of their own commanders and work for achieving definite objectives. Hence, the requisitioning of the Armed Forces should specify identified tasks rather than the quantum of tasks. If the district and state plans incorporate the role expected from the Armed Forces and there is continuous flow of information, the procedure for deploying the forces would be quick and smooth. This would also enable the appropriate local military authorities to obtain in advance, the necessary sanction for using certain critical resources. Apart from clarifying these in the planning stage itself, the specific administrative authority that should contact the appropriate military commander should have a clear understanding of these. The civil authorities have to ensure that when an Army column is requisitioned, their role is clarified and the relief stores for distribution or special stores for a given task are provided to them. If the scale of disaster so warrants, all available resources must be requisitioned immediately for a speedy response.

The troops send a report to their service headquarters immediately on being requisitioned and also on withdrawal of aid to civil authorities. When further aid is not required, the report submitted include details of number and type of troops employed, equipment used, duration of employment, brief report of work done, result achieved and all additional expenses incurred. All expenditure for the employment of Armed Forces for the maintenance of law and order will be borne by the Central Government. For aid in the maintenance of essential services, assistance during natural calamities or execution of development projects, all costs except those for normal pay and allowances, rations, clothing, equipment and supervision charges, will be met by the State Government. These are generally in respect of the following:

(a) Consumable stores including fuel and lubricants.
(b) Non-consumable stores and equipment including depreciation assessed on the replacement cost on account of wear and tear,

cost of repairs and maintenance and cost of replacement of equipment rendered unserviceable as a result of such use.

(c) Incidental expenditure necessitated by the work, e.g., cost of moving personnel and equipment to and from the site of work, extra allowance, rations, clothing and amenities provided, etc.

(d) Hospitalisation and treatment of the service personnel injured during the discharge of duties.

(e) Any damage to crops or any other compensation payable.

CAPABILITIES AND RESOURCES

The three wings of the Armed Forces, with the Army in the forefront, have always been the lead agencies in providing relief during many disasters. As already mentioned, the Army being the largest in terms of strength and being located all over the country including remote inaccessible areas, is invariably involved with its expertise and various resources in virtually all aspects of disaster relief ranging from logistic support and medical help to repair of roads and bridges. The Air Force is in the lead for providing aircraft and helicopters needed during a disaster primarily for the transportation of personnel, stores, and equipment, for air dropping food, water and essential stores and for undertaking aerial survey for damage assessment. The Navy is called upon for providing diving assistance for search and rescue at sea, for salvaging sunken ships and boats, for clearing harbour areas, for transporting relief material and equipment by sea to the coastal areas, for medical assistance and for aerial reconnaissance of coastal areas. Besides their own men and equipment, the Armed Forces are also equipped with local work forces they voluntarily train to encounter emergencies, as part of their general welfare programmes. Involvement in disaster risk reduction would be an important aspect for them.

Even though the equipment with the Armed Forces is primarily meant for military operations, its resources are made available and are extremely useful during a crisis situation. However, any other special equipment required should be planned for and provided by the civil authorities. An over view of the capabilities and resources of the Armed Forces will give an indication of the diverse nature of tasks that they are capable of performing.

Planning. Planning is carried out by the Armed Forces at various levels i.e. at the national, state and local levels. The Ministry of Defence and the Service Headquarters are involved at the highest level. At the State government level, the Command/Area Headquarters of the Services interact through periodic civil-military conferences with the local administration, police and others organisations. Once contingency plans for disaster management have been drawn incorporating the various actors, regular rehearsals are carried out and the contingency plans are also periodically updated. These plans are also generally graded at different levels for catering to different types and intensity of disasters that call for different approaches. The planning process is an important aspect in disaster risk reduction as it also creates awareness, cooperation, mutual understanding and an integrated approach with preventive and precautionary measures incorporated. As the Armed Forces are employed in all emergency support functions, it would be worthwhile to incorporate them in the planning process. Their expertise in terrain analysis, planning and implementation and other skills could prove to be useful in national disaster management planning.

Internal Security Schemes are prepared on a Command basis and are framed to meet such situations as are most likely to develop within a particular area. Disaster management is incorporated in these plans. Each Sub Area Headquarters, military station and other formations has a local Internal Security Scheme. These are drawn up in close collaboration with local civil authorities and have their approval. Special schemes are designed by the local formation Commander to deal with situations not covered by local Internal Security Schemes. Typically, these would address specific disaster threats like floods, cyclones, earthquakes and landslides in their area of jurisdiction.

Organisation. The structure of the Armed Forces is based on a brick or layered system of units and sub units that enable a force of any size to function and adapt to any situation. There are specialist teams that are trained to assist local governments and departments in coordination and liaison with outside agencies. An Operations Headquarters, usually set up in the affected area, coordinates the operations in a that area. The organisational strength of the Armed Forces and their disciplined and systematic approach, make their services apt to deal with large-scale disasters. The army has different echelons of headquarters to control subordinate formations which are composed of functional units. Generally the smallest independent organisation is a Battalion or a Regiment of different arms (such as infantry, armour, artillery, engineers, signals and aviation) and services (such as supplies, ordnance, workshops, medical, etc). The main component of the Navy is a fleet of various types of fighting and support ships and a complementary component of naval aircraft. There are also on-shore establishments for support of its fleet. In addition to maritime warfare, some of the peacetime tasks where the Navy indirectly gets involved are fisheries protection, anti-smuggling, rescue and assistance at sea and protection of agencies exploiting the sea-bed. Marine survey, oceanography and meteorology are some of the fields to which the Navy contributes substantially. They also provide assistance to shipping firms in distress. The very fact that warships are fully self-contained as regards food, medical and other stores and can provide landing parties makes it possible for them to be despatched to troubled areas at very short notice, to render assistance or relief until such time when other forces arrive and establish themselves. Air Force units are called Wings, each having a particular type of aircraft squadron, viz. fighter, transport or helicopter. It is an Air Force Station that controls the various components within its jurisdiction. For disaster management, the Air Force Stations should be contacted at the local level. The use of transport aircraft and helicopters for the movement of stores and personnel, aerial reconnaissance, air dropping of life-saving medicine, food and water and such tasks is usual during disasters.

Communication. The Armed Forces own a wide variety of equipment for short- and long-range communication including satellite-based systems, which immediately provide for the necessary communications in a disaster area. Besides, the expertise of the engineering and supervisory specialists can be availed to restore quickly, the essential local communications. The sophisticated communication network of the Armed Forces can be utilised if there is a breakdown of the communication systems at the disposal of the civilian administration. Long distance radio and satellite connectivity for communication may be established with military assistance. Local line and radio communication could be established between the key elements in an area and also a small telephone exchange set up for internal line communication of the control room. Army signallers can provide valuable assistance in the re-establishment of the city telephone network and the restoration of long distance trunk routes.

Engineering. Roads and bridges invariably get damaged during a disaster. The engineering wing of the Army helps in repairing or in constructing temporary bridges to facilitate the movement of goods and people. Earth moving and bridging equipment and the heavy recovery vehicles available with the Army are useful in most disaster situations for clearing debris and restoring road communications. Trained people are available to handle explosives and carry out demolition tasks. Clearing, construction or repair of roads is often needed to enable relief teams and material to reach the affected area. The Army can also provide water supply, electrification, and the equipment needed to restore emergency services. Assistance for the restoration of damaged roads and bridges, clearance of debris and such tasks is provided by the Army in the form of man power, expertise and equipment. Generating sets are available with various defence units that could be temporarily used for providing electricity to essential installations during an emergency. Assistance is also offered in the restoration of power lines in an area.

Manpower. Trained and disciplined personnel is the basic strength of the Armed Forces. Their very presence creates a sense of security in the minds of the people, which in turn gives them the confidence to handle traumatic situations. In the case of large-scale disasters, the entire force in an area is galvanised into action and everything is done on a war footing, with relief operations as the top priority and all other activities in the Command taking a back seat. In addition, their various technical specialists in the areas of communication, power, engineering, medical and veterinary services, diving and hydro graphic survey, etc., can be used for damage assessment, general administration, liaison, control, clearing debris, repair work, general management or any other activity for which manpower is required. The large number of disciplined ex-servicemen spread all over the country can play a vital role in consolidating community efforts towards disaster management.

Search and rescue. The capability of the Armed Forces for aerial search and rescue is often indispensable in the event of disaster in remote areas. They invariably undertake this task as a priority in all disaster situations and coordinate it with local agencies. The soldiers are well-trained in this aspect in their normal course of duty. Depending on the nature of a calamity, specially equipped columns could be tasked for search and rescue in a given area, e.g. in a flood situation divers of the Navy could be grouped with the columns along with army boats while after an earthquake, Army engineers and earth moving equipment would be required for debris clearance. Helicopters are also very useful for conducting aerial search and rescuing marooned survivors.

Transport. Vehicles, ships and large transport aircraft of the Armed Forces are useful for transporting large amount of relief supplies and for evacuation of affected people in addition to moving their own men and material. Air transport, especially helicopters, are of immense use in the the evacuation of critical areas and for reconnaissance. Army motorboats and naval motorised rubber boats are extensively employed in flood situations. Aircraft, both fixed wing and helicopters, are also used extensively for aerial assessment and dropping food, water and supplies. Naval ships are also used for carrying stores to cyclone hit coastal areas. Helicopters are pressed into service often to evacuate the injured requiring medical treatment. The Armed Forces can provide emergency transportation by air, sea and land. Military helicopters can drop emergency supplies to marooned people. Boats can be used for rescuing people from flooded areas.

Medical and health. The Army Medical Corps has often rendered signal service in providing medical relief to people affected by disaster. Field hospitals can be established in any area for importing emergency treatment to the injured and the sick before their evacuation to civil hospitals. Sanitary engineering, decontamination and disease control can also be undertaken by the Armed Forces on a war footing. A point to note is that the medical teams of the Armed Forces are self-contained units and can provide immediate assistance in any area without depending on the local area for accommodation, food, water and other requirements. A Division has two Field Ambulance units, which offer the services of an advance dressing station, forward treatment centre, medical aid post, staging post, and dental unit besides detachment at helipad/airfield/railway station. In addition there are various military hospitals. The troops are trained in first aid and all columns are capable of providing immediate assistance to disaster victims. Mobile medical teams equipped with the essential equipment and medicines are organised for administering on the spot treatment to seriously injured people. Field medical camps could be organised to handle large number of casualties before shifting them to hospitals. Military hospitals at various locations can also temporarily handle influx of a large number of casualties.

Protection of life and property. The Army assists the civil administration in maintaining law and order on specific request by the civil administration. This happens when the state police is over-stretched and is not able to cope up with the problems following a crisis. Troops would be used for effectively cordoning vulnerable areas and for the protection of vulnerable assets.

Shelter and camp. Tented camps for emergency shelter can be provided along with arrangements for food, water, power supply and other essentials. Management of such camps can be effectively undertaken by the Armed Forces. The engineering units can construct temporary shelters with provision for water supply and electricity. One of the columns could be assigned the management of field kitchens, hygiene, sanitation, etc. Medical resources would be utilised for the people seeking shelter in these camps. The construction material and daily supplies should be provided by the civil administration. The Army may be employed for the transportation of materials, distribution of food and water and for organising camp kitchens.

Territorial Army. Territorial Army is a voluntary part-time civilian force having departmental units (raised from employees of government departments and public sector undertakings) and non-departmental units. The units are Infantry Battalions, Railway Engineer Regiments, General Hospital Units, Indian Oil Corporation/Oil and Natural Gas Commission Units, Dock Companies, ASC Battalions, Ecological Battalions and Wasteland Development Units. In addition to providing units for the regular Army when required, the role of the Territorial Army is to relieve the regular Army of static duties, provide aid to the civil authorities in dealing with natural calamities, maintain essential services in situations where the life of the community is affected or the security of the country is threatened and to provide support to all national endeavous. The Territorial Army units are like a reserve force which can be employed to meet special situations. In times of need, their Railway units help in running essential rail traffic. Similarly, there are medical units of the Territorial Army for supplementing the military hospitals personnel. The Ecological Battalions are tasked on project basis with restoring the ecological balance of a degraded area through afforestation, arboriculture and similar works. Specialised Territorial Army units could be raised for disaster management, which can play a significant role in enhancing disaster preparedness in addition to combating disasters.

Border Roads Organisation. The Border Roads Organisation, staffed largely by personnel from General Engineer Reserve Force with some representation from the Army Engineers, has been set up for accelerated development of infrastructure in the remote areas along the Eastern, Northern and Western borders to support army deployment. The Director General, Border Roads, at Delhi, works through Project Chief Engineers for various geographical areas, which further have a number of Border Road Task Forces. A Border Road Task Force consists of one or more road construction companies with specialised platoons and labour force provided by pioneer companies, and also has other units, like a Stores Company, field workshop and medical units. These units are located in remote areas and handle large amounts of plant and equipment, their responsibility being the construction and maintenance of roads. They are experienced in dealing with land slides and other natural hazards which frequently occur in their areas of responsibility and hence they must be incorporated in all local disaster management plans.

Veterinary. The Army veterinary services come to the aid of the population by saving their animal wealth through treatment, zoonotic disease control, food inspection and evacuation. They are also involved in research to improve the breed of the animals. Their specific role in disaster management aspects concerning livestock is invaluable.

Bomb disposal. Army also undertakes disposal of bombs and improvised explosive devices and anti-terrorist measures. Bomb Disposal Unit is a unit specially meant for this purpose. They work in close liaison with the Civil Defence and Police organisations and also undergo special training.

Nuclear, biological and chemical threats. The Armed Forces have capability to meet nuclear, biological and chemical threats. They have specially trained personnel and specialised equipment for monitoring and decontaminating in case of such requirements.

DEPLOYMENT FOR DISASTER MANAGEMENT

Each wing within the Armed Forces would be deployed for specific tasks as per its capabilities, with the Army being predominantly engaged for assistance. Each service will be independently responsible for rendering aid for the maintenance of essential services during natural and other calamities (e.g. floods and fires) and for other type of assistance (e.g. for development projects) when required by the civil authorities.

Coordination will be done in joint meetings of the representatives of the Services and the civil authorities concerned when resources of more than one service are required.

For maintenance of essential services, the assistance provided by the Armed Forces will pertain to supplying technical personnel and specialised equipment. The Corps of Engineers, Corps of Signals, Corps of Electrical and Mechanical Engineers and the Army Medical Corps are likely to be called upon to provide such assistance. The Air Force would mainly be used for the transportation of stores and personnel by fixed and rotary wing aircraft while the Navy would best be employed for all activities at sea and for the provision of divers. Generally, the tasks undertaken would be the provisioning of machinery and equipment which is not readily available with civil authorities, provisioning of technical personnel, and supervisory staff who should coordinate and guide the work of civil agencies.

Army assistance should not be called for work mainly involving unskilled or manual labour that can be obtained from other sources like voluntary organisations, paid labour, Home Guards, the police or the civilian government staff. Even if troops have to be detailed to save life or on humanitarian grounds at short notice, they should be withdrawn as soon as possible.

In the country a dedicated force for emergency management does not exist as such. The Armed Forces, with their organisation, infrastructure and dedication are uniquely suited and could be entrusted with the overall responsibility for disaster management. However, this would entail meeting the cost of their primary operational role and additional budgetary support to them besides an acknowledgement of the fact that the civil authorities are not the primary agency for disaster management. The potent resources available with the Armed Forces in the form of trained personnel, equipment and supplies can be availed to consolidate the national response to major disasters. The committees existing at the national and state level should have representative of the Armed Forces for formulating disaster management plans and advise on the suitability of employing the military and for coordinating the military aspect of the support the Armed Forces can offer for disaster management. This Military Coordinating Officer could be a senior operations staff officer from the nearest headquarters who should be part of the disaster management team at the national or state level. As part of the control room, the Military Coordinating Officer should set up a small liaison cell with mobile communication system. With this, all information of the military relief effort would be available in real time to control room. Any requirement for the Armed Forces can be immediately communicated and executed.

Typical Army Column. Generally the Army is employed in self-contained, self-sufficient and mobile columns capable of action whether it is for managing a natural calamity or for restoring law and order. A column is built around an administratively self-contained Company of about a hundred men with additional resources attached, such as a signal detachment, a medical team, the repair and recovery element, and other functional requirements like boats needed during floods. Each column may function independently upto a limited role with a number of them functioning under a common, designated headquarters. The role assigned to each column could be just one or a combination of many, like evacuation, medical, distribution, security, search and rescue operations, clearance of debris, providing shelter, or any other. In any case, the material required to be expended for the affected population should be provided by the local administrative authority.

EX-SERVICEMEN

A large number of disciplined and trained men who have retired from the Armed Forces are available in all parts of the country. Some of these men are trained and experienced in the different aspects of disaster management and could be the key persons at the local level for disaster management. Ex-servicemen have qualities of leadership, skills, expertise and dedication for community work. The Director General, Resettlement, formulates and implements schemes for the resettlement of ex-servicemen in collaboration with The Soldiers, Sailors and Airmen Board which functions under it and has branches in all states and districts of the country. The ex-servicemen must be involved at the district level in the planning and implementation of local disaster management activities. These persons can be organised into specialised teams for disaster mitigation and prevention and assigned the responsibility of imparting training and awareness and formulating plans for disaster management at the village and district level.

As already discussed, during natural disasters, the Armed Forces participate in relief, search and rescue operations, aid in the movement of men and material and render other types of assistance to the affected population. During man-made disasters, the Armed Forces are mainly called in to restore law and order and to help the injured and protect life and property. Not withstanding the difficulties involved, the Armed Forces can always be relied upon to help their fellow citizens. However, the civil administration must use the Armed Forces as the last resort available to them and never out of panic, as once, reportedly, it was contemplated to call in the Army for arresting a former Chief Minister. The involvement of the Armed Forces in disasters is the maximum during floods, which are the most frequent and often, the most devastating. However, the Armed Forces have been involved during earthquakes, landslides and cyclones. The scale of response from the Armed Forces depends upon the gravity of the calamity and the amount of assistance the civil administration requires.

The civic action projects in some areas like Assam and Jammu and Kashmir have been monumental in upgrading the civic amenities; encompassing social welfare, rural upliftment, education, health, medicine, sports, culture, environment, local administration and generation of employment. In Nagaland and Manipur, the Army launched 'Operation Good Samaritan' for creating the infrastructure to cater to the basic needs of the insurgency stricken people. Various development activities undertaken included building of roads, schools and community-related structures, human resource development and vocational training, agricultural related development programmes, veterinary assistance and training, and medical assistance. Along with this, various activities for disaster risk reduction could be accomplished involving the local population. The mere process of preparation for integrating the Armed Forces at various levels of disaster management would go a great deal towards disaster risk reduction.

For effective disaster risk reduction it is prudent to clearly consider the following aspects in the context of the defence forces:

- National development planning
- National disaster management policy
- Disaster legislation
- Counter disaster planning
- Special disaster related circumstances in which national assets may be at risk.

Assistance can be provided to include:

- Seminars and workshops aimed at enhancing knowledge of disaster management and encouraging the development of counter-disaster capability.
- Provision of support and expertise for formulation of plans.
- Provision of key systems and facilities, such as emergency communications and warning networks.
- Provision of training for key officials and others.

Some of the issues that could be considered for making an optimum use of the expertise of the defence forces in disaster management at national level are as follows:

- Support from the Armed Forces should be sought only when other resources are unavailable and it does not interfere with its primary operational role.
- Requisitioning of the Armed Forces should invariably have prior concurrence of the Central Government. The task expected from them should be clearly specified.
- The Armed Forces when employed for disaster response must be de-requisitioned at the earliest.
- Use of Territorial Army needs to be incorporated in disaster management plans. In highly disaster prone states, it could be

considered raising specialised Disaster Management Battalions similar to Ecological Battalions.

- Boarder Road Organisation, if available should be suitably incorporated in disaster management plans.
- The potential of ex-servicemen available throughout the country should be tapped for disaster management.
- A Military Coordinating Officer should be part of the disaster management team at the national and state level.

The Armed Forces have always been involved with selfless dedication in assisting people in distress. It is important to highlight that at the planning stage, all the concerned agencies should be involved to formulate a joint disaster management plan and the Armed Forces should be a prominent constituent of this system. Although various agencies are nodal in different types of disaster, they all rely on the Armed Forces to play a major role in one way or the other. Exchange of information, cooperation and mutual understanding of the needs and peculiarities of each one among all those involved goes a long way in better disaster management. It is surprising to note that the Armed Forces are not formally part of the national or state crisis management committees. The Armed Forces are well trained and organised, but there is very little training specifically for natural disaster management as the major emphasis is on internal security duties apart from operational responsibilities. However, they do possess the awareness required for preparing themselves for meeting the needs of disaster management and they do plan and rehearse their role. This aspect is now being incorporated in various training institutions of the services. The Armed Forces must be utilised only as a last resort and only in a supportive role after being requisitioned for specific tasks to ensure a focused response. Their efficiency might tempt the administration to requisition them even before the government's resources are fully utilised. In major calamities, the Army has been the first intervener highlighting the deficiency at national level of an alternative to it. There could also be a high-powered federal disaster management agency and under it, similar agencies at the states and also in highly disaster prone areas where the experts including specialists from the Armed Forces are pooled so that all aspects of disaster management are integrated.

The Armed Forces, as in the past, expect to continue to be involved in coordination with other governmental agencies in disaster relief owing to the special features and characteristics of the forces, outlined earlier. Hence it is important that the organisation, capabilities and limitations of the Armed Forces must be understood by all agencies both, Government organisations and NGOs. This would ensure that the resources of the Armed Forces are optimally utilised and there is no adverse effect or dilution of their primary function. However, in view of its vast experience in disaster relief dispensation and its well-oiled organisation, it is only appropriate that the proposed revamped structure of disaster management, primarily the provision of relief, at all levels in the country (National/State/District) include appropriate components of the Armed Forces. Yet, with every disaster management operation involving them, the Armed Forces must be allowed to withdraw as soon as the initial and the most urgent phase of relief operation is completed and the other agencies have arrived and consolidated their resources. This is important so that the other agencies fulfill their responsibilities and as subsequent recovery and rehabilitation processes are time consuming and often not so urgent or ardous as to expect the Armed Forces to contribute. It is appropriate that the involvement of the Armed Forces in disaster risk reduction be suitably incorporated at all levels.

19

Pilot Demonstration Project for Drought Risk Reduction

— **Kulwant Singh**

INTRODUCTION

Disaster is an event, natural or man-made, sudden or progressive, which impacts with sudden severity that the affected community has to respond by taking exceptional measures. Frequency as well as intensity of natural disasters is increasing globally, including in India. Disaster impacts are felt more in developing countries in comparison to developed countries due to borderline economic status of the vulnerable populations with no or inadequate adjustment capacity.

India, with a wide range of climatic and topographic conditions is subject to various types of natural disasters like floods, drought, cyclones, earthquakes, etc. in varying degree. Here, we discuss drought, the various risk factors and the alternative livelihood options for risk reduction in the state of Rajasthan.

DROUGHT

Drought is not dramatic like other disasters as stated in its definition. There is a slow and creeping onset of this kind of disaster but it involves a much larger population and lasts for a much longer period of time. It means a long drawn battle for the vulnerable communities who despite all their coping mechanisms, need external assistance and inputs for their survival. We tend to focus on drought only when it is upon us. We are then forced to react, to respond to immediate needs, to provide what are often more costly remedies, and to attempt to balance competing interests in a charged atmosphere.

Droughts are not confined to a particular topographic setting and their impact can extend over very large regions. Famine is the most serious outcome of drought although causal links, are not always clearly recognisable. The atmospheric factors that initiate drought are not always well understood and, consequently, drought is often defined in terms of its effects rather than its causes.

To define drought as any unusual dry period which results in a shortage of water suggests that rainfall deficiency is the trigger. It is, however, the shortage of useful water in soil, rivers, ponds and reservoirs that create the hazard. It is important to view this water shortage in relative rather than absolute terms, although droughts are often associated with semi-arid climates, they also occur in areas that normally enjoy abundant rainfall.

In India, the Meteorological Department declares a drought when annual rainfall falls below 75 per cent of that which is expected and severe drought at 50 percent. Recently it has been recognised that the problem is aggravated by irresponsible human interference.

Drought Classification

Acute water shortages arising from lack of rain or its proper distribution is the prime cause of drought. However, drought means different things to different people and there are probably as many definitions as there are users of water. To the meteorologist it is the absence of rain, to the agriculturist it is the absence of soil moisture to support plant growth and to the hydrologist it is the absence of water in the storage reservoirs and canal systems for irrigation and power.

According to the National Commission of Agriculture (1976) drought can be defined under three categories.

(i) **Meteorological drought.** It is a situation when there is a significant (more than 25 per cent) decrease of rainfall from the normal value for the area.

(ii) **Hydrological drought.** Meteorological drought, when prolonged, results in hydrological drought with a marked depletion of surface water and consequent drying up of reservoirs, lakes streams and rivers, cessation of spring flows and also fall in ground water level.

(iii) **Agricultural drought.** It occurs when soil moisture and rainfall during the growing season are inadequate to support a healthy crop growth to maturity and therefore result in crop stress and wilting.

DROUGHT IN RAJASTHAN

Drought in Rajasthan in the year 2000–2001 may be categorised under all the three categories i.e. Meteorological, Hydrological and Agricultural.

Drought in Rajasthan is a recurrent phenomenon as stated in an old saying, *'Sat Kal, Satias Jamana, Trisath Kabakacha Baki Tinghesan'* meaning that in every 100 years there are seven years of acute drought, 27 years of scarcity, 63 manageable years and 3 normal years. *However, in last 51 years there has been 44 years of drought and scarcity.*

Mr. La Touche (later Sir Diggs La Touche), who made the first regular settlement of Ajmer-Merwara in 1874 A.D., has observed, *Ajmer-Merwara can never rely on two good harvests in succession.* Famines, which disfigure the history of this otherwise beautiful province distinguish it from all other provinces of India; for there is no province or district in the country where famines occur so frequently.

Impact of Drought

During a period of a year in 2000–2001,

- Drought conditions were prevalent in 31 out of the 32 districts.
- Out of a total of 41,529 villages, 30,583 have been declared as scarcity affected.
- There were high temperatures and inadequate rainfall with long spells of dry weather during the months of August and September.
- There was a skewed distribution of rainfall in Rajasthan; in the two districts of Rajsamand and Sikar it was 60% less than the normal.
- In another 5 districts, the rainfall was deficient by 40–60% of the normal.
- As many as 16 districts out of the total of 32 districts had a rainfall deficient of 20–40% of the normal distribution.

Effects on Population

- The affected human population was nearly 33 million whereas livestock affected was nearly 40 million. This accounted for 3.3% of the total human population of 1,000 million of the country which was fighting a losing battle against a disaster for the third year in succession when all the reserves of finance and food grains stood practically exhausted.
- The human population along with their cattle migrated from southern and western Rajasthan as the fodder and drinking water became scarce.
- There was considerable reduction in the hemoglobin level of pregnant women and the weight of children below the age of 5.
- Vulnerable sections of the society were severely affected. The vulnerable population include:
    - Women, especially pregnant and lactating mothers.

- Children, especially infants and new born babies
- Old and infirm people
- Specially abled and destitute
- Landless labourers and small and marginal farmers.

Effect on Crops

- During the year 2000–2001, the damage to crops was more than 75% in 19,817 villages between 50–74% in the remaining 10,766 villages.
- Due to hostile conditions, 8.95 million hectares of Kharif crops out of the total cropped area of 11.6 million hectares got damaged severely causing an estimated financial loss of $ 763 million, apart from the loss of $ 400 million in Rabi crops.

Effect on Water Sources

- Most of the rivers and other natural water sources had gone dry.
- The major impounding reservoirs like Rajsamand, Ramgarh, Meja, Fatehsagar and Pichhola were nearly dried up.
- Out of 2,370 tanks, only 131 tanks had over flown in 1999–2000. Most of the tanks, by and large, remained empty.
- The recharge of ground water aquifers, wells, tanks and other water bodies were adversely impacted.
- There was acute shortage of drinking water supply to major cities, towns and villages. Already in some of the villages, the condition has gone so bad that water had been supplied through tankers.

Causes

- The average rainfall of Rajasthan i.e. 575.1 mm is as such quite low in comparison to that in other parts of the world.
- The state did not receive rainfall at the right time i.e. during critical crop growth and flowering stages, which led to massive crop loss.
- The rainfall distribution was uneven with most of the rain occurring in the month of July and the rest of the year totally dry.
- The average total number of rainy days was just 34.3 in eastern Rajasthan and 16.8 in western Rajasthan.
- There is lack of widespread watershed management approach for water and soil conservation. The total area till date covered under watershed development programme is just 190,000 hectares out of a total of 2,440,000 hectares. This is just 8% of the total area that could be covered under watershed development programmes.

- Due to a very limited use of water conservation and harvesting practices, 84.6% of rain water is lost through surface run-off and evaporation.
- The impact of drought is more severe in the areas where assured irrigation is unavailable.
- Poor availability of certified seeds of drought resistant traditional crops like moth, guar, etc. is another aggravating factor.
- The excessive number of cattle (that exceeds the human population) leads to over grazing making vegetation scarce resulting in soil and water erosion.

The Much-overlooked Factor

One of the primary causes attributed to drought is inadequate rainfall but on a closer look one finds that there are:

- Only 5 districts which receive less than 300 mm rainfall.
- 7 districts receive rainfall between 300–500 mm rainfall.
- The remaining 20 districts receive more than 500 mm of average rainfall.

Of the 5 districts receiving less than 300 mm of annual rainfall, the two districts of Ganganagar and Hanumangarh do not suffer the impact of drought because of the alternate and additional sources of irrigation made available through the canal system. The land productivity of these districts is the highest and is comparable to that of the highest food grain producing state of Punjab.

The other factor that aggravate the severity of drought is that, most of the vulnerable populations of small and marginal farmers, women and landless labourers are dependant on agriculture which itself is dependant on the monsoon and its variations. The traditional arts and crafts industry has been suffering for long and in absence of effective drought-proofing mechanisms, the impact is more pronounced. There is underdevelopment in the secondary and tertiary sectors like agro-based industries and processing units in rural areas. Awareness regarding dry farming with improved agricultural practices is still lacking among the farming communities making their stakes in agriculture quite high.

ROLE OF THE GOVERNMENT

Over the years, the major actor in mitigating the drought situation has been the Government, which is now getting more into action by taking long-term initiatives for drought management on a sustainable and long-term basis unlike the previous years when every year millions of rupees were spent in immediate relief activities. The Government has spent about $ 910 millions since 1951 in 44 years of drought and now it has realised the need to take long-term drought risk reduction measures and put suitable mechanisms and initiatives in place.

The UNDP has suggested to the Government of Rajasthan to formulate a drought management plan; and suitable actions have already been initiated to draw up draft proposals. Rajasthan would not be far behind if it reduces its over dependence on the age-old scarcity and famine codes where relief measures were considered as responsibility of the State. There is a need to have a rights based approach to drought, especially in a country like India, which has been declared as a welfare state after independence.

DROUGHT RISK REDUCTION AND PREPAREDNESS—NEED OF THE HOUR

Drought risk reduction, that is the solution to this problem, can only be taken up through the communities, which can replicate and sustain water harvesting and management methods. Mass mobilisation of communities is necessary if this work has to be done in the shortest timeframe in a cost-effective manner. This should be followed up by at least one Food For Work programme to construct a water harvesting structure in the village since mobilisation of communities and the advocacy of these concepts have to be accompanied by a 'seeing is believing approach' so that the village communities get motivated for its adoption. This structure would reduce the risk involved in facing the recurring drought by providing alternatives to the rural communities for increased water table and stored water in their villages.

The mobilisation of the communities at a large scale is possible only through effective advocacy and awareness campaigns accompanied by development of model water harvesting and drought proofing structures by communities themselves as per the potential and needs of their villages. The total area covered by water management programmes in the last 51 years is just 4.63% of the total potential area. Massive resources are required if it is to cover the remaining 95% which is spread over more than 40,000 villages out of which more than 30,000 are acutely affected by the drought in the year 2000–2001.

The Pilot Plant in Lalwadi is part of an extensive exercise. It shall lead to a bigger project in the first phase of which it is proposed to create 'Islands of Excellence' in 500 villages in the 25 severely affected districts of the state. This would have a 'demonstration effect' on the communities in the neighbouring villages that shall also adopt these measures on seeing their beneficial effects in the model villages. In the subsequent years, 1,000 villages may be selected every year to cover 10,000 villages in 10 years.

Lalwadi village has been selected as a Pilot Project universe subject to the following:

- Extent of scarcity: The Government of Rajasthan has carried out the entire exercise for evaluating the scarcity in the state. Other tools like reports of media, field visits and people's feedback has also been taken into consideration.

- The village has been selected to have minimum intervention and maximum impact; for this, the project would encourage rejuvenation of traditional rain water harvesting techniques and also cost-effective new ones.
- Absence of any other player: UNDP Jaipur is networking with all the leading players in drought and ensuring there is no other player so as to avoid duplication of efforts.

Expected-Project Goal

The goal of the project is enhancement of livelihood options for drought risk reduction, which shall especially be sensitive to vulnerable sections primarily, women and the marginalised, by strengthening the capacities of village community. The project implementation is to ultimately contribute to (1) long-term self-reliance among the community to face future disasters, especially drought, which is a recurrent phenomenon, and (2) serve as a replicable model.

Objectives of the Project

The main objectives of the Project are to:

- Generate awareness among the communities regarding long-term coping mechanisms for drought.
- Create model drought proof villages in various parts of the state by constructing traditional as well as modern water and soil conservation structures.
- Organise the communities at the village level for long-term water management.
- Rejuvenate traditional drought coping mechanisms depending on local conditions and felt needs.
- Make an impact on policy matters of the government by exemplifying model drought proof villages.

Following special drought proofing and water harvesting activities have been proposed to be undertaken:

1. Mass awareness programmes for community mobilisation

- In the village, a Drought Mitigation Group (DMG) shall be formed which shall comprise of members of different social strata of the village and age groups.
- There shall be an initial orientation programme for these groups which shall be the entry point for all the members of the village.
- Phase-wise training programmes will be implemented. The DMG shall be trained in different aspects of risk and vulnerability reduction. The broad areas to be covered shall include:

- Dry land farming as the area is basically in the arid and semi arid zones.
- Animal Husbandry
- Fodder Management
- Soil and water conservation techniques that shall also include efficient use of water
- Farm forestry
- Pasture Land Management
- Improved agricultural practices which shall include use of bio-fertilisers, bio-pesticides, vermi-compost techniques, and other techniques which demand less external inputs.

2. Programmes for water management

- Construction of model rain water harvesting structure (RWHS): It is proposed that in the village one water conservation structure having larger impact shall be chosen and constructed with the involvement of communities.
- Attempts shall be made for the following activities to be initiated for deriving the maximum benefit:
    - Earthen embankments.
    - Shelter belt plantations
    - Bunding
    - Loose stone and permanent check dams
    - Nadis, johads
    - Khadins
    - Tankas
    - Kuis, baories.

Advocacy

An effective advocacy campaign is being launched in collaboration with leading media groups, academic institutions and other organisations to advocate the cause of water conservation and harvesting practices throughout the state.

The efforts of this project would be brought to the notice of all leading players periodically via meetings and workshops to advocate the cause of effective water management for drought risk reduction. UNDP/ UNDMT, Jaipur is already networking with all the major players including the Government and hence an effort is being made to include drought risk reduction in normal development programmes. The success of each experiment in the villages shall be highlighted for replication in other parts of the state. Government being the leading player shall essentially be made part of these workshops and meetings to advocate establishment of linkages between the drought risk reduction and normal development programmes.

UNDP/UNDMT, Rajasthan is also working with the Government to formulate a draft drought policy/act to provide legal framework to drought management, which is still dependent upon age-old famine codes.

Implementation Strategies

The project shall be executed by NISTHA, an NGO, in coordination with the PRI of the village, the local people, the Government and UNDP. The Departments of Agriculture, Animal Husbandry, Forestry, Revenue, and Community Development blocks, etc. shall facilitate, train and orient the village community for their empowerment. The orientation and training programmes shall be conducted with resource persons from the Government and other institutions. A close interaction shall be maintained with the Government departments for the successful implementation of the project. Special training materials depicting various aspects of drought proofing shall be distributed in the village.

For the construction of water harvesting structures, Food For Work programmes shall be launched to provide immediate livelihood support to the villagers. These structures would also address the long-term drought problem in the villages.

Indicators/Output

Following indicators shall be used to monitor and evaluate the project and shall also be used during the implementation phase for effective execution:

- 50% of the rain water would be saved out of a total of 84% of rain water, which is other wise lost as surface run-off.
- Increase the irrigated area from the present 33% to at least 80%. Efforts shall also be made for bringing improvement in ground water table.
- The cropping intensity would double from 119% to more than 250%. Consequently, the agricultural production would be doubled and also losses due to recurrent droughts would be minimised.
- Adoption of improved agricultural practices by community to improve the land productivity and to increase the number of working days on farms. This shall lead to increased income with less of risks leading to better livelihood opportunities in agriculture.
- The purchasing power of the community shall also improve as people shall adopt farming of cash crops like vegetables and less water consuming crops like Kear, Sangri, etc. which are not only suitable to local conditions but also have a very high market value.

- The animal husbandry that is the primary alternate livelihood source in most of the villages of Rajasthan shall get a boost with increased fodder production and better water availability. The milk yield and also meat production shall double.
- Improved capacity of communities to mitigate drought will be effected through a well functioning village drought proofing committee.
- Optimal utilisation of village resources by a well aware and technically equipped village.
- Immediate relief to the drought struck villagers who are on the brink of starvation.
- Better utilisation of fodder and saving of around 40% of the fodder that is being wasted in the villages now.
- A model for the entire state to combat the problem of drought.
- Community asset creation in terms of water harvesting structures.
- Improved livelihood opportunities for the people in their villages itself which shall reduce labour migration.
- Mass mobilisation of communities through out the state for the replication of these methods of drought proofing.

A look at the enclosed Annexures reveal the prevailing conditions in the State of Rajasthan. Annexure-I indicates the extent of scarcity in the year 2000. The expenditure incurred by the Government under relief operations from 1950–1990 in stated in Annexure-II. The extent of famine in Rajasthan during 1998–2000 is mentioned in Annexure-III. The estimates of losses to Kharif and Rabi crops in 2000–2001 are indicated in Annexure-IV and V respectively. An analysis of the conditions that existed in the past calls for the concerned stakeholders to come out with simplistic, workable but innovative schemes with clearly laid down objectives, approach and strategies for improving the conditions of the communities through drought risk reduction measures.

ANNEXURE I

Extent of Scarcity Year-2000

(*Source*: Department of Relief, Government of Rajasthan)

DISTRICT	*Total villages*	*Affected villages*			*Population affected* (Lakhs)	*Sown area hect.* (Lakhs)	*Affected area hect.* (Lakhs)	*Affected cattle* (Lakhs)	*Value of damaged crop* (Lakhs)
		50–74 (%)	*75–100* (%)	*Total*					
Ajmer	1,056	143	906	1,049	11.30	3.42	3.07	8.68	27,371.03
Alwar	1,987	240	182	422	4.29	0.66	0.44	5.07	1,965.08
Banswara	1,472	0	1,455	1,455	11.55	2.25	1.84	11.82	11,153.00
Baran	1,216	199	204	403	9.82	0.92	0.62	2.73	4,538.60
Barmer	1,941	1,185	586	1,771	13.20	15.21	10.13	38.36	16,556.00
Bharatpur	1,392	1	21	22	0.17	2.29	0.09	0.15	98.30
Bhilwara	1,780	191	1,586	1,777	15.81	3.20	2.87	21.60	12,389.44
Bikaner	783	373	261	634	9.74	11.02	7.24	17.93	7,237.25
Bundi	863	364	312	676	8.19	1.45	0.88	9.12	5,763.41
Chittore	2,415	1,688	658	2,346	14.99	3.83	2.70	15.45	19,623.00
Churu	990	358	600	958	14.58	7.05	5.92	21.33	27,337.44
Dausa	1,061	489	479	968	8.77	1.38	1.02	6.29	3,597.95
Dholpur	584	No Damage							
Dungarpur	871	0	871	871	8.74	1.14	0.99	9.58	986.09
Ganganagar	3,026	10	51	61	0.33	0.26	0.22	0.65	1,124.05
Hanumangarh	1,912	76	330	406	4.22	1.52	1.16	4.43	19,726.00
Jaipur	2,312	68	2,158	2,226	23.42	4.42	4.02	20.35	28,628.69
Jaisalmer	637	16	567	583	3.41	4.12	3.74	12.43	653.28
Jalore	712	46	666	712	11.43	5.74	5.33	6.26	28,406.03
Jhalawar	1,613	1,364	221	1,585	9.61	2.84	1.93	8.60	9,291.14
Jhunjhunu	865	534	331	865	16.00	4.01	2.95	12.43	1,812.94
Jodhpur	1,080	191	857	1,048	21.73	11.04	9.67	31.90	28,341.11
Karauli	799	696	76	772	9.04	1.36	0.85	6.95	4,984.00
Kota	937	339	22	361	2.55	0.65	0.39	2.01	2,987.69
Nagaur	1,502	772	373	1,145	13.77	9.47	6.57	18.12	30,185.34
Pali	968	30	916	946	14.76	4.92	4.64	28.04	8,516.00
Rajsamand	1,004	31	973	1,004	8.03	0.85	0.80	11.93	8,071.57
S. Madhopur	800	501	76	577	6.07	1.03	0.63	5.05	3,654.50
Sikar	1,000	585	414	999	18.31	4.64	3.38	16.63	5,487.00
Sirohi	474	18	447	465	6.42	1.34	1.22	9.97	11,571.98
Tonk	1,102	7	1,095	1,102	9.38	2.40	2.28	9.56	10,195.53
Udaipur	2,375	251	2,123	2,374	20.81	2.30	1.86	26.28	8,923.93
Total	**41,529**	**10,766**	**19,817**	**30,583**	**330.41**	**116.70**	**89.47**	**399.69**	**3,51,177.37**

ANNEXURE II

Expenditure Incurred Under Relief Operations
in Rajasthan Since 1950–51 to 1989–90
(*Source*: Department of Relief, Government of Rajasthan)

Year	*Expenditure* (Rs. in Lakhs)	*Year*	*Expenditure* (Rs. in Lakhs)
1950–1951	36.06	1977–1978	265.68
1951–1952	30.63	1978–1979	430.54
1952–1953	104.80	1979–1980	1,662.30
1953–1954	49.57	1980–1981	3,331.47
1954–1955	15.72	1981–1982	14,015.01
1955–1956	25.57	1982–1983	11,396.75
1956–1957	6.13	1983–1984	7,629.94
1957–1958	17.58	1984–1985	1,043.64
1958–1959	24.70	1985–1986	9,133.07
1959–1960	3.96	1986–1987	16,840.91
1960–1961	14.72	1987–1988	62,231.32
1961–1962	40.29	1988–1989	32,440.71
1962–1963	11.01	1989–1990	3,186.73
1963–1964	126.95	1990–1991	3,842.69
1964–1965	416.58	1991–1992	574.61
1965–1966	112.55	1992–1993	13,166.65
1966–1967	1,141.98	1993–1994	3,421.39
1967–1968	793.47	1994–1995	16,658.96
1968–1969	1,542.07	1995–1996	4,568.86
1969–1970	6,307.36	1996–1997	20,364.56
1970–1971	4,210.81	1997–1998	524.85
1971–1972	222.68	1998–1999	18,729.86
1972–1973	1,179.24	1999–2000	27,697.00
1973–1974	5,109.68	2000–2001	1,40,200.00
1974–1975	704.16	**Total**	**4,27,656.62**
1975–1976	1,025.85		
1976–1977	207.41		

ANNEXURE III

Extent of Famine

(*Source*: Department of Relief, Government of Rajasthan)

Activities	Year 1998	1999	2000
1. No of villages affected	20,069	23,406	30,583
2. No. of Population affected (in Lakhs)	215.07	261.79	330.41
3. No. of Cattle affected (in Lakhs)	295.78	345.60	399.69
1. Damage to Crop			
A. Area (in Lakh Hectares)	64.96	78.18	89.47
B. Value (in Lakhs Rs.)	2,28,348.96	3,40,701.61	3,51,177.37
2. Rainfall Deficit	-3%	-16%	-29%
3. No. of Overflown Irrigation Reservoirs	277	308	131

ANNEXURE IV

The Estimates of Losses to Kharif 2000–2001 Crops

(Prod. '000 Tonnes)

Crops	*Actual 1999–2000 Prod.*	*Production* Original target	Estt. prod.	Loss in prod.	*M.S.P.* Rs./Ton	*Loss in* Rs. Crores
(a) Cereals:						
Rice	252	155	168.4	13.4	5,100	7
Jowar	173	300	129.5	-170.5	4,450	-76
Baira	1,299	2,225	1,294.0	-931.0	4,450	-414
Maize	964	1,160	569.7	-590.3	4,450	-263
S.Millets	1	5	2.2	-2.8	4,450	-1
Total	**2,689**	**3,845**	**2,163.8**	**-1,681.2**		**-747**
(b) Pulses:						
Kh. Pulses	118	530	276.6	-253.4	12,000	-304
Arhar	16	10	15.1	5.1	12,000	6
Total	**134**	**540**	**291.7**	**-248.3**		**-298**
Total Foodgrains (a + b)	**2,823**	**4,385**	**2,455.5**	**-1,929.5**		**-1045**
(c) Oilseeds:						
Sesamum	16	85	44.4	-40.6	13,000	-53
Groundnut	264	250	170.1	-79.9	12,200	-97
Soyabean	601	775	546.1	-228.9	8,650	-198
Castorseed	40	60	99.7	39.7	10,000	40
Total	**921**	**1,170**	**860.3**	**-309.7**		**-309**
(d) Others:						
Sugarcane	787	1,100	764.1	-335.9	561	-19
Cotton	502	765	502.8	-262.2	16,250	-426
Guar	232	550	547.3	-2.7	12,000	-3
Total						**-1802**

ANNEXURE V

The Losses in Rabi 2000–2001 Crops

(*Source*: Department of Agriculture, Government of Rajasthan)
(Prod. '000 Tonnes)

Crops	*Actual*	Production				
	1999–2000 Prod.	Original target	Revised target	Loss in prod.	*M.S.P.* Rs./Ton	*Loss in* Rs. Crores
Wheat	6,732	6,850	5,462	–1,388	5,800	–805
Barley	365	400	483	83	4,300	36
Total Cereals (a)	**7,097**	**7,250**	**5,945**	**–1,305**		**–769**
Gram	678	1,200	736	–464	10,150	–471
Rabi Pulses	87	40	34	–6	10,150	–6
Total Pulses (b)	**765**	**1,240**	**770**	**–470**		**–477**
Total Foodgrains (a + b)	**7,862**	**8,490**	**6,715**	**–1,775**		**–1,246**
Rape and Mustard	2,459	2,450	1,940	–510	11,000	–561
Taramira	19	75	40	–35	10,000	–35
Linseed	5	12	6	–6	8,000	–5
Total Oilseeds	**2,483**	**2,537**	**1,986**	**–551**		**–601**
Total						**–1,847**

20

Timely Intervention for Disaster Risk Reduction

— **Dhanushi Senanayake**

INTRODUCTION

Landslides pose a severe threat to life and could cause extensive damage to natural resources and economic infrastructure. Often landslides disrupt livelihoods and thereby inflict a strain on development.

In Sri Lanka, sharp increases in the instances of landslides were reported during recent years. Landslides have been a constant occurrence in the central and southwestern regions since early 1980s. In the recent past, landslides have occurred in 1986, 1989 and 1993. The lives lost in the 1989 and 1993 landslides were 300 and 351 respectively. A number of families became homeless in an extensive part of the hill country.

According the Department of Social Services statistics on the distribution of relief to those affected by natural disasters during the last two decades in Sri Lanka, 40% has gone to landslide and flood victims. Further, during the period 1987–1996 out of the reported number of deaths from natural disasters, 95% deaths were caused by landslides.

Landslides are caused by natural and human activities. Heavy rain and certain soil varieties in areas surrounded by mountains contributes to the occurrence of landslides. Further, indiscriminate land digging for construction purposes, poor drainage facilities, incorrect land use policies and high-density population settlements too lead to the occurrence of landslides.

The landslide mapping process conducted by the National Building Research Organisation identified seven districts in and around the central hills as landslide prone. Landslides affect about 30% of the population occupying 20% of the land area of the country. Hill country of the island is most subjected to the landslide hazard every rainy season.

NAWALAPITIYA—A CITY IN THE GRIP OF LANDSLIDES

In the Nawalapitiya district several landslides had taken place causing severe destruction to lives and human settlements. The Nawalapitiya district is located 142 km away from the capital city of Colombo. Nawalapitiya town receives an annual average rainfall of 300–400 mm. Out of the total land area, 17% consists of steep slopes. Natural resources are not available within the Nawalapitiya city area.

The economic base of the town is determined by other sectors of the economy, such as industries and services. The economic activities of the Nawalapitiya town are based mainly on the railway service sector.

The available usable land area in the jurisdiction of the township has been almost fully utilised. Further, expansion of the city is hindered due to the geographical location of the area. This has resulted congesting the town area with public, private and religious dwellings.

Most of the paddy lands of the town are categorised as abandoned lands. To accommodate the increasing population of the town; it is anticipated that these lands will be filled. Filling of this land requires careful consideration. Even the tea plantation land area is less, which is about 1% of the main land area of the town. Many of the recent constructions according to a government official is unauthorised and on encroached land. The source of livelihood of people in Nawalapitiya area is mostly confined to the railway and services sector. Opportunities for self-employment are much limited as majority of the labour force is unskilled. Due to unavailability of suitable land, the opportunities for agricultural activities or animal husbandry are limited. Living standards of the people living in this area could only be improved by expanding the industries and services sectors.

COMMUNITIES IN PRE- AND POST-LANDSLIDE SITUATIONS

ITDG South Asia in collaboration with the National Building Research Organisation (NBRO) under the Livelihood Options for Disaster Risk Reduction project conducted a research study in two settlements in the jurisdiction of the Nawalapitiya Municipality Council. Two communities were selected to represent pre- and post-landslide situations and to understand the impact of landslides on the livelihood patterns of the landslide prone communities. Gondennawa area was selected to represent pre-disaster community and Soysakele area was selected to represent post-disaster community.

Both the communities are placed at different strata in terms of socio and economic conditions. The post-disaster community at Soysakele is low in terms of economic conditions such as income levels, employment status, living standards assets and resources.

In Soysakele, the majority of the income earners are casual labourers, with an average monthly earning of Rs. 7,700 per household. Most of them are occupying land without proper authorisation. The population density of the area is high. This has resulted in blocking the city's drainage system and further aggravating the threat from landslides. The education level of the community is low in comparison to Gondennawa. About 83% of income earners had less than G.C.E. (O/L) education. Out of the total income earners, about 27% never had any formal education.

Gondennawa area is placed between a large rock and an old landslide prone area. This area is threatened especially with rock fall threats. The community in Gondennawa is subjected to constant fears and the threat of rock falls and landslide. About 60% of this community occupies the government railway quarters, which are condemned due to landslide threats. Majority of the income earners in Gondennawa are white collar workers with a monthly income of Rs 8,500 per house hold. In this community, although about 75% of people owned between 6–20 perches of land, the property value of their holdings are very low due to the prevailing threat from landslides. The level of educational attainment in Gondennawa was high compared to Soysakele. Only 42% had education less than G.C.E. ordinary level.

Both these communities have very limited alternatives available to the besides their existing occupation.

An interesting feature of both the communities is the peoples' perception of landslide threat. People consider the extent of severity according to the number of lives lost. In these communities though people continue to live with fear, they expect that landslides will not occur, despite the threat being severe. Further, they do not have economic means to settle elsewhere.

As a result of the risk reduction measures adopted at the government and non-governmental levels, the interventions by such organisations have shown some impact on the communities.

The National Disaster Management Centre (NDMC) has identified the following key agencies that are involved in disaster management.

1. Department of Social Services
2. Department of Health
3. Department of Irrigation
4. Department of Meteorology
5. Police and Armed Forces
6. Urban Development Authority
7. National Building Research Organisation.

The Government of Sri Lanka in 1989, with the assistance from UNDP launched a capacity building project on landslides. As a result of the number of awareness generation and capacity building measures, the number of deaths got reduced. It happened because of the fact that people were evacuated before a landslide.

In Nawalapitiya a pioneering organisation in landslide risk reduction activities was the Centre for Housing Planning and Building. In 1997 the Centre for Housing Planning and Building in coordination with the Nawalapitiya Urban Council initiated the 'Work book' documenting information about the Nawalapitiya Urban Council Related Information. A hazard zone mapping process, which has enabled easy access to the information about the district to be used for further developmental planning initiatives, followed it.

The initiative taken by the Government of Sri Lanka in 1990 to assess and map landslide hazards in two districts included ascertaining socio-economic issues of resettlements, creating awareness among resident communities in hazardous areas about the adverse impacts of improper land use, and other preparedness measures. This initiative was extended and replicated in other areas. It resulted in both immediate and long-term positive impacts. Landslides were earlier considered an area of technical expertise, but is now recognised as a multidisciplinary issue with much focus on social aspects. This initiative has also verified that it is the encroachers or illegal squatters who are the most vulnerable to landslides. Such information help resolve many issues related to town and country planning. The government organisations responsible for town planning, building and settlement now take a more systematic approach and get clearance from the NBRO.

A risk assessment methodology and a methodology to determine priority areas for detailed landslide mapping have been developed for use in future plans.

Further steps undertaken in this direction include awareness creation campaigns involving the heads of various government departments such as the Urban Development Authority, National Building Research Organisation, Department of Social Services, Telecommunications Department, etc. At the same time, there were on going awareness creation programmes for school children too.

Awareness building and capacity strengthening process for architects and draughtsman, construction sector workers such as electricians, contractors, etc., has also been initiated.

The awareness building activities followed a preparation of an action plan involving the National level governmental organisations like the National Building Research Organisation, Urban Development Authority, Social Services Department, etc. Tasks were identified for and responsibilities assigned for each organisation in pre, during and post-landslide situations.

The immediate impact of such awareness generation exercises resulted in an increased interest in reporting and reacting at inappropriate constructions by community members particularly in Nawalapitiya. The extension to a school Anurruddha Maha Vidyalaya, in an area where a landslide had occurred previously, had to be immediately stayed by the Nawalapitiya Urban Council on the demand made by the public.

The Centre for Housing, Planning and Building has included landslide hazard assessment in their curriculum in the training courses on building construction.

In Sri Lanka, up to 1996, disaster management was merely limited to the provision of relief to affected people. Even today, the Social Services Department is handling this function. Some forms of landslide disasters were referred to NBRO. Apart from this, Sri Lanka Urban Multi-hazard Disaster Mitigation Project (SLUMP) has been formulated as a joint project of Urban Development Authority, National Building Research Organisation and Centre for Housing Planning and Building.

In 1994, a committee was appointed to review the environmental laws in Sri Lanka. This committee considered land use, agriculture and social conservation as the subject of land use planning. The recommendations of this committee led to the establishment of Land Use Policy Planning Division (LUPPD) of the Ministry of Agriculture.

Gaps in the Institutional Measures

Although a number of agencies are involved in disaster management and for enacting Counter-Disaster Measures Bill, the importance of disaster mitigation has not been well recognised.

In order to reduce the risk of disasters, the most essential thing is to mitigate it at the planning stage. Therefore, National Disaster Management Council should act as a coordinating agency between various developmental agencies.

One of the most notable gaps in institutional set up is the non-availability of disaster control or management agency at the national level. It eventually leads to gaps in regional as well as local level management. Though NDMC or the Department of Social Services are handling disasters in the relief activities stage, a key agency which has technical expertise is needed at the national level to take care of the pre-disaster phase.

Adhocism prevails in listing out priorities and in managing problems arising from uncertainties relating to investigations and analysis. Absence of focus on Landslide Impact Assessment as an essential element of the planning process was felt greatly in developmental activities. Probably, it was due to the non-availability of landslide hazard maps.

Public awareness about occurrence of landslides and community participation in the landslide disaster mitigation are of a low order.

In Nawalapitiya, the Urban Council presently dumps their solid waste into the riverbanks at two places. The site affected by landslide in Soysakele is filled up with waste from the town dumped without taking any precaution. Further, the sewerage treatment plant in the hospital is not functioning at present and sewage flows into the Mahaweli River adding to the possibility of landslide.

Coping Mechanisms Demonstrated by Affected Communities

It has been observed that both the communities demonstrated a high degree of awareness on landslides.

These two communities live under constant fear from the threat of landslide and rock fall situations. During heavy rainy situations they take turns in keeping vigil. There have been instances when they could alert others by shouting out loudly about the impending danger. The only method of evacuation they could think of was to flee from the vicinity. Children were asked to stay in rooms perceived by the parents to be hit last by the falling rocks.

However, the Pilot demonstration project initiated by ITDG South Asia in collaboration with the NBRO involved community participation at different levels. The initial awareness building activity demonstrated a high degree of participation from both communities involving equal participation of men and women. In recent times, a few new community-based organisations have emerged. ITDG's pilot project is implemented by involving the community members from both locations as representatives. Their contributions are sought for needs assessment, formulating pilot project plans, and for implementation and maintenance.

Community-based Project Initiative

The community-based project initiative is implemented by ITDG South Asia in collaboration with the National Building Research Organisation. The project is coordinated by the National Building Research Organisation. This project lays emphasis mainly on knowledge and capacity building activities, structural measures, advocacy and policy influencing.

The knowledge and capacity building exercise is taken care of by:

- Raising awareness and conduct training on suitable engineering techniques for build on the existing surface that to avoid heavy soil excavation.
- Developing early warning systems.
- Mobilising the community to face the risk of landslides in a collective manner rather than individually.
- Strengthening linkages of the community with the relevant local government authorities and thereby increasing and strengthening the capacities of the community to cope better when faced with a landslide situation.
- Improving and increase awareness of women in the households on risk reduction from landslides. In both these communities women remain in their homesteads, while their husbands go to work. Therefore women very often have noticed signs of landslides and rock falls.

The structural measures are strengthened by:

- Reconstructing the drainage system in order to avoid disrupting and blocking waterways.
- Building a proper retaining wall.
- Mobilising the community members to plant a belt of trees to prevent and minimise the damage from falling rocks and boulders.

The advocacy and policy influencing process is put in place through:

- Impressing upon the local authority Nawalapitiya Urban Council to effectively implement construction codes for any new construction in identified landslide prone areas.
- Generating awareness through poster campaigns.
- Raising awareness among school children.

Lessons and Best Practices

The high degree of awareness, capacity and commitment of the local government authorities of the Nawalapitiya urban Council is quite praiseworthy. The Centre identified the Chairman of the urban Council for Housing Planning and Building for training on Community-based Disaster Management, which was conducted by the Asian Disaster Preparedness Centre, Bangkok. The chairman himself was a pioneer in taking forward the awareness creation campaigns and obtaining the necessary support from other government departments. However, many local government authority heads involved are without adequate training on disaster management. This implies that identifying and conducting training on community-based disaster management for local authorities is essential for effective implementation of any mitigation or disaster management activity.

Community members from Gondennawa and Soysakele were not willing to consider relocation. It was mainly because no alternative places for relocation were offered by authorised local government sources. The local government had not taken any measures against unauthorised construction or occupation of state land. Further, limited availability of income generation opportunities for the community members outside Nawalapitiya, non-availability of vocational training centres in close proximity, lack of expansion in terms of commercial activities were also reasons for the reluctance of the community members to move into another location. The local authority officials have not taken action against illegal/unauthorised settlers. These unauthorised settlers have caused drainage problems, which add to landslides in Soysakele. Therefore, it is important that the relocation areas provide means of income generation and employment opportunities.

Another observation was that, it was the low-income settlers who were most prone to the risk of landslides. Therefore, urban and any development planning activity at local level needs to pay attention on resettling low-income dwellers in safe areas.

RECOMMENDATIONS AND SUGGESTIONS

Nawalapitiya is located at a central point, and can be reached by bus or train from the capital city of Colombo and Kandy. Therefore, improving the accessibility to Kandy can attract investors and entrepreneurs into this area from other areas of the island. It is also important to improve the infrastructure facilities in the town.

Considering the threat, the local authorities need to search and identify proper residential locations to relocate the communities at risk from landslide and falling rocks.

Nawalapitiya with its picturesque surroundings located in the central hills has the potential to be developed as a tourist attraction for both foreigners and locals. For this, it is important to develop the present road network and the infrastructure facilities.

The low-income unauthorised settlers are at high risk to landslide threat. They have no alternative places to settle and their employment activities are limited to labour work. Their housing conditions are also inadequate and unsafe.

The urgency of this problem should be considered by the local housing authority and the urban council and immediate measures taken for the resettlement and for the construction of residential apartments for these communities.

Both communities do not have adequate access to systematic credit schemes. This could be one reason why these communities do not go in for any income generating self-employment activities. It is therefore necessary to establish a micro-credit scheme to provide financial facilities for self-employment programmes.

Setting up of an industrial base might prove to be beneficial to both the community and entrepreneurs considering the availability of wage labour in this area for a comparatively low rate.

In the direction of reducing the risk from landslide and working towards achieving preparedness, an information dissemination network that will enable free flow of information can be established with assistance from relevant government and non governmental organisations. For landslide risk reduction measures, masons, carpenters, electricians and plumbers need to be involved in programmes for specific skills development. It is also important to involve staff from the base hospital, Police and divisional Secretariat offices for capacity strengthening activities during an emergency situation.

Considering the high degree of awareness of the communities, the Urban Council's prevention measures such as planting of appropriate trees to prevent the falling of rocks and boulders and public awareness campaigns will demonstrate better results.

CONCLUSION

Nawalapitiya town is situated away from the main road network between two major links to the Capital City, Colombo. The physical fabric of the town is obsolete mainly due to lack of investment by the successive governments during past two decades.

Infrastructure facilities such as water supply and sanitation, solid waste disposal and drainage system are not adequate. Most of the housing units are old but the type of colonial setting with various facilities resist people to abandon them. Closeness to the town and the existence of a high percentage of Moor and Hindu communities has also led to the concentration of settlements.

Some houses in both these communities do not have a 'well-engineered' construction. This further increases problems in landslides prone areas. There are instances of houses being constructed on encroached land or without proper authorisation from the local authority. The local authority officials have not taken action against illegal/unauthorised settlers. These unauthorised settlers have caused drainage problems, which lead to landslides in Soysakele. This in turn, results in the minimal development of housing and land. Besides, obtaining loans from lending institutions is not possible due to land ownership problems. The Nawalapitiya city has also limited space for any expansion.

22 families in Gondennawa and 13 families in Soysakele have moved in to the area after the landslides which occurred in 1994. In Gondennawa, nearly 35% of the migrant households have moved into the area after 1997. The migration pattern indicates an increasing trend, which is contrary to the awareness created by the local authority officials. This further confirms that non-occupation of state lands by any confirms form of activity leads to unauthorised occupation. Moreover, the study has revealed that nearly 50% of the households migrated to the areas under investigation by the local authority.

The analysis of the factors brings to light the fact that lower land values may have attracted them to the area. Availability of employment in and around Nawalapitiya and closeness to the Kandy, the second largest city in the country, may also have encouraged them to migrate to this area.

Nawalapitiya particularly does not seem to be attracting investors for any new commercial activity, probably due to the low level of infrastructure facilities and the unskilled labour force. The town has little scope for industrial or agriculture development apart from some possibilities for plantation industry.

However, the recent initiatives have demonstrated a high level of enthusiasm of the community to counter the threat of landslides. This can be utilised to implement the community-based pilot project initiatives successfully. Further, the local authority head, i.e., the chairman of the Urban Council himself has demonstrated high degree of knowledge about the problems of the affected communities. The community members' easy accessibility to the local government authorities is a noteworthy feature.

The high degree of awareness and commitment demonstrated by the officials of the local government authority can be made use of to convince other national government organisations. This in turn will be of assistance in obtaining land and other infrastructural measures for relocation and for improving the living conditions of the landslide affected communities.

Considering the unskilled labour force and the increasing unemployment rate in both the communities, local authorities, with the support of other government departments such as the National Apprentices and Industries Training Authority, Ministry of Industries and Youth Affairs Ministry, a Vocational Training Centre needs to be established in the town.

Considering the ideas presented in the research and the general observations of the ITDG team, it is strongly felt that efforts need to be made for undertaking detailed, data based vulnerability assessment and risk analysis in all areas prone to different disasters so that timely actions are initiated for disaster risk reduction.

21

Disaster Risk Reduction: Livelihood Options in Drought Situations (A Case Study of Tharparkar, Sindh, Pakistan)

—Pir Abdul Haque Jan Sirhindi

INTRODUCTION

Tharparkar district is a sandy dune desert in Sindh Province of Pakistan and is part of great Rajisthan desert of Indo-Pakistan. The lower Thar comprises Tharparkar and Umerkot districts of Sindh Province. Tharparkar district is spread over 19,638 square kilometres in the south-east part of Pakistan and includes the ecological zones of the sandy dune area called the *Thar area* and the hilly tract and marshy lands of *Parkar area*. Administratively, it is divided into the four sub-districts of Mithi, Diplo, Chachro and Nagarparkar. It has a population of 9,14,000 with a population density of 46.17 per square kilometre. Tharparkar has a rain dependent agricultural economy. Rainfall pattern is not uniform and ranges from 50 mm to 300 mm per year received mostly in the monsoon season (June–September). Drought usually occurs in every four to five years and generally, during such periods, half of the population migrates to the plains of Indus in the north which would support human life and livestock in such times.

The most severe drought that affected Thar in the last hundred years were in the years of 1899, 1939, 1948, 1952–53, 1985–87 and 1999. The drought that hit the area in 1999–2000 resulted in the migration of forty per cent of the human population and eighty per cent of population livestock. Both the Government and Non-Government Organisations undertook an extensive relief operation. Owing to the capacity and the commitment of the organisation towards the people of Thar, the Thardeep Rural Development Programme (TRDP) was one of the key stakeholders in the process. Along with relief work, Thardeep also conducted and coordinated studies for combating the challenges posed by drought.

One of such research studies was on 'Livelihood Options for Disaster Risk Reduction', conducted by Intermediate Technology Development Group (ITDG) in February 2000. The aim of the study was to establish the links between drought and socio-economic conditions in relation to the livelihoods of communities and to identify approaches that can support, create and sustain options/opportunities for the livelihood of people living in drought-prone areas.

The research report concluded with some short-term and some long-term recommendations. In the light of these recommendations, Thardeep, in collaboration with ITDG, facilitated a community project in the village, Mithrio Charan, in the sub-district Chachro of Tharparkar district. The implementation phase of the project has been completed. An attempt has been made in this paper to present a critical analysis of the project along with a few suggestions for extending the scheme to the wider area of the 'Arid Zone' of Tharparkar.

SITUATIONAL ANALYSIS AND CRITICAL REVIEW OF IMPLEMENTATION PROCESS

The Project was undertaken in village Mithrio Charan at the distance of 45 kilometres in the north-east of Chachro. Mithrio Charan comprises four hamlets with a total population of 936 individuals. Rain-fed agriculture is the major source of income for 80 per cent of the population. Wage labour and remittances are the additional resources. The villagers have two options available for meeting their drinking water needs; harvesting the rain water in small traditional reservoirs and constructing wells to get access to deep aquifers (found at depths of 250 to 300 ft). Both the options have seldom been used for irrigation even in a small scale in the recent past. Ostensibly it is due to lack of any joint, organised and well-planned venture but intrinsically, the crucial elements of poverty and meagre resources at the disposal of the community were the key constraints that prevented harvesting the water for a livelihood option instead of using it for the sole purpose of drinking.

The project was based upon the idea of offering support to the community to bridge the gap between the technical and managerial fronts and thereby facilitating to extend the scope of available water to small-scale irrigation. The initiative, as was conceived, would generate opportunities of subsistence and income generation. The process analysis aims at defining and delineating the whole approach of creating an alternative risk reduction strategy in times of natural disasters, particularly drought.

Major activities included constructing rain water harvesting tanks in every household and cultivating 4 acres of land to be irrigated by utilising the ground water from a dug well besides grafting the indigenous bush of Ber (Sisyphus Jojaba) to produce its improved variety and establishing 'kitchen gardening plots'. Smokeless stoves were to be constructed as a measure to discourage the use of wood as fuel. A fattening and stall-feeding practice for herds would be initiated for increasing production and profit. The project plan clearly defined the terms of references between the three partners involved—the Community, Thardeep and ITDG. It also explained the work plan, budget allocation and the implementation time frame and also stated explicitly, the outlines for monitoring the process.

The visit to Project universe followed by a discussion with the community members and the Thardeep team revealed that to a large extent, the desired outputs were achieved. However, the long-term implications in terms of the nature and the amount of disaster mitigation that such a project can ensure in times of drought remains to be seen.

The analysis of the key strengths that contributed to the smooth execution of the project points to their long-term efficacy, durability and compatibility. Some of the major strengths include the individual capacity of the partners, the low scale of activity, the clear framework for community involvement, and the favourable geophysical fabric.

Capacity of the Partners

The Project could be a success primarily due to an integrated structure, sharing of resources, and due to the Project activities complementing the other ongoing Projects in the area. These could be attributed to the capacity of the partners.

The community organisations were already in place, with trained social workers. The implementation of the Project was supervised by a qualified and dedicated leader whose primary focus was on 'doing and learning'. These factors however, raise the question whether these would always be available for similar Projects to be implemented successfully elsewhere.

Low Scale of Activity and Clear Framework for Community Involvement

The project has indeed resulted in an innovative model. It did demonstrate a means to produce fresh, green vegetables and desert fruits during drought, not only for consumption but also for deriving income. The communities in the proximity of the site benefited considerably and in some instances were motivated to replicate it. The location has turned out to be a training station that teaches an alternative mode of small irrigation. The community has exhibited practically whatever could be contributed and shared for the success of a collective venture. The choice of village organisations has proved to be apt as it succeeded in harnessing and using the community's potential. Of course, it is a low scale activity.

Favourable Geophysical Fabric

Factors like soil texture and quality as well as quantity of water need to be taken stock of for making the decision regarding use of the appropriate technology to combat drought situation in an area since the success of any venture is largely dictated by these factors. The favourable geographical features which existed in the Project universe, went a long way in making the results encouraging. The project can be successfully implemented in other areas also by making suitable alterations to the methodology depending upon the existing geophysical traits of the areas concerned.

RECOMMENDATION AND SUGGESTIONS

(a) The project may be extended to 25 villages in Chachro sub-district involving community organisations and Thardeep with the same terms and conditions applied. In the near future, the district Government and ITDG may be the real partners with clearly defined roles and responsibilities.

(b) The project may be used as a training and demonstration station for government functionaries with the assistance of the district Government particularly of the Agriculture department. They can use the results wholly or in piece-meal as best applies to their context to prepare effective schemes for all the villages of Tharparkar district.

(c) ITDG and Thardeep may support the district Government in identifying the partners that can facilitate in exploring the resources and in suggesting methods to evolve structures at the grassroot level that can secure community participation on one hand and cooperation and assistance on the other hand to make the schemes successful.

(d) ITDG, the district Government and Thardeep may organise a comprehensive technical study to clarify the feasibility and the appropriateness of such a project with regard to the other parts of Tharparkar district and Sindh.

22

Livelihood Strategies in Disaster Risk Reduction in Bangladesh

— Hafiza Khatun

INTRODUCTION

The term 'environment' is usually defined as the system of biological and physical resources, and their process of interaction, that affect life and livelihood. These biophysical systems are in fact, in constant interaction with human beings and their social system. The physical development activities for human benefit use and interact with these biological and physical resources i.e., these components of the environment in a complex manner. The development projects formulated and assessed according to technical, economic and political criteria without adequate consideration of their environmental and social impacts have resulted in unforeseen harmful consequences that adversely affect the desired benefits and the sustainability of the projects. In the same way, attempts for appropriate environmental management without integration of human behavior and social forces is likely to be ineffective (Ahmed, et al., 1994). In addition to this, over exploitation of natural resources also leads the environment to a deplorable condition. Natural phenomena such as cyclone, earth-quakes, volcanic eruptions, tsunamis, wildfires, floods, landslides, droughts and river bank erosion are threatening the environment more than ever. So do their basic man-made counter parts such as accidents, communal riots, war, etc. These disasters continue to cause grievous damage to the environment (Carter, 1991).

South Asia is a very disaster-prone region. It is not to be perceived in a pejorative sense, rather we all should take pride in our resilience and sustained efforts to mitigate losses and cope with natural hazards and eventually contribute in environmental management. South Asia is exposed to nearly half a dozen types of natural hazards. The Asian Disaster Preparedness Center in Bangkok has rated the intensities of various types of hazards for South Asian countries. Bangladesh is severely vulnerable to floods and cyclones, and Pakistan to floods, earthquakes, and droughts. India is highly vulnerable to floods, cyclones, droughts and earthquakes, while Nepal is prone to earthquakes, landslides, droughts and Flood (Khan, 2001). Each type of these disasters differs in its nature and impact. The policy makers in each of these countries have to design unique disaster risk reduction measures for each type of disaster in environmental management planning.

Despite a rising awareness about the needs for protecting the environment, environmental degradation has already cut a swathe through South Asia especially through Bangladesh quite fast during the last few decades (Chowdhury, 2001). Bangladesh is a densely populated, relatively poor country facing a number of natural and technical challenges. It is a country where disaster and the natural environment are linked, although the nature and mechanism of these bonds are not always clear. There are several factors that make Bangladesh particularly vulnerable to environmental degradation. The major ecological and environmental problems in Bangladesh are: natural disasters such as floods, cyclones, river bank erosion, drought, arsenic contamination in ground water and others like earthquakes, tornados, landslides, etc.; besides deforestation and forest degradation; salinity intrusion in the coastal area; pollution from agrochemicals and vehicles, urban and industrial waste; lack of safe water and sanitation, drying up of wetlands and water logging; loss of bio-diversity; and poorly planned development projects. These natural disasters are often costly in terms of lives lost and their impact on the economy. On an average, disasters caused 530 deaths and loss of 23.614 million Taka per year in the period from 1996 to 2001 (Kelly and Chowdhury, 2001). The effect of disasters is much more severe on women than it is on men because of biological and psychological factors. On the other hand, women virtually shoulder the whole responsibility of looking after the children and protecting their families. Even in normal times, women lack social power and their condition is being aggravated by the increased demands on them. The welfare of women constitute an issue of predominant concern for a country in ensuring a sustainable environment.

Disasters are usually classed into two; those which are caused by the forces of nature and those which are due to the interaction of human actions and natural processes and resources. Social scientists have added a third category; that of social disasters. This chapter will focus only on the livelihood strategies of coping with disasters caused by nature in Bangladesh.

NATURAL DISASTERS IN BANGLADESH

Flood

Bangladesh lies on the flat alluvial plains in the combined delta of three mighty rivers; the Ganges, Brahmaputra and Meghna. Flood is almost a regular annual affair in Bangladesh. One fifth to one third of the country is flooded every year during June through October when nearly two thirds of the food grain (mainly rice) is produced. Floods encountered by Bangladesh are of four categories:

(i) normal monsoon floods, due to over topping the river banks and the excess water submerging the adjoining areas. This is the normal type of flood (with variation in extent and duration) which the country has to bear with year to year.

(ii) floods due to on rush of rain water down the hill slopes overland and its ultimate ponding up in the low lying areas.

(iii) flash floods in the eastern and northern rivers with sharp rise of river stages within a matter of days only, followed by sharp fall.

(iv) tidal surge because of tidal fluctuations and water level set-up due to cyclones (Khatun and Ali, 1990).

The country has been hit by many catastrophic floods of a single category or a combination of various categories of various durations resulting in huge loss to human lives and property. For generations, flood plain inhabitants of Bangladesh have adapted to the annual floods through numerous indigenous strategies so much so as to reap benefits from this recurring natural phenomenon. However, floods become a major public concern when catastrophic floods which surpass their adjustment ability occurred in the year 1954, 1955, 1956, 1974, 1984, 1987, 1988, 1998 and 2000.

With increase in population and growth of physical infrastructures, vulnerability of the society to floods has increased; consecutive floods at times (viz. 1987 and 1988 floods) drastically reduce the percentage growth rate of GDP. The growth rate of GDP dropped down from 4.4 per cent in the fiscal year 1985–86 to 3.9, 2.9 and 2.5 per cent during the fiscal years 1986–87, 1987–88 and 1988–89 respectively (Kelly and Chowdhury, 2001). The flood of 1998 which affected two thirds of the country and continued for 75 days caused enormous damage to the country's economy. The loss in the agricultural sector alone amounted to Taka 43 thousand million. The loss included crops, livestock and poultry, fisheries and forestry (Ali and Khatun, 2001). In the year 2000, the usually flood free area i.e. the south-western part of Bangladesh was affected by a flash flood which caused damage of more than Taka 800 billion (BDPF, 2001). About 98 per cent of the earthen houses were damaged, households lost their livestock and poultry and other durable assets. Standing crops like Amon paddy, vegetables, and tree resources were lost (CARE Bangladesh, 2000).

Since the unprecedented flood of 1998, it was felt that flood forecasting and dissemination services to help the community at the local level to prepare against such events are urgently required. Recently, the government has undertaken a project, namely, 'Consolidation and Strengthening of Flood Forecasting and Warning Service' that began in

January 2000 for capacity building, development of equipment and tools for improved data collection and flood forecasting and for effecting improvement in the dissemination of services to communities through the Comprehensive Disaster Management Programme (of UNDP). It is hoped that a community-based approach to flood forecasting would empower local people to correlate the danger levels with past experience, subject to any changes brought about by new infrastructure.

Traditionally people have developed different kinds of coping strategies related to their livelihood. Of the indigenous responses to flooding in Bangladesh, a significant number relate to agriculture coping measures including selecting appropriate variety of rice and other crops—depending on the timing—and water level as well as type of soil. People also tend to reduce the magnitude of economic loss through crop loss from flood and erosion by cultivating low cost varieties. Housing techniques are also adapted according to the risks posed by floods and erosion. Houses are built on raised lands or earthen platforms so water can not reach the plinth during normal floods. They also try not to use any housing material susceptible to flooding (like mud) but preferably corrugated iron sheet, cemented pillar and walls (if possible). Poorer people use thatch, bamboo and corrugated iron sheets. Plantation of water friendly plants/trees like bamboo, banana, Hogla, Kolmi and others beside the homestead is very common to protect the house from erosion as well as for the use of these plants/trees during floods and after recession of the flood. During the dry season, the kitchen is made in the open courtyard, but during floods they make bamboo platforms that can be raised when the water level is increasing and use portable stoves for cooking. Food, household items and crops are also stored on raised platforms. Selling animals, grains and other assets like jewellery, sale and lease of land, advance sale of labor and fishing are the common coping mechanism to survive during the crisis period.

Cyclone

Bangladesh lies in the path of tropical cyclones and has a wide and shallow continental shelf which is particularly vulnerable to cyclone. As a result, the coastal area of Bangladesh have been ravaged by tropical cyclones associated with tidal surge almost every year, particularly in the pre-monsoon months of April–May and post-monsoon months of October and November. Among the many natural disasters afflicting Bangladesh, cyclones are surely the worst, because they cause intense damage and there are no mitigation features as such to counter them. It is estimated that about ten thousand lives have been lost due to floods over the past forty years, whereas the loss of lives from cyclone has been close to half a million in the same period. Damage of livestock and other assets (trees, houses, etc.) have been that much greater (Rashid, 2001).

As regards the impact of cyclones, out of its three accompanying effects of heavy rain, strong wind and storm surge, it is the storm surge of tremendous height and force, which causes most of the damage including loss of human lives and livestock, destruction of property and agricultural and industrial production. During the period from 1960 to 1998, a total of 47 cyclones of different magnitude struck the coastal area of Bangladesh. Of these the cyclones of 1970, 1985, 1988 and 1991 registered storm surges of height 10.0, 4.5, 4.4, 7.6 m respectively and have made worst history in respect of human casualties and property destruction. The cyclones of 1970 and 1991 claimed 300,000, people and one million cattle heads; 150,000 people and 70,000 cattle heads respectively and there were wide spread damage to agricultural production and property. The estimated loss in monetary terms was Taka 42 billion (Kelly and Chowdhury, 2001).

Cyclones kill people and livestock destroy crops, damage infrastructure, houses and vital installations and cause wide spread health hazards. Apart from short-term problems, storm-surge creates long-term problems because the salt water makes the soil unproductive. The impact of cyclones, with frequent occurrences and substantial magnitude, on the marginal landowners and landless people happens to be tremendous. These affect livelihood assets and seriously challenge the efforts of the country towards self-reliance.

Cyclones in the coastal region have occurred from time immemorial, but it is only recently i.e. from early 1960s, that some efforts are being made to mitigate the sufferings and loss of life and property caused by them. The major efforts are prediction and dissemination of warning systems, construction of infrastructures like shelters and killas, coastal forestation, creating awareness among the people to use the given facilities at the time of emergency, immediate post disaster relief and educating the vulnerable people to take necessary measures initiating from own and community level in saving life during cyclone and facing the post disaster emergency crisis and salvaging the livelihood.

From the experience after 1991 cyclone, Akhter (1992) mentioned that during the post-cyclone period, free distribution of relief and reconstruction undertaken by organisations, tended to make the community irresponsible and unwilling to participate in the rehabilitation process. She also emphasised that it is important to ensure a proper work attitude rather than create dependencies. After the 1991 cyclone, literacy was found to be inversely related to death rates. Since literacy is a proxy for socio-economic status, this observation confirms the contention that it is always the poor in the society and the economically disadvantaged who suffer more, even in a natural disaster.

Experience from the 1995 cyclone shows that different community-based disaster preparedness programmes implemented by incorporating various indigenous practices like storing food, water and other necessary items organised by the Bangladesh Red Crescent Society, donors and

NGOs for realising facilities like shelter, killas and proper dissemination of warning on time showed a more effective positive impact on the community in saving life and rejuvenating the livelihood in the post-disaster period than providing only relief and infrastructure.

River Bank Erosion

In Bangladesh, river bank erosion is one of the most important natural hazards which causes widespread damage to man and his habitat. A substantial area of Bangladesh is prone to be affected by river bank erosion. The major factors responsible for river bank erosion are: rapid rise and fall in the water level, high variation in the maximum and minimum discharge, high rate of sedimentation and scouring in the bed material, formation and movement of large bed forms, soil condition of bank materials and flow pattern and deflection currents towards the bank line. Besides submerged islets, discharge capacity of the channels and waves are also responsible for river bank erosion (Ahmed, et al., 1994).

Land loss due to river bank erosion is the highest in the Jamuna; the erosion rate is estimated to be between 139 and 353 hectares per year. Channel shifting is a major outcome of river erosion. During the last 200 years, the Jamuna has shifted laterally about 19 km and the Teesta by 17 km. The Ganges-Padma has migrated eastward during the last 400 years leaving a number of right bank distributaries. Land erosion in the coastal region caused by rivers, sea waves and tide is also significant. Bhola has suffered a net land loss of about 227 sq. km in the last 50 years. Hatiya was reduced from over 1,000 to only 21 sq. km in 350 years and Sandwip has lost 180 sq. km in the last 100 years (Ahmed, et al., 1994). As per the recorded report, 125,935 acres of agricultural land and 4,084 houses fell into the rivers due to erosion in 1997 (Disaster Forum, 1997).

As per BWDB, erosion takes place at 242 identified sites along the banks of 77 rivers in 210 upazilas of 58 out of 64 districts. About 1200 km of river bank erodes and about 8,700 hectares of land are washed away every year. About one million people are directly or indirectly affected by river bank erosion every year in Bangladesh. A field-based research shows that within the period of July 1998 to June 1999 about 18,000 people from 39 villages of Chilmari upazila alone, living along the river Jamuna were directly affected by river bank erosion (Islam, et al., 2001). Erosion affects, the rich as well as the poor. The poor, however, are affected more. The marginalised victims of riverbank erosion lose their homes, land and livelihood and become helpless with little resources. The majority of them find some place in the neighborhoods to live but an increasing number are migrating to the cities and ending up in the urban slums and squatters.

With persistent erosion, the loss and damage to properties, roads and human habitation in general, are remarkably high. The material consequences are short range as economic recovery is possible within a predictable time. But the socio-economic impact and demographic dislocation due to bank erosion are mostly permanent and most often long-term (Halli, 1991; Rogge, 1991). Although the affected people try to make an adjustment to the new (often hostile and unfavorable) environment, their standard of living falls far behind from what they had enjoyed before.

Response by the government to address the problem of erosion has always been confined to the construction of embankments. Over the time, embankments have proved to be ineffective in withstanding river bank erosions. However, since long, the vulnerable people have been adopting various community-based coping strategies to survive after being affected by this inevitable disaster. The coping strategies adopted by the people are mainly considered as corrective, with an aim to modify the event and to minimise the loss. Islam, et al. (2001) have identified some of the coping measures adopted by the affected people at Chilmari as part of their livelihood strategies. These are change of location, use of movable housing materials so that in case of emergency, the structure can be easily dismantled and moved away within a short time; living in clustered villages which helps the settlers to mobilise necessary manpower and assistance at the time of emergency and also provides for moral and emotional recovery from the hazard effects; formation of samaz/gushti based social organisations; possession of a means of transport, preferably a country boat and one of the family members (if possible) adopting a non land based occupation (may be secondary) so as to be, supportive for the family at the time of crisis.

Drought

A natural disaster that is seldom frequently mention is drought. This is probably because it is difficult to diagnose the disaster until it strikes. Bangladesh is less prone to drought, and rainfall variability is the main cause of drought especially in the pre-monsoon and post-monsoon periods. Drought is a temporary, spatially irregular and non-periodic phenomenon. It affects regions with low annual rainfall amount i.e. the western and northwestern parts of Bangladesh, particularly the Barind tract in Rajshahi division. About six per cent of the total land area comes under the most drought prone zone in the country. Drought may however, affect quite wider areas of the country. Bangladesh experienced two serious droughts (1979 and 1989) in the last 25 years, which affected almost the entire country in varying degrees (Ahmed, et al., 1994).

Drought generally affects farmers by reducing their crop productions drastically. The continued drought of 1994 and 1995 (started in October 1994 and ended in July 1995) in the northwestern districts of Bangladesh led to a shortfall of 3.5 million tons in rice production. The total loss of rice production due to the 1982 drought was 52,896 metric tons. The amount was about 41% of the total damage caused by natural hazards that affected Bangladesh in that year. Tangible losses from drought can be estimated in terms of monetary value but not intangible losses like distress and miseries. The impact of drought in Bangladesh is manifested in rural unemployment, population displacement, water level depletion and price increase.

Local communities of drought prone areas have, over the years, developed adaptation strategies to cope with the adverse effects of drought. In the 1994–95 droughts, the affected households adapted various remedial adjustments at the household level. As Bangladesh is an agrarian country, crop adjustments are the main focus to avert any risk. As part of agricultural adjustment, people re-sow crops to compensate for the loss of crop production and others like application of irrigation water for crop yield increase. Under intense drought conditions when domestic food stocks become exhausted or very low, drought affected people, not finding any other means of survival, go for non-agriculture adjustments as part of livelihood strategies to cope the disaster.

Arsenic Contamination of Ground Water

Arsenic dissolved in fresh water is common all over the world. It becomes a serious problem when there is so much that it adversely affects the human physiology. Bangladesh has recently acquired a new form of calamity: dangerous level of arsenic in ground water, the principal source of water used for drinking and cooking in the country. Arsenic is a toxic and carcinogenic substance that attacks internal organs, produces gangrene and a number of concerns affecting the skin, lungs, liver and bladder. WHO considers 10 parts per billion to be the maximum allowable amount of arsenic in drinking water. Bangladesh authorities consider this limit to be 50 parts per billion (Rashid, 2001).

The oxidation of arsenopyrite or ferrous hydroxides rich in arsenic present in the Bengal Delta sediments may be responsible for the release of arsenic oxides in solution to the groundwater. The subsequent migration of this arsenic contaminated groundwater through these deltaic sediments may be one of the principal causes of arsenic poisoning in Bangladesh. Arsenopyrite and ferrous hydroxides would be stable in the reducing environment below the groundwater table. If the groundwater table were lowered by over pumping of water for drinking and cooking purpose and increased irrigation during the dry season and the sediments exposed to the oxygen of the atmosphere, these arsenic rich minerals would oxidise releasing arsenic (Bridge and Husain, 2000).

The problem of arsenic contamination in the ground water of Bangladesh is a crisis of unprecedented proportion. Millions of people in rural Bangladesh are now exposed to the risk of arsenic poisoning. So far known, out of 64 districts, 59 districts covering an area of about 65,000 sq. km have been found to be affected by arsenic pollution in the tube-well water, particularly in the shallow aquifers on which about 97 per cent of the people are dependent for drinking water. About 75 million people are at risk of arsenic disaster (Alam, 2000 and Karim, 2001; Chowdhury, 2001).

Ironically, the arsenic contamination in drinking water in Bangladesh is the result of the safer water-drinking programme mounted by the UNICEF. The villagers were weaned away from the traditional use of surface water for drinking and cooking as surface water was declared to be the source of deadly microbes and bacteria conducive to life-threatening pathogenic diseases. The UNICEF and other donors assisted the government to sink pipes into underground aquifers for clean groundwater. Presently, out of 5 million such tube-wells, 3 million may be contaminated with arsenic (Karim, 2001). Excessive use of ground water for drinking, cooking and irrigation by installing thousands of shallow and deep tube-wells led to withdrawal of enormous quantity of groundwater from various underground aquifers without allowing the chance to refill, causing gradual dewatering of the basal sand, which caused air to enter. In addition, dropping of water level facilitated further exposure of the arsenic rich beds to air. Collectively, these happenings have caused the underground aquifer to become aerated and have changed the anaerobic environment to aerobic one.

Arsenicosis is a silent killer in the rural areas of Bangladesh and has emerged as a major public health problem of the country. It is very hard to estimate the total number of arsenicosis patients in Bangladesh. However, about 1.2 lakhs (0.001 per cent of the population) are clinically exposed to arsenicosis disease (Bhuiyan, 2001) and sub clinically the figure may be 10 times higher. The figure may rise gradually with respect to time and spatial dimensions. So public health is in jeopardy in areas where arsenic contamination is extensive.

Ministry of Health and LGRD have undertaken a number of short-term and long-term programmes to mitigate the arsenic contamination problem. Many national and international NGOs and Agencies have also undertaken many programmes and activities to mitigate the problem. The government launched a campaign for creating awareness among the people about the hazards of drinking arsenic contaminated water by issuing warning through public media like radio and television. However, the massive job of screening majority of the tube-well for the contamination is yet to be undertaken. Measures should be undertaken to find out the alternate arsenic safe water options, which would be sustainable, acceptable, feasible and affordable by the community.

DISASTER AND GENDER PERSPECTIVE

When disasters strike, the most vulnerable are women and children, especially those women who are household heads and those children who live in single parent families. Disaster affects both men and women but it has a gender dimension. The disaster related problems affect rural women more severely than men because of the wider responsibilities they have for their households and the fact that those responsibilities keep them tied to their households more strictly than the male members. Saleheen and Huda (2001) have identified three principal reasons for this vulnerability: low status of women in the family and society, lack of awareness and the physical nature of women. The 1991 cyclone surge immediately killed 40–59 per cent of the children, 25–30 per cent of the women and 15–20 per cent of the old people in the affected area. In Kutubdia, about 85 per cent of all dead bodies were of women and children (BCAS, 1991).

Although women are usually at greater risk than men, it is the women who make it possible for the community to cope with disaster. The role of women is absolutely central to the management of disaster-coping strategy. Now, there is general agreement on that women are active environmental stewards, particularly in South Asia. Women constitute a rich depository of traditional and indigenous knowledge regarding the environment and their skills in coping with disasters are often remarkably efficient. The vast wealth of women's knowledge of traditional and local methods of coping with disasters should not be undermined or ignored. This is being used by the disaster stricken society as livelihood strategies for surviving disasters.

The woman, as the manager of the household, irrespective of socio-economic status, play a significant role in the management and preparedness for pre-, during and post-disaster situations. They take various measures to save lives of the family members, specially the children, and the elderly and sick persons. They keep themselves prepared with the materials required to save the lives of the family members and protect household items. Often, they make arrangements to save children and elderly persons during disasters. Pregnant women are at great risk during a disaster. They need to get special care not only from the family members but also from the other women of the community. These women should also remain prepared with their first aid equipment and materials including the oral saline and other necessary materials.

Women traditionally are responsible for preparation and management of food for the household members. As part of this responsibility they take preparation long ahead of disasters, not only in view of the family members but also the poultry and livestock. They preserve dry readymade food for emergency and also make arrangements for cooking by preparing movable burners and collecting and preserving fuels. They preserve cereals, vegetables and other dry food items too. Women also preserve kerosene, matches or lighters, lamp and hurricanes for emergency lighting.

Preparation is also taken for post-disaster rehabilitation by preserving seeds for field crops and preparing seedlings for homestead gardening. Women also remain prepared to process the emergency harvest. They carry their poultry and livestock if necessary and they keep themselves ready for that by preparing baskets for poultry and rafts for livestock.

Women try to serve clean drinking water to the family members either by collecting it from far away tube-wells or by boiling the available water or using purifying tablets. However, women suffer most for many reasons, among them for lack of public conveniences during disasters, especially during floods. Pregnant women and lactating mothers and babies suffer most. Sometimes their suffering continues long after the disaster.

As the manager of the homestead she plants many types of plants and trees around the homestead, which are useful in normal times, during, disasters as well as in post-disaster periods. As long-term planning for repairing the homestead, they collect and stock clay soil from the fields during the dry season and do the repairing work and cleaning of the homestead in the post-disaster period.

LIVELIHOOD OPTION FOR DISASTER RISK REDUCTION IN BANGLADESH (ITDG STUDY)

The study conducted by the Intermediate Technology Development Group (ITDG), Dhaka, explored the livelihood strategies of the disaster stricken people of Bangladesh. This study covered three types of disasters including flood, river bank erosion and arsenic pollution of ground water. The Sustainable Livelihood (SL) approach was the conceptual frame work of the study covering two aspects: how do disasters affect the life and livelihood and, what impacts do everyday livelihood strategies have in reducing the risks of disasters. The greater Faridpur district covering Faridpur, Madaripur and Rajbari districts located in the central part of Bangladesh was selected as the study area. Historically, this area is very much exposed to these three types of disasters. Qualitative data collected from 30 villages and part of urban area through FGI and PRA exercises and secondary information were the sources of data for the study. The field's team consisted of four facilitators, two women and two men.

Some of the villages selected for the study are situated on low lands. Large areas are permanently submerged. One quarter of the villages have flood refuges. Study villages of Faridpur district have good road communication which facilitated the marketing facilities of the local products. This opportunity has led many households to shift from traditional rice cultivation to high priced vegetable cultivation. However, the embankment and local roads have created some waterlogging problem. Fisheries have received a boost with intense aquaculture among the people of the submerged area. A significant percentage of the population is engaged in transport related occupation. Availability of electricity also facilitated some people to be involved in secondary type of activities. The study area perceived to be improving day by day in respect of the life and livelihood of the people irrespective of the huge loss inflicted by disasters like flood and river bank erosion. The only drawback of the area is that more than 90 per cent of the tube well contains arsenic above acceptable level.

Other study villages are also found to submerged during the monsoon period. The settlement pattern of these villages is linear on the natural levee along the river, which is comparatively higher than the surrounding areas. The people who live on the higher land along the river are comparatively wealthier than those who live in the inland areas. The proximity to the river and the roads along the river places these people in an advantageous position for involving in trading mostly related to fishing and transport in the region. On the other hand, the inland households are handicapped by the separation from the waters almost round the year. However, these people also earn enough by fishing in the beels. However, with the trading opportunity in the locality improving, these people also involve themselves in the business in the locality.

NGOs and Christian missionaries in the locality facilitated to develop flood refuge in the locality by providing free tube wells and latrines. People from the surrounding villages collect drinking water from these tube wells. However, the refuge serves only a limited number of people compared to the huge demand at the time of disaster. Literacy rate is comparatively higher in this locality than the surrounding areas, which facilitated them to become aware about the disaster and mitigation measures. By constructing houses on high land above the normal flood level, engaging in diversified occupation and taking advantage of flood refuges, these people are facing disasters like flood and river bank erosion with courage and recovering their economic status very quickly after the disasters strike. However, they do not know yet what to do with the arsenic menace. Many of them are not yet aware of this disaster either.

Sudden loss of human lives due to flood, river bank erosion or arsenic is not as much pronounced as compared to that in the other disasters like cyclones or fire hazards. However, the flood-affected households do suffer in different ways by loosing or damaging different kinds of assets and properties including housing, equipments and supplies. The magnitude of these losses is again related to the characteristics of flood like the rate of increase in water level, duration of flood and the timing of flood occurrence. Households obviously take some measures for self and livelihood maintenance before, during and after floods. The common measures are: Preservation of emergency food especially dry food, storing fuel and fodder, preparing portable stoves, making earth/ water hyacinth barrier around the homestead, storing household items, selling grain for want of secure storage, building and moving the family to high platform dwellings, using gang-plank made of bamboo or preparing raft with banana tree trunk, reducing food intake quantity, placing livestock on raft, saving the equipments, preserving seeds for plantation after recession of flood water, preparing seedlings (if possible) of winter vegetable, keeping arrangement of emergency transport, etc.

It is evident from a study that a number of communities and households receive assistance either from the government or NGOs during disasters, but this is not enough as per the requirement. The vast majority of the communities and households are forced to cope with disasters on their own with limited assistance from government organisations and NGOs. The study also reveals that the coping capacity of the households and the community increases many fold with time, even with little technical assistance from government organisations and NGOs. There is an urgent need to expand the reaction measures to include programmes by the government and other agencies for stepped-up security, rescue, emergency transport and refuge during disasters.

Among the three types of disasters under study, attention is mostly focused on floods. Programmes that directly address the problems arising out of river erosion are practically non-existent. A list of the immediate needs of communities prone to erosion would include disaster forecasting and warning, improved security during times of crisis, the institution and enforcement of regulations making it mandatory for households to move in the face of danger and the development of a system for personal identification, which would enable people to establish their identity, regardless of location, and thereby access loans from banks to recover from a disaster.

While there is considerable discussion about the arsenic threat there is weak political will to accord it the seriousness it deserves. Community level interventions are mostly limited to providing information about the dangers of arsenic polluted water but next to nothing is being done to enable communities to acquire facilities for arsenic-removal or the treatment of surface water.

DISASTER RISK REDUCTION MEASURES

As Bangladesh is a densely populated, relatively poor country, the impact of any disaster on lives and livelihood always appears to be extremely grave. It is a country where disaster (both natural and man-made) not only lead to a heavy drain on domestic resources, but also trigger poverty, put the environment in a fragile condition and ultimately put a brake on the pace of sustainable economic development. Long-term disaster management needs to be undertaken for sustainable economic development of Bangladesh. On the other hand, the concept of environmental management entails a holistic approach in taking measures in risk reduction from disasters as well as in monitoring and auditing the environment through laws and customs relating to use and protection of natural resources. Long-term planning in reducing the disaster impact, both in terms of casualty and economic loss is a necessary element for the over all development of Bangladesh. However, livelihood strategies adopted by the vulnerable people of Bangladesh from time immemorial to live with disaster should be evaluated and considered with due importance in this regard.

Disaster and poverty are mutually reinforcing, in any disaster, the poorest are the hardest hit. So the poor need special attention in this regard. Now a days it is observed that poverty alleviation i.e. socio-economic development of the poor has a positive impact on reducing the disaster impact and ultimately helps in minimising environmental degradation. Kelly and Chowdhury (2001) have identified strong linkages between poverty, disaster and environment. Yadmani (2001) has supported these linkages. According to these scholars, disasters are no longer seen as extreme events of natural phenomena but as manifestation of unresolved problems of development. However, it is observed that disaster risk management becomes more effective when the issues of long-term risk reduction for the poor are addressed rather than emphasising relief approaches to response the hazards. The researchers used DFID's Sustainable Livelihood (SL) model as base to establish the linkages between disaster, livelihood and environment.

Government of Bangladesh has identified Natural Disaster as one of the major constraints to development in the successive plan documents of the country. The Fourth Five Year Plan (1990–95) was mostly concerned with flood control measures. However, the Fifth Five Year Plan (1997–2002) emphasised the need for disaster preparedness of the people as opposed to the earlier concepts of reacting after a disaster. A massive programme to train and equip people living in disaster prone areas for improving their capability to cope with natural disasters was highlighted. Preparedness, response and rehabilitation were accorded priority in the government approach. The Fifth Five Year Plan also recognised a direct linkage between natural disasters and the poverty alleviation programmes of the country. Given the risk and vulnerability to natural hazards that are likely to continue as a serious threat to national development efforts, macro level policies of the government are being adapted in view of the past experiences. Different government agencies as well as local and international NGOs and some donors have taken numbers of programmes in this respect in different hazard prone areas of Bangladesh.

The National Water Policy (NWPo) has recognised the importance of implementing effective non-structural measures to reduce the impact of floods and erosion. As opposed to the structural measures against floods and river erosion, the recent policies and plans have recognised the importance of community capacity building that emphasises on their ability to survive the natural shocks 'with dignity and stamina'. An important trend in the approach is the recognition that past plans were driven by technology rather than by demand; and, that there is a need for 'community-based planning for disaster response'.

Researchers from various disciplines are also lending their efforts in identifying the appropriate risk reduction measures by evaluating the causal factors, the behavioral pattern of the disaster stricken people of different parts of Bangladesh and the activities of different agencies involved in disaster management aspects. Through frequent management of disasters, both, the NGO sector and the Government have gained adequate experience and expertise and the capacity to deal with large scale disasters. In this regard, perhaps Bangladesh is way ahead of other countries in the region, yet needs to effect wider use of this expertise so as to cover larger population facing disaster.

Various research work and previous work experience have identified varieties of adaptive suggestions for long-term planning in reducing risk and surviving disasters. Some of the key suggestions are as follows:

- Enhancing the literacy level.
- Strengthening of different poverty alleviation programmes.
- Developing infrastructure in the locality for assisting economic development and creating employment opportunity as well as for minimising disaster risk.
- Raising awareness in the community regarding the preparedness and coping strategies of the disasters by adopting available knowledge and opportunities provided by different government organisation and NGOs and other agencies.
- Activating efficient disaster forecasting and dissemination system, enhance immediate reaction capabilities like rescue, emergency life saving measure, emergency transport, refuge, step up security during disaster, provide incentive for the rehabilitation of livelihood, create employment opportunity, etc.
- Providing special attention to gender perspective in all the aspects.

Three categories of primary actors including government organisations, NGOs and international agencies would be responsible to adopt and implement these suggestions by incorporating the community. Ministry of Disaster Management and Relief (MDMR) under the Government of Bangladesh would play a vital role in this respect. All activities would take place through inter-ministerial coordination and in collaboration with different relevant agencies including government organisations, NGOs and international agencies at different levels. Setting up of some demonstration field projects by these agencies and researchers bear critical evidence for the long-term management plans.

CONCLUSION

Adoption of various kinds of livelihood strategies to survive in the disaster prone environment is part of the tradition of the Bangladeshi people. However, some times the environment comes under threat due to abuse or over exploitation of the natural resources to survive the disaster inflicted situation. Disasters not only damage lives and property but also take toll of the domestic resources, trigger poverty, put the environment in a fragile condition and ultimately disrupt all sustainable economic development. It is observed that the marginal and landless households are the most vulnerable to any disaster and are pushed to below the poverty line. In this situation either they themselves overexploit the natural resources or are abused by the others to overexploit the natural resources, ultimately leading to an increase in environmental degradation. Taking appropriate measures for reducing the risk of damage from disasters can mitigate this. Disaster risk can also be reduced by adopting appropriate sustainable livelihood strategies aided with institutional support from government and other agencies before, during and after the disaster.

REFERENCES

Ahmed, Q.K., N. Ahmad, and K.B.S. Rasheed, *Resources, Environment and Development in Bangladesh with Particular Reference to the Ganges, Brahmaputra and Meghna Basins,* Bangladesh Unnayan Parishad, Academic Publishers, 1994.

Akhter, F., *Women are not only victims,* in H. Hossain, C.P. Didge, and F.H. Abed (Eds.) *From Crisis to Development, Coping with Disasters in Bangladesh,* University Press Limited, 1992.

Alam, M.K., *Arsenic Problem in Bangladesh: An Overview,* in M.F. Ahmed (Ed.) *Bangladesh Environment 2000,* BAPA Bangladesh Poribesh Andolon, pp. 84–99.

Ali, M.H. and H. Khatun, *Impact of 1998 Flood on the Agricultural Sector and its Mitigation,* in N. Ahmed and H. Khatun (Eds.) *Disaster: Issues and Gender Perspectives,* Conference Proceedings, 2001, Bangladesh Geographical Society, pp. 254–262.

Bangladesh Centre for Advanced Studies (BCAS), Cyclone '91; An Environmental and Perceptional Study, Dhaka, 1991.

Bhuiyan, R.H., *Social and Psychological Scenario of Arsenicosis Patients: A gender Perspective,* in N. Ahmed and H. Khatun (Eds.) *Disaster: Issues and Gender Perspectives,* Conference Proceedings, 2001, Bangladesh Geographical Society, pp. 101–121.

Bridge, T.E. and M.T. Husain, *Groundwater Arsenic Poisoning and Solution to the Arsenic Disaster in Bangladesh,* in M.F. Ahmed (Ed.) *Bangladesh Environment 2000,* BAPA Bangladesh Poribesh Andolon, pp. 100–111.

Carter, W.N., *Disaster Management, A disaster Management Handbook,* Asian Development Bank, 1991.

CARE, Bangladesh 2000, Rapid Household Livelihood Security Assessment Report on Bangladesh Flood in Southwest Region, Disaster Management Project.

Chowdhury, A.M., *Natural Disaster and National Security,* in *National Security of Bangladesh,* Dhaka.

Chowdhury, Q.I., *Bangladesh: The Environmental Scenario*, in Q.I. Chowdhury (Ed.) *Bangladesh, State of Environment Report, 2000*, 2001, Forum of Environmental Journalists of Bangladesh (FEJB).

Disaster Forum Bangladesh 1997, Disaster Report, Bangladesh, Dhaka.

Halli, S.S., *Economic Impact of Riverbank Erosion in Kazipur Upazila*, in K.M.E. Elahi, K.S. Ahmed, and M. Mafizuddin (Eds.) *Riverbank Erosion, Flood and Population Displacement in Bangladesh*, Dhaka REIS-JU, 1991.

Islam, A.S.M.M., S.T. Islam, and M.S. Alam, *Riverbank Erosion and Associated Consequences: A Case Study of Chilmari Thana*, in N. Ahmed and H. Khatun (Eds.) *Disaster: Issues and Gender Perspectives*, Conference Proceedings, Bangladesh Geographical Society, pp. 35–46, 2000.

Karim, N.R., *Arsenic—A Catastrophe Around the Corner*, Holiday, December 14, 2001.

Kelly, C. and M.H.K. Chowdhury, Issues Paper Poverty, *Disaster and Environment in Bangladesh: A Quantitative and Qualitative Assessment of Causal Linkages*, Department for International Development Bangladesh, Government of the United Kingdom and Northern Ireland, Dhaka, 2001.

Khan, F.K., P*reliminary Words and Greetings*, in N. Ahmed and H. Khatun (Eds.) *Disaster: Issues and Gender Perspectives*, Conference Proceedings, Bangladesh Geographical Society, pp. xix–xxii.

Khatun, H. and M.H. Ali, *Impact of Flash Flood on the Environment of Foothill Areas of Bangladesh: A Case Study*, The Journal of BNGA, Vol. 18 (1 & 2), 1990.

Rashid, H., *Disaster: Issues and Gender Perspective*, Keynote Address, in N. Ahmed and H. Khatun (Eds.) *Disaster: Issues and Gender Perspectives*, Conference Proceedings, 2001, Bangladesh Geographical Society, pp. xxiii–xxxvii.

Rogge, J.R., *Individual and Institutional Responses to Riverbank Erosion Hazards*, in K.M.E. Elahi, K.S. Ahmed, and M. Mafizuddin (Eds.) *Riverbank Erosion, Flood and Population Displacement in Bangladesh,* Dhaka REIS-JU, 1991.

Saleheen, M. and K.M.S. Huda, *The Vulnerability of Women in Disaster Prone Areas in Bangladesh*, in N. Ahmed and H. Khatun (Eds.) *Disaster: Issues and Gender Perspectives,* Conference Proceedings, 2001, Bangladesh Geographical Society, pp. 96–100.

Yadmani, S., *Disaster Risk Management and Vulnerability Reduction: Protecting the Poor*, Presented at the Asia and Pacific Forum on Poverty, Organised by the Asian Development Bank.

23

Impact of Flood on Gender: A Case Study

— Balkrishna Lohani
— Harish Chandra Acharya

INTRODUCTION

Floods have been a recurrent natural disaster in Nepal. These have been destructive to life not only because of the casualities and physical injury owing to drowning they cause but also because of the associated diseases and livelihood disruption. But there is a dearth of information obtained directly from the affected households on the socio-economic effects of floods, the coping mechanisms adopted at its occurrence and the hardships faced by the affected people in earning livelihood and managing the household. Natural disasters affect the entire family but the vulnerability of women and children is much greater because of their subordinate position in the family, weaker physical structure and dress materials (such sarees) used by them. Studies undertaken to analyse the condition of women in rural parts of South Asian countries reveal the worse condition of women relatively to men within households even under normal livelihood condition due to patriarchal and traditionally embedded cultural values. These studies reveal higher female mortality owing to women's high maternal mortality rates, inequalities in food intake relative to men, unequal work burden due to productive as well as reproductive responsibility, lack of control over the means of production, limited facilities for training, lack of employment, etc. The hardships faced by the women of poorer households become worse when the households are victims of natural disasters.

Objectives of the Study

A study was conducted in Chitwan, Nepal, to analyse the gender issues and concerns in the context of flood in Nepal and to suggest ways of incorporating gender based capacities into disaster mitigation and management planning.

The specific objectives were to:

(i) Analyse the economic and non-economic activities performed by men and women in managing the family during normal time, at the time of flood, and in post-flood situation.
(ii) Examine the differential impact of disaster on different occupational and age groups of women.
(iii) Identify issues and concerns of women of various age groups.
(iv) Analyse whether gender based needs and concerns are taken care of in rescue/relief and rehabilitation programmes.
(v) Suggest measures to make flood mitigation and relief planning more effective and gender sensitive.

Methodology

The study was based on a field survey of five Village Development Committees (VDCs) of Chitwan district. Chitwan was chosen for field survey because it was one of the hardest hit districts by the devastating flood of 1993. The district is at a road distance of 145 Km from Kathmandu and is also connected by air. The East West Highway passes through this district. The district is situated in between the Trishuli River at Mungling and Narayani River in the north south side and the Lother, Manohari and Rapti rivers in the east west side. The latter three rivers confluence at different points and join the Narayani River. All these rivers of the district experienced severe flood owing to extremely intense rainfall during July 20–21 and August 10–11, 1993 in the Mahabharat Ranges. Rapti River broke the fragile embankment at Sungurmara of Piple VDC and gushed into the settlements. The settlements lying along the lower sides of these rivers were fully or partially swept away with lesser impact on the farther as well as interior settlements from east to west. The Narayani River also inundated some areas. In Madi area, River Riu caused the flood. A total of 20 VDCs of the district was estimated to have been affected by the flood of 1993. Of these, five VDCs were selected for the study to cover the most severely affected areas and the families. These were Piple, Bhandara, Kathar, Kumrose and Bachhauli, in order of the severity of the flood in terms of losses of lives, property and infrastructure.

A total of 100 flood-affected households with about 20 from each of the selected VDCs were surveyed for gathering data and information. Households were used as the sampling unit rather than individual women because the problems of women are linked with the characteristics and the condition of the household. The sexual or gender division of labour in the households, women's livelihood options, and the relative position of women to men within the household under different stress situations cannot be visualised in isolation from the economic, social, and organisational set up to which they belong. Variation in the condition of women because of class, age, assets, differences in the forms of work organisation and modes of production were seen as the main features that distinguished the sampled households. The interviews focused on probing deep into the issues and concerns of women of the households to find out their needs and concerns.

The fieldwork was carried out between July 24 and August 10, 2000. A team of field researchers consisting principally of a team leader, a sociologist, a gender studies specialist, a development economist, a communication expert and two graduates involved in field research. A snap survey was conducted earlier for identifying the seriously affected VDCs. The survey team met the Chairmen of the respective VDCs and prepared the list of households. The team also met the District Development Committee (DDC) Chairman, District Police, and Army Chiefs and NGOs involved in rescue/and rehabilitation programmes. A second visit was made to each of the five VDCs to carry out the household level interviews, in-depth interviews with the Chairman of VDCs, interviews with the local clubs and to focus group discussions. The household interviews were made in places of resettlement and initial places of residence. A large majority of households in Piple, Bhandara and Bachhaulli are resettled in new places, while in Kathar, and Kumrose families are staying in their old houses after the removal of debris and repair of damages caused by the flood. Some families in these VDCs have also built new houses in nearby places. The households were selected using stratified random sampling technique in order to ensure representation of all castes and ethnic communities.

Role of Local Authorities and NGOs

Royal Nepal Army helicopters started rescuing people from the next day of the flood. The people were kept in schools, Sajha go down and even in the homes of other people. In the evening, sugar and beaten rice (Chiura) was distributed to the people by the Red Cross. All party Flood Victims Relief Committees were formed at the VDC and DDC levels under the convenorship of the Chairmen. Construction work of Rapti dam was started in cooperation with the Royal Nepalese Army and the Nepal Police.

Several sub-committees were also formed in different VDCs to carry out the removal and restoration of the belongings of the victims, to erect temporary huts and tents, to dump the swept away and dead cattle, to spray medicines, to conduct health camps and to distribute foodstuffs. On the occasion, RRN made available clothes along with lodging and food utensils to the people according to the size of their families. WFP provided rice and lentils to all victims for two months. RRN provided rice, as food for work, to level up the affected land area under its supervision. Later, UNDP provided the fund to control the flood by erecting spurs. From the second year, East Rapti Irrigation Project reconstructed all damaged irrigation canals, culverts and bridges whereas the VDC repaired the roads and the land surface. Similarly, the VDC constructed 3 primary and 1 lower secondary school, but the construction of high school has not been done yet. The work is expected to be completed by this year with the funds granted by the DDC and the VDC.

The 1990 flood affected families presently stay in temporary dwellings in the government forest area in the northern part of Piple VDC known as Kupon Tole and in Chauki Dada of Bhandara VDC. The 1993 flood affected families live in the government forest area of Piple-Bhandara VDCs. About 55 families have been resettled at Chainpur VDC following the issue of landownership certificates with the purchase of land from the government through the DDC and with the support from Red Barna.

General Impact of the Flood

In the VDCs covered by the survey, the flood affected about 1,000 families. The death toll was 15, of which 6 were women, 4 were men and 5 were children. Many flood-devastated settlements were veritable hell for several days due to the bad odor of decaying animal carcass. Community properties such as schools, drinking water systems, and irrigation systems were also destroyed. In Kumrose VDC, the unique community canal of 572 metres, constructed with the support of CARE, Nepal in 1986 was severely damaged and rendered useless. The canal had brought about a significant change in the socio-economic condition of people residing in the VDC. Among the household covered by the survey, 5 households were bereaved by the death of household members in the flood. The dead persons were 2 girls, 2 men and 2 children, of whom one adult and one child were swept away by the flood. A baby of 13 days died of pneumonia 7 days after the flood. The total land damaged by the flood was 29.84 hectares. About 74 houses were completely damaged and the many, partially. Household assets such as utensils, bed, etc., were also swept away in large quantity. Some 70 households lost livestock and poultry. The number of livestock and poultry lost was estimated as 39 cows, 16 oxen, 45 goats, 6 pigs and 136 chickens. Community resources such as the drinking water system, irrigation canal, several hand pumps, and 2 schools were also damaged.

Impact on Women

The flood of 1993 made women's life more difficult. There was diverse impact on different age groups of women. Before the flood, the housewives managed the households with the produce of the land and the earnings of their husbands and their own. They were used to the habitat of their house. They also used to get seasonal employment as wage labourers in agricultural season. But after the flood they have been exposed to various problems and difficulties. The main common problem of women and young girls is the lack of employment opportunity for earning a livelihood. The women who lost all their meagre belongings and their life long savings have not been able to compensate their losses even after 7 years.

A large majority of mothers (40 out of 110) are faced with the problem of sending their children to school owing to lack of income to provide for books, stationery, and clothes. They said that spending on children's education is out of question, when the family was hovering on the verge of starvation.

A large majority of school going age girls is out of school. The reasons behind this are the economic hardship and the lack of basic facilities inflicted by the flood. Forty percent of girls are not in a position to attend school due to extreme poverty.

Mothers as well as young girls are also facing problems in gathering fuel, fodder and water in the present resettlement area. The resettlement area is a squatter with limited access to safe water, electricity and sanitation facilities. There do not even have a tube well. Family latrine in the home stead is not possible because of the very small piece of land (0.017 hectare) provided to each family for housing purpose. The lack of public conveniences has affected women and adolescent girls more than men because of the need of the former for privacy.

CASE STUDIES

Seven case studies were made in order to probe deeper into the condition of women who were either lactating mothers of a few days or had lost their primary support due to the flood. These were conducted mainly on families of the Rai/Danuwar and Chaudhary/Tharu ethnic clans. All the seven case studies showed that women belonging to indigenous and lower solid status suffered the most. Such women were not given any special consideration in the relief and rehabilitation programmes. They are still agonised by the trauma of the death of their close relatives and have severe difficulty in managing livelihood.

CONCLUSION AND RECOMMENDATIONS

The findings of the study, when critically analysed, led to various important conclusions and recommendations. The on-going and future plans of action for flood control and management should take serious view of these conclusions and recommendations.

Conclusions

An overwhelming majority of households in the flood affected five VDCs covered by this study belong to the poorer sections. The meagre land they own reflects their poverty. In rural areas, land is the major determinant of a peasant household's welfare situation. The more land they have, the better off their socio-economic position is. Their condition worsen further when floods wash away the meagre land areas they own. The flood of 1993 has made the poor more vulnerable, marginalized them further and also uprooted some of them from their land. There has also been damages to the natural environment and community infrastructure such as schools, health posts, growing land, etc. Although these losses did affect the entire family, the traditionally disadvantaged position of women intensified their sufferings in many ways.

Shortage of food has compelled women to allow themselves only low intake of micronutrients. Generally women are the last ones to eat in the family and hence in a situation of shortage they have to remain content with left-overs or miss some meals altogether. This has severe implications on the health of pregnant and lactating women.

Recommendations

The findings of the study reveal the need for a multi-pronged approach to disaster management, particularly in the context of flood. In this regard it seems necessary to initiate a long-term programme towards control and management of floods in Chitwan district. Activities to be undertaken for the control of flood should focus on protective measures and awareness programmes, while the activities relating to management of flood should concentrate on capacity building of the local community for disaster preparedness and mitigation. Since women are the most vulnerable sections of the society and since they have needs and concerns different from those of men, it is important to adequately incorporate women issues in disaster preparedness, mitigation and management plans.

Flood Control Plan

The risk of flood in the areas surveyed has increased in recent years due to the accumulation of sediments in the Lothar and Rapti rivers which has elevated the riverbeds. As a consequence, the settlement areas outside the embankments have been left to remain at relatively lower level than the riverbeds. In case of heavy rainfall, the river may break the weaker part of the embankments and gush into the settlement areas carrying huge amounts of silt and debris. The risk of flood has also been accentuated due to soil erosion caused by the depletion of forest for various purposes related to agriculture farming and for the collection of forest products. Nevertheless, the risk of flood could be minimised, if not controlled by initiating various measures. Incorporating the following measures in an integrated framework could develop an extensive flood control plan:

1. Implementation of afforestation programmes in the upper parts of the rivers and alongside at rivers and rivulets.
2. Control of depletion of forest by providing alternatives for those who depend on forest products or practice slash-and-burn farming in the catchment areas of the rivers.
3. Raising up the embankments of the river.
4. Lowering the river beds by allowing extraction of stone and sand (this would also generate income to the local bodes).
5. Continuous protection of deck (embankment) through reinforcement of gabion wires and stones in vulnerable parts of the embankments and the spurs.
6. Installation of early warning systems.
7. Conducting adequate inspection while constructing canals and dams.
8. Diversion of mainstream of the river immediately after the monsoon starts.
9. Check dams must be constructed in Hardekhola from Churiya region to control floods. This will solve 90 per cent of the flood problem, by diverting Herdekhola to Sungurmara from where Rapti take its course.
10. Households in flood prone areas should be provided secured residence.
11. Installation of rain gauges (hydrometer) in various catchments of the river.

Capability Building of Local Communities

The following measures may be suggested for capability building of local communities:

(i) Making local people as well as school/college students of the area aware of the causes and effects of flood.
(ii) Training local people in various aspects of disaster management.
(iii) Involving local people in various stages of flood management.
(iv) Imparting navigation training to selected youths of the communities in flood-prone areas.
(v) Conducting seminars, rallies, etc. to highlight such slogans as 'Let's Protect the Deck (embankment)'. 'Let's save Ourselves from Floods' in the source areas and plain areas of the river routes.

Capability Building of Local Institutions

Following measures are suggested for capability building of local institutions in flood management:

(i) Setting up of a rapid rescue force in DDC.
(ii) Training the officials of local institutions and village people in disaster management.
(iii) Granting the provision of rescue materials such as, rafting boats, dinghy, etc., to the VDCs of flood prone areas.

Women-centered Flood Management Plan

(i) Pregnant and lactating women are allocated extra food, clothing and other materials in addition to ration quantity.
(ii) Particular attention is paid to their health and childcare.
(iii) Women who have lost their husbands or children in the flood are provided emotional support and psychosocial care.

The rehabilitation resettlement programme should focus on:

(i) Adequate provision of dwelling space so that construction of toilets is possible.
(ii) Provision of drinking water facility (e.g. hand pump) within a reasonable distance.
(iii) Skill training to women, supplemented with support services such as easy credit, training on business practices, market linkages, etc.
(iv) Support for schooling of girl children of ultra-poor households.
(v) Provision of seed money and training in office management to Women's Saving Groups.
(vi) Promotion of local NGOs to act as social mobilisers of women for awareness build-up and organised group action directed towards self-reliance.

24

Towards Risk Reduction—A Perspective of Disaster Management in Nepal: Challenges and Opportunities

—Kul Chandra Shreshtha
—Meen Bahadur Poudyal

INTRODUCTION

Remote, rural, rugged and fragile geophysical structure of the country, unplanned settlement, population pressure, weak economy, low literacy rate and lack of public awareness are the main contributing factors of natural disasters in Nepal. Every year, landslides, floods, fire, avalanches, storm, heavy rains, epidemics and various other natural (and man-made) disasters cause the loss of thousands of human lives and the destruction of physical property worth billions of rupees. Thus, Nepal is facing an acute problem in the form of natural disasters, which adversely affects her development efforts. Albeit, natural disasters cannot be stopped, the magnitude of their impact can be considerably reduced by adopting adequate preventive measures. Proper, prompt and efficient response with anticipatory approach to natural disaster and timely rescue, relief and rehabilitation operations to mitigate its effect can help in disaster risk reduction.

Natural disaster may be defined as a sudden or progressive calamity or misfortune causing destruction or damage to physical construction, and leading to the disruption of normal pattern of life. As a result, the affected people need help to redress the situation in one way or the other.

According to the Natural Disaster Relief Act (NDRA), 1982, Natural Disasters include earthquakes, fires, storm, floods, landslides, heavy rains, drought, famine, epidemics, etc. It also includes industrial incidents or accidents caused by explosions, poisoning and such other kinds of disasters.

The Act defines Natural Disaster Relief Work as any relief work to be carried out in the area affected or likely to be affected by the natural disaster in order to remove the grief and inconvenience caused to the people, to rehabilitate the victims of the natural disaster, to protect the public property and life and property of the people, to control and prevent the natural disaster and to make advance preparation thereof.

Major Natural Disasters in Nepal

A brief description of some past major natural disasters in Nepal are as following:

Earthquakes

The high mountains and the Himalayan range of Nepal, which is a young mountain chain that stretches almost 2,500 kilometres in the east-west directions fall under the seismically active zone mainly due to the subduction of the Indian plate under the Tibetan plate. The seismic record of Nepal goes back to 1255 A.D. Since then, a series of earthquakes have occurred in 1408, 1681, 1810, 1833, and 1866. Among these, the earthquake of 1833 was a major one. After that, an earthquake with a tremor of 8.4 on the Richter scale magnitudes hit Nepal in 1934. Its epicentre was in Kathmandu. This disaster claimed the life of 16,875 people and destroyed 318,139 houses. Nepal next experienced two major earthquakes, one in 1980 and another in 1988. The earthquake of 1980 recorded 6.5 on the Richter scale with the epicentre lying in Bajhang district. In this event, 178 people lost their lives and about 40 thousand houses were destroyed. The earthquake of 1988 measured 6.6 on the Richter scale, with epicentre in Udayapur district. It took toll of 721 people, 1,566 cattle heads and destroyed about 64,467 houses.

Flood and Landslides

Floods and landslides are the most destructive types of disasters in Nepal. Three quarter of the total land area of Nepal is hilly and many villages are situated on or adjacent to unstable hill slopes. As a result, frequent landslides and floods with debris flow occur. Unplanned settlements and physical constructions without due consideration to the natural hazards considerably aggravate the mountain environment. On the other hand, the landslides add enormous load to the streams and rivers causing flood and debris flow downstream. Every year, such types of disasters cause the loss of a number of human life and immense damages to agricultural land, crops, human settlements and other physical property. In July 1993, Nepal experienced a devastating flood in the Tarai region, which took the life of 1,336 people and left 487,534 people homeless. In 1999, flood and landslides killed 113 while 47 people were reported missing and 91 seriously injured. 8,844 families were affected, 3,507 houses and cattle sheds were destroyed and 177.32 hectares of land and agricultural crops were ruined in that year's flood and landslide. The disaster caused a total loss of NRs. 3.6 million.

Fire

Fire disasters take place mostly in the rural areas of the Tarai and Middle Hills region of Nepal. As 90.8 per cent of the total population lives in the rural areas in poor housing conditions, fire hazards are very common. The houses of the rural areas, especially in the Tarai areas, are usually very close to each other and are made up of straw or reeds and timber, which are easily affected by fire. In the year 1999–2000, fire killed 39 people and injured 10. The number of families affected by this disaster was 1,065. Besides, 1,035 houses and 52 cattle sheds were destroyed. The number of livestock loss was 48. The total loss has been estimated to be NRs. 4.5 million.

Epidemic

In most cases, the epidemic of cholera, gastroenteritis, encephalitis, meningitis, typhoid, jaundice, malaria and so on occur during the summer and rainy season. This type of disaster occurs mainly due to the lack of proper health care and sanitation. In 1999–2000, 1,207 people died of epidemics and 6,119 families were affected in various parts of the country.

Avalanche

As the northern part of the country is covered with high mountains, the Himalayas, avalanche is very common and sometimes it claims the life of human beings as well. The avalanche of November 1995 killed 43 people including some foreign trekkers at Khumbu and Kanchanjungha areas.

Glacier Lake Outburst Flood (GLOF)

Glacier lakes are common in the Himalayan region of Nepal. A total of 159 glacial lakes have been found in Koshi basin and 229 in Tibetan Arun basin. Among them 24 are potentially dangerous. The areas like Upper Barun, Lower Barun, Chamlangtsho, Tsho Rolpa, Sabou, Dudh Kunda, Majang, Inja, Thulari have potentially dangerous glacier lakes. These lakes contain huge volume of water and remain in unstable condition. They can burst any time and a natural catastrophe may cause loss of life and physical property. About 14 such glacier lake outburst floods have already been experienced from 1935 to 1991.

Windstorm, Thunderbolt and Hailstorm

Windstorm occurs mainly during the dry season between March to May. Thunderbolt occurs during the monsoon and hailstorm takes place during the beginning and end of the monsoon. Hailstorm causes heavy losses of agricultural crops though human life loss is seldom. Windstorm and thunderbolts causes the loss of human life as well as physical property. Windstorm and thunderbolts killed 22 people in 1999 A.D. in the whole kingdom.

Drought

Some parts of the country face the problem of drought. Uneven and irregular monsoon rainfall is the main cause of drought. The mountainous region (the northern belt) of Nepal is generally dry. The lack of irrigation facilities makes the problem even more serious as prolonged drought condition has adverse effect in crop production. The drought of 1994 affected 35 districts of the country.

Considering the physical size of the country and its population, the magnitude of the natural disasters is as serious as in other disaster prone countries.

The Ministry of Home Affairs is the nodal agency in relation to disaster management in Nepal. The vital functions of the Ministry are the formulation of national policies and their implementation, preparedness and mitigation of disaster, immediate rescue and relief works, data collection and dissemination, collection and distribution of funds and resources. The Ministry has its network throughout the country to cope with the natural disasters. Thus, the Ministry is the main agency for immediate response in the time of natural disasters. Despite limited funds and resources, the Ministry of Home Affairs is managing natural disasters in all respects though there are a number of problems and enough room for improvement.

AN OVER VIEW OF THE NATURAL DISASTER MANAGEMENT IN NEPAL

Natural Disaster Relief Act, 1982 has constituted the following organisational structure under the chairmanship of the Home Minister through which rescue, relief, rehabilitation and resettlement programmes are being carried out effectively and efficiently (see Figure 24.1).

According to the NDR Act, Central Natural Disaster Relief Committee (CNDRC) has been constituted under the chairmanship of the Home Minister in order to formulate and implement the policies and programmes relating to the natural disaster relief work and to undertake other necessary measures related thereof. Moreover, the Central Committee prepares specific norms of relief assistance to be given to

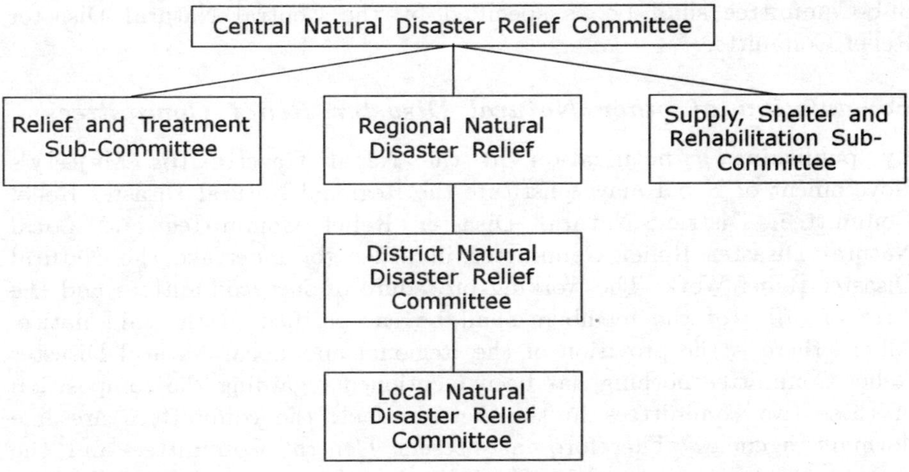

FIGURE 24.1 Flow Chart for Disaster Management in Nepal.

the natural disaster victims of the affected area in cash and/or in kind. The Central Committee itself may manage the working procedures of the Central Committee.

The Central Committee may constitute Relief and Treatment Sub-Committee (RTSC) and Supply, Shelter and Rehabilitation Sub-Committee (SSRSC) which provides necessary advice and suggestions to the Central Committee, helps to execute policies and directives of the Central Committee and operates effectively the rescue, relief and rehabilitation work during a very serious natural disaster.

Functions and Duties of the Central Natural Disaster Relief Committee

The main functions of the Central Natural Disaster Relief Committee are to recommend to His Majesty's Government to declare the area affected by natural disaster as disaster area, to formulate national policies regarding the rescue, relief, rehabilitation and reconstruction works, to prepare plans and programmes for the control and prevention of natural disaster, to collect cash and kinds and disburse them in the affected areas, to involve and coordinate social organisations in relief works, to form groups and send them to affected areas and to give necessary direction to the District Natural Disaster Relief Committees and Local Natural Disaster Relief Committees as and when necessary.

This committee itself may manage the working procedures of the Relief and Treatment Sub-Committee. However, the functions, duties and responsibilities of the Sub-Committee shall be as specified by the Central Natural Disaster Relief Committee. Similarly, the working procedures of the Supply, Shelter and Rehabilitation Sub-Committee may be managed by itself and the functions, duties and responsibilities of the

Sub-Committee shall be as specified by the Central Natural Disaster Relief Committee.

Formulation of other Natural Disaster Relief Committees

By publishing a notification in the Nepal Gazette His Majesty's Government of Nepal may constitute the Regional Natural Disaster Relief Committee, District Natural Disaster Relief Committee and Local Natural Disaster Relief Committee in order to undertake the Natural Disaster Relief Work. The working procedure of such committees and the term of office of the members shall be as specified in the said notice. Albeit, there is the provision of the Regional and Local Natural Disaster Relief Committee nothing has been mentioned regarding the composition of these two committees in the Act. In fact, the committees are like dormant agencies. Therefore, at present Central Committee and the District Committees are active. In fact, the present system has helped to expedite rescue and relief works, as there are only two tiers.

Functions and duties of the Regional Natural Disaster Relief Committee

Regional Natural Disaster Relief Committee provides necessary suggestions and information to the Central Natural Disaster Relief Committee regarding the natural disaster relief works carried out in the affected area. It coordinates with the District Natural Disaster Relief committees and implements the directives of the Central Natural Disaster Relief Committee. It goes without saying that, at present, this Committee is not in operation. But, it will be activated at the time of a very serious natural disaster.

Functions and duties of the District Natural Disaster Relief Committee

The main functions and duties of the District Natural Disaster Relief Committee are to coordinate with Local Natural Disaster Relief Committee regarding natural disaster relief work, to prepare district level plans on natural disaster relief works and submit them to the Regional Natural Disaster Relief Committee, to support and supervise the natural disaster relief work conducted by the Local Natural Disaster Relief Committee, to provide information frequently to the Regional Natural Disaster Relief Committee regarding the ongoing relief work and to work according to the directives of the Central and Regional Natural Disaster Relief Committees.

Functions and duties of the Local Natural Disaster Relief Committee

The functions and duties of the Local Natural Disaster Relief Committee are to prepare detailed description of the loss caused by natural disaster, to estimate the means and resources required for the relief and rehabilitation of the victims of natural disaster and to submit them to the District Committee, to organise volunteer's teams as and when necessary and conduct the relief work, to take the injured people to the nearest hospitals and health posts as soon as possible, to evacuate the victims to a safe place, to distribute the cash and kind to the families of the victims of natural disaster, to raise public awareness programmes and to hand over the goods, cash balance and the accounts to the district committee upon the completion of the relief works. However, like Regional Natural Disaster Relief Committees this committee also is not in operation. But, it will also be activated at the time of a very serious natural disaster.

Types of Rescue and Relief Assistance

The Central Natural Disaster Relief Committee has formulated some norms for immediate relief assistance to the victims of natural disaster in cash and/or in kind. The Central Natural Disaster Relief Committee may revise the amount and quantity of such assistance from time to time. According to the present norms specified by the Central Committee, Rs. 10,000 (ten thousand) shall be provided as relief assistance to the family of a dead victim. Injured persons shall be provided the facility of treatment in the nearest hospitals or health posts. If an injured person is very serious, he/she will even be airlifted and taken to the well-equipped hospital. If a house has been destroyed in a natural disaster then up to Rs. 3,000 shall be made available to the victim i.e. owner of the house. Albeit, the house has not been destroyed, but, if there is the threat of disaster or the house is not safe to live, then up to Rs. 2,000 shall be provided to the victim for the arrangement of a temporary settlement. Homeless families shall get 7 kilos of rice or Rs. 125 per head as food grain assistance and Rs. 500 shall be provided for the clothing and utensils. A victim whose land and crops have been completely destroyed and has nothing to eat shall also get an amount of Rs. 500 as an immediate relief. Victims of the Natural Disaster will also get 50 cu. ft. timber at concessional rate. Generally, all the above relief assistance is to be provided through the District Natural Disaster Relief Committee.

Fund Channel Process

Provision of a Central Natural Disaster Relief Fund has been made under the control of the Central Natural Disaster Relief Committee. The fund consists of the cash and kind provided by His Majesty's Government of Nepal, fund received from the Prime Minister Aid Fund, cash and kind assistance received from foreign countries, agencies and individuals and from other sources. In addition, provision has also been made of a Regional Natural Disaster Relief fund, a District Natural Disaster Relief Fund and a Local Natural Disaster Relief Fund. These funds are composed of the cash and kind provided by His Majesty's Government of Nepal, the cash and kind received from Central Natural Disaster Aid Fund and from other sources. Whatsoever, at present, only the Prime Minister Aid Fund, Central Natural Disaster Relief Fund and District Natural Disaster Relief Fund are into operation. In other words, the Regional and Local Natural Disaster Relief Funds are not into operation. The Central Fund releases budget to the District Natural Disaster Relief Fund according to the need and justification for immediate relief assistance to the victims of the natural disaster.

Agencies Involved in Disaster Management

At present, various agencies of His Majesty's Government of Nepal, e.g. the Ministry of Home Affairs, the Ministry of Water Resources, the Housing and Physical Planning, the Ministry of Health, the Ministry of Finance, Ministry of Defence, the Ministry of External Affairs, the Ministry of Public Works and Transport, the Ministry of Information and Communication, the Ministry of Forest and Environment, the Ministry of Agriculture, the Ministry of Education, the Ministry of Science and Technology, the Ministry of Women and Social Welfare, the Ministry of Supplies, the Secretariat of National Planning Commission, the Royal Nepal Army, Neal Police, Nepal Red Cross Society, Nepal Scout, Mining and Geological Department, Department of Meteorology, etc. are involved in disaster prevention and mitigation works in close cooperation with various international agencies such as: Japan International Cooperation Agency (JICA), Asian Disaster Reduction Centre (ADRC), Asian Disaster Preparedness Centre (ADPC), United Nations Development Programme (UNDP), International Centre for Integrated Mountain Development (ICIMOD), International Red Cross Society (IRCS), United States Agency for International Development Mission to Nepal (USAIDMN), United Mission to Nepal (UMN), Cooperation for American Relief Everywhere (CARE), World Food Programme (WFP), Save the Children Fund (SCF), Technical Cooperation of the Federal Republic of Germany (GTZ), Lutheran World Service (LWS), etc. Besides, professional and non-governmental organisations of Nepal are also providing highly valuable support at the time of disasters.

Concept of Regional Cooperation

While talking about Regional Cooperation Nepal, India, Bangladesh, Bhutan and Pakistan have common problems and thus a country's disaster situation affects the others as well. For instance, if it rains heavily in Nepal, India and Bangladesh may be flooded. And if a cyclone hits any part of India or Pakistan or Bangladesh, it might rain heavily in Nepal and may cause flood and landslides. Therefore, all these countries have to work together to solve the problems by protecting the environments. It is very necessary to establish a Regional Information Centre to share the information so as to cope with the natural disasters.

PROBLEMS IN DISASTER MANAGEMENT IN NEPAL

Disaster management is a difficult task. Disaster happens all of a sudden. Thus, the suddenness of a disaster and its destruction, especially during a very serious disaster, becomes very difficult to cope with a normal administrative set up. In view of the suddenness of the disaster and limited fund and resources the Government is facing a number of severe challenges. Issues like whether or not the disaster management policies and regulations are compatible to present situation, what should be the role, duties and responsibilities of the various disaster management related agencies, how to accumulate, manage, disburse and distribute the fund and resources, whether or not the present resource mobilisation is compatible to the needs of the victims, how could the resource mobilisation be simpler and faster, how could the duplication of services be minimised, how could the cooperation and coordination between the various disaster management agencies be established and how could mass public awareness be raised are some of the main challenges. Problems of behaviour indifference, the lack of cooperation and coordination among various disaster management related agencies, inadequate funds and resources, the lack of roads, transportation and communication facilities and the lack of modern technology especially early warning system are other severe problems in the disaster management in Nepal. In addition, in Nepal where only 39.6 per cent of the total population is literate and where the people accept natural disasters as an act of God, it is very difficult to carry out awareness programmes.

Combining the above features with the unyielding geographical nature of Nepal, the set of challenges faced here with respect to disaster management are complex and difficult to overcome.

Measures to Solve the Problems

Despite the above problems and hard choices associated with the disaster management of Nepal, selection of proper strategies could help to solve the problems, which will simultaneously help to prevent and reduce the natural disasters. In this context the need to amend the Natural Disaster Relief Act, 1982 and the formulation of Natural Disaster Relief Regulations is very necessary whereby the role, functions, duties and responsibilities of all the disaster management related agencies should be specified so that no agency could ignore or shift their responsibilities. For want of mutual understanding and dialogue between the disaster management related agencies, duplication of work and delays in rescue and relief works has been experienced in the past. Therefore, there is the need to work in close cooperation and mutual understanding with all the concerned agencies.

As public awareness is one of the vital problems in managing disaster in Nepal, it is felt necessary to work at increasing the literacy rate. Moreover, disaster management course should be included in the school and university curriculum. It is also necessary to train school teachers, selected students, women leaders, health workers and other social workers to educate others in measures to prevent or mitigate the natural disasters. For all this, active people's participation is necessary. Such types of programmes may convince people to believe that natural disasters are not an act of God.

There is an urgent need to improve road infrastructure, transportation and communication facilities. In order to prevent inappropriate construction of buildings, the building code should be strictly implemented.

Political determination and quick decision making is needed to carry out disaster management activities efficiently and effectively. It will be better to include disaster management component in the development plans and programmes of concerned agencies for the effective implementation of disaster mitigation programmes.

CONCLUSION

In view of the complexities and diversities of the disaster management, a concrete, effective, practicable and proactive policy is needed. Thus, if above-mentioned measures could be adopted and implemented by all concerned agencies it goes without saying that disaster management will be more effective and efficient. The lack of coordination, insufficient funds and resources, poor public awareness and problems on count of resource mobilisation, etc., have to be addressed through the effective implementation of aforementioned measures. Disaster mitigation, early warning system, emergency rescue and relief operation, rehabilitation and recovery plans should involve activities such as training, post-disaster evaluation, monitoring of relief works, review and cooperation and coordination of central, district and local level preparedness and research works. Recovery planning should involve immediate recovery as well as long-term programmes. Political determination, effective and efficient policy implementation and people's participation is of great importance to achieve the above goals.

25

Drought Profile, Management and Risk Reduction in India

— **Pardeep Sahni**

INTRODUCTION

Disaster is an event, natural or man-made, sudden or progressive, which impacts with sudden severity that the affected community has to respond by taking exceptional measures. Frequencies as well as intensity of natural disasters are increasing globally including in India. Disaster impacts are felt more in developing countries due to borderline economic status of the vulnerable population, which has inadequate adjustment capacity. India with a wide range of climatic and topographic conditions is subject to various types of natural disasters like floods, cyclones, drought, earthquake, etc., in various degrees. Drought is a slow creeping disaster. It can be defined as lack or shortage of water for an unusually long period. Drought can occur at any place causing anything from inconvenience to deaths through famine. Drought connotes a situation of scarcity and distress usually caused by prolonged failure of rains, affecting agricultural activities adversely, leading to loss of production and employment, drinking water shortages, deficiency in water supply, etc. Consequently, there is migration of people in search of an alternative employment and livelihood.

Drought Conditions in India

The problem associated with drought is a recurrent feature in India. About 107 million hectares of the country spread over administrative districts in several states, are affected by drought. We now discuss the common causes for drought in India.

Meteorology	*Water resources*	*Agriculture-crop yield*	*Population*
• Inadequate monsoon rainfall. • High temperature and evaporation, wind speed. • Unseasoned rains and fog/ snowfall.	• Inadequate water availability, high water loss in storage and distribution, utilities. • Over exploitation of surface and ground water.	• Shift in agricultural practices (low to moderate water demand crops to high crops). • Crop damage due to rain and snow/pest.	• High greater rate of human and animals. • Location of high water consuming milestones at semi arid/arid regions.

The administrative districts frequently affected by drought in different states are shown in the table below:

States	*Districts*
Andhra Pradesh	Anantapur, Chittoor, Cuddapah, Hyderabad, Kurnool, Mehaboobnagar, Nalgonda, Prakasam
Bihar	Munger, Nawadah, Palamau, Rphtas, Bhojpur, Aurangabad, Gaya
Gujarat	Ahmedabad, Amrely, Banaskanta, Bhavanagar, Bharuch, Jamnagar, Kheda, Kutch, Meshana, Panchmahal, Rajkot, Surendranagar
Haryana	Bhiwani, Gurgao, Mahendragarh, Rohtak
Jammu & Kashmir	Doda, Udhampur
Karnataka	Bangalore, Belgaum, Bellary, Bijapur, Chitradurga, Chickmangalur, Dharwad, Gulbarga, Hassan, Kolar, Mandya, Mysore, Raichur, Tumkur
Madhya Pradesh	Betul, Datia, Dewas, Dhar, Jhabuva, Khandak, Khargaon, Shahdol, Shahjapur, Sidhi, Ujjain
Maharashtra	Ahmednagar, Aurangabad, Beed, Nanded, Nashik, Osmanabad, Pune, Parbhani, Sangli, Satara, Sholapur
Orissa	Phulbani, Kalakhandi, Bolangir, Kendrapada
Rajasthan	Ajmer, Banaswada, Barmer, Chiru, Dungarpur, Jaisalmeer, Jalore, Jhunjhunu, Jodhpur, Nagaur, Pali, Udaipur
Tamilnadu	Coimbator, Dharmapuri, Madurai, Ramanathapuram, Salem, Tiruchirapalli, Tirunelveli, Kanyakumari
Uttar Pradesh	Allahabad, Banda, Hamirpur, Jalan, Mirzapur, Varanasi
West Bengal	Bankura, Midnapur, Purulia

The Government of India's Meteorological Department has declared droughts in meteorological subdivisions where annual rainfall has been less than 75% of the normal. Severe droughts are declared in years when rainfall has been 50% below normal. The primary cause for the occurrence of drought is the deficiency of precipitation either in the form of rain or snow. In this context it is relevant to consider that major part of the country except Tamilnadu receives bulk of the annual precipitation during the south-west monsoon period, from June to September. October–December constitutes the main rainy season for Tamilnadu. Winter precipitation is significant in Jammu & Kashmir, Himachal Pradesh and West Uttar Pradesh hills and its abundance or deficiency constitutes the level of stream flow in the following season. The average annual rainfall in India is shown in the following map.

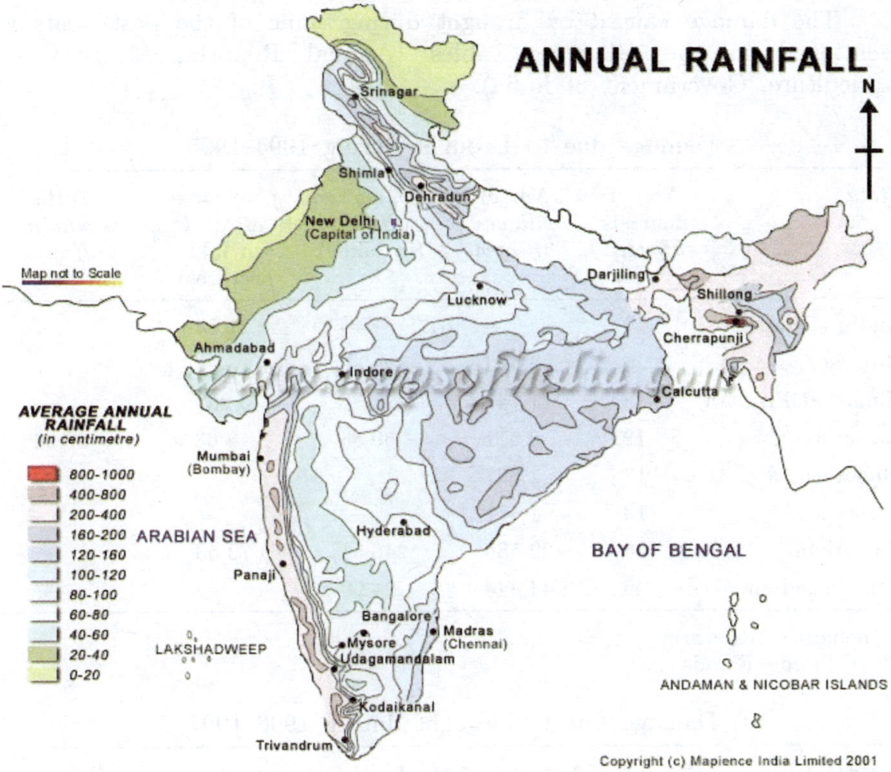

The following table reveals the periodicity of drought in different meteorological subdivisions:

Meteorological subdivision	*Recurrence of very deficient rainfall*
Assam	Once in 15 years
West Bengal, MP, Coastal AP, Kerala, Bihar, Orissa	Once in 5 years
North Karnataka	Once in 4 years
Eastern UP, Vidarbha, Gujarat, Eastern Rajasthan, Western UP, TN, Kashmir, Rayalseema, Telengana	Once in 3 years
Western Rajasthan	Once in 2.5 years

Impacts of Drought

In India, 191 districts covering 174 blocks of 12.03 lakh sq. km. are identified as drought prone. Drought has socio-economic impact on the society. The economic impact includes reduced income of farm lands, loss of live stock, fall in industrial output, unemployment, shortage of essential goods, inflation in higher price, decrease in agricultural output, poor purchasing power, etc. The social impact on drought is evidenced through poor hygiene, migration, malnutrition, in health, increased stress and mobility, etc.

The damage caused by drought during some of the past years is mentioned in the following tables (Annual Reports, Ministry of Agriculture, Government of India).

Damage due to Drought during 1994–1995

State	*No. of districts affected*	*No. of villages affected*	*Population affected* (in lakhs)	*Crop area affected* (in lakh hectares)	*Cattle population affected* (in lakhs)
Andhra Pradesh	14	—	—	21.24	—
Gujarat	5	2,562	—	—	—
Himachal Pradesh*	12	11,741	29.458	3.269	—
Karnataka**	12	6,571	59.55	8.68	22.64
Maharashtra	13	2,367	—	—	—
Orissa	13	1,471	—	—	—
Rajasthan	25	22,586	246.81	72.53	—
Uttar Pradesh	40	44,344	24.90	—	—

* Including Hailstorm.
** Including Floods.

Damage due to Drought during 1996–1997

State/UT	*No. of total districts*	*No. of districts affected*	*No. of villages affected*	*Population affected* (in lakhs)	*Crop area affected* (in lakh hectares)	*Estimated value of crop loss* (in lakhs)
Gujarat	19	1	593	5.10	—	—
Madhya Pradesh	45	9	4,233	31.63	8.48	41,912.63
Maharashtra	31	7	1,650	—	—	—
Orissa	30	26	27,081	256.75	—	—
Total	**125**	**43**	**33,557**	**293.48**	**4.48**	**41,912.63**

Damage due to Drought during 1999–2000

State	*No. of total districts*	*No. of villages affected*	*Cropped area affected* (in lakh hectares)	*Estimated value of damaged crop* (in thousands)	*Population affected* (in lakhs)	*Cattle population affected* (in lakhs)
Andhra Pradesh	22	512*	15.22	25,669,100	—	—
Gujarat	15	8,000	—	—	—	—
Himachal Pradesh	12	—	2.87	2,348,007	—	—
Jammu & Kashmir	6	—	2.96	5,504,600	—	—
Karnataka	18	11,078	22.84	—	55.38	—
Madhya Pradesh	4	5,010	9.53	3,514,800	26.64	—
Manipur	5	—	0.71	612,000	—	—
Mizoram	3	—	0.51	263,400	—	—
Rajasthan	26	23,406	78.18	21,552,000	261.79	345.60
Tripura	4	—	0.20	556,700	0.82	—
West Bengal	10	—	1.20	4,389,400	25.25	—
Total	**125**	—	**134.22**	**64,410,007**	**369.88**	**345.60**

Damage due to Drought during 2000–2001

State/UT	*No. of districts affected*	*No. of villages affected*	*Population affected* (in lakhs)	*Cattle population affected* (in lakhs)	*Cropped area affected* (in lakh hectares)	*Estimated value of damaged crop* (in thousands)
Chattisgarh	16	—	69	36.06	—	113,600
Gujarat	17	9,449	250	71.33	NR	NR
Madhya Pradesh	22	14,851	26.64	34.28	9.53	3,514,800
Orissa	24	NR	NR	NR	270.00	90,300
Rajasthan	31	30,583	32.5	400	87.49	—

GOVERNMENT POLICY AND PROGRAMMES

National Water Policy

India's National Water Policy (NWP) was adopted in September 1987. The National Water Resources Council (NWRC) under the Chairmanship of the Prime Minister lays down the NWP, reviews development plans and advises on implementation. The policy envisages strategies covering ground water development, water allocation priorities, drinking water, irrigation, water quality, water zoning, water conservation, flood control and management. In the context of water use, the main issues are the pricing of water for various end uses including drinking, irrigation and industrial use. The NWP of the Government of India accords highest priority to drinking water supply. The State Governments in India make their water policies within the overall framework of the NWP. The National Water Policy, Government of India, 1987, also deals with drought management. It states:

1. Drought prone areas should be made less vulnerable to drought associated problems through soil-moisture conservation measures, water harvesting practices, the minimisation of evaporation losses, the development of the ground water potential and the transfer of surface water from surplus areas, where feasible and appropriate. Pastures, forestry or other modes of development which are relatively less water demanding should be encouraged. In planning water resource projects, the needs of drought-prone areas should be given priority.
2. Relief works undertaken for providing employment to drought stricken population should preferably be for drought proofing.

Sustainable Agriculture and Rural Development Policy

The Agricultural Development Strategy of 1999 focused on Sustainable Agriculture and Rural Development (SARD). The strategy is essentially based on the policy on food security and alleviation of hunger. A regionally differentiated strategy, based on agro climatic regional planning which takes into account agronomic, climatic and environmental conditions, is adopted to realise the potential of growth in every region of the country. The thrust is on ecologically sustainable use of basic resources such as land, water, and vegetation; in such a way that it serves the objectives of accelerated growth, employment and alleviation of hunger.

Major activities to implement the SARD policy are as follows:

1. Development of crops based on regionally differentiated strategy
2. Development of horticultural crops
3. Adequate and timely delivery of core inputs
4. Integrated pest management
5. Greater use of bio-fertilisers and bio-technology
6. National Agricultural Technology Project
7. Rained farming and watershed management
8. Soil and water conservation
9. Animal husbandry and dairying
10. Development of fisheries
11. Agricultural research and education
12. Development of human resources

Approaches for Land Use

A wide range of approaches have been employed in India to address problems of land degradation, some of which include:

- Prevention of soil loss from the catchments.
- Promotion of multi-disciplinary integrated approach to catchment treatment.
- Improvement of land capability and moisture regime in the watersheds.
- Promotion of land use to match land capability.
- Reduction of run-off from the catchments to reduce peak flow into the river system.
- Upgrading of skills in the planning and execution of watershed development programme.
- Increase of productivity of land affected by alkalinity for increasing sustainable agriculture production.
- Identification of critical degraded areas.
- Generation of data on land suitability and capability for regulating land use.
- Preparation of soil resource map and inventory of soil and land resources.
- Development of technical skills in soil and water conservation.
- Building up and strengthening of land capability of State Land Use Boards.
- Promotion and implementation of land use policy relates to land base programme.

The following Soil and Water Conservation Programmes have been launched in response to the need for conservation and rehabilitation of degraded land including:

- Strengthening of State Land Use Boards (SLUBS).
- National Land Use & Conservation Board (NLCB).
- Soil Conservation Training Centre DVC Hazaribagh (Plan 7 Non-Plan).
- Centrally Sponsored Scheme of Soil Conservation for Enhancing Productivity of Degraded Lands in the Catchments of River Valley Projects.
- Centrally Sponsored Scheme of Soil Conservation in the Catchments of Flood Prone Rivers.
- Centrally Sponsored Scheme for Reclamation of Alkali (User) Soils.
- EFC Assisted Project for Reclamation and Development of Alkali land in Bihar and U.P.
- Uttar Pradesh Sodic Land Reclamation Project with World Bank assistance.
- Watershed Development Project in shifting Cultivation Areas of North Eastern States (WDPSCA).

- Indo-German Bilateral Project on Watershed Management.
- Reclamation of Marginal and shallow ravines in the states of Uttar Pradesh, Madhya Pradesh, Gujarat and Rajasthan.
- Centrally Sponsored Scheme for Reclamation of Saline Soils including Coastal Saline and Sandy Areas.
- Centrally Sponsored Scheme for Amelioration of Acid Soils.

In order to integrate and intensify the activities aimed at combating desertification, a comprehensive plan for control of desertification under the National Forestry Action Programme has been proposed. The programme aims at evaluating the present status of deserts in the country, assessing the implementation of ongoing programmes for the development of deserts and desert prone areas, formulating broad policy guidelines and action plans for implementation aiming at control of desertification, developing strategies involving people in desert control through various means, and including appropriate measures relating to research and training in desert control.

The integrated Wastelands Development Project has been started to facilitate pilot projects using an integrated approach to wasteland development by initiating area-specific projects taking into account land capabilities, site condition and local needs, and ultimately aiming to promote optimal land use for both ecological and socio-economic needs. The different types of problem lands for which projects are prepared include saline/alkaline lands, arid/sandy areas, ravine areas, and the Aravallis. The activities cover soil and water conservation, afforestation, silvi-pasture development, grazing management, etc.

The main objective of the Afforestation Project for the Aravalli (Rajasthan) is to check desertification and restore ecological status by re-afforestation and also to increase the production of fuel wood, fodder, timber, and non-wood forest products to meet local needs. Rehabilitation of common lands in Aravallis (Haryana) is being implemented in the four southern districts of Haryana that is Bhiwani, Mahendragarh, Gurgaon, and Faridabad since 1990. The project outlay is Rs. 480 million and covers environmental protection, restoration of green cover in the semi-arid Aravalli Hills, and improvement in the living conditions of the local people through meeting their biomass needs.

These programmes target specially the poor inhabitants of rural areas and attempt to combine development with drought management.

The Agricultural Policy of the country announced in July 2000 is the product of extensive consultation with all stakeholders, including State Governments. The policy seeks to actualise the vast untapped growth potential of Indian agriculture, strengthen rural infrastructure to support faster agricultural development, promote value addition, excel growth rate of agro-business, etc.

The Panchayati Raj Institutions (PRIs) have been mandated to function as effective institutions of local self-governance and to prepare plans for economic development and social justice and implement them.

The PRIs are the umbrellas for the integration of sectoral programmes with poverty alleviation and rural development programmes. The Council for Advancement of People's Action and Rural Technology (CAPART) continues to provide projected financial assistance to voluntary organisations, which are to play a more dynamic role in empowering the poor through advocacy, awareness generation and formation of Self-Help Groups (SHGs). In order to promote people's participation and create awareness, the practicing farmers, village youth and school dropouts are working as focal points for dissemination of information, for example, on low cost technology and producing plant material for conservation measures. Stress is being laid on organising SHGs to institutionalise people's participation to improve household production systems (cattle rearing, mushroom cultivation, sericulture, bee-keeping, etc.)

DROUGHT MANAGEMENT IN INDIA

Drought Monitoring and Warning

In India, India Meteorological Department (IMD), National Remote Sensing Agency (NRSA), Central Water Commission (CWC), and Ministry of Agriculture undertake drought monitoring. IMD prepares the aridity anomaly charts on a fortnightly basis using observations from agro-met stations. NRSA uses vegetation index derived from IRS and NOAA data monitors drought at district and block level, under the National Agricultural Drought Assessment and Monitoring System Programme. The monitoring of water level of all major rivers and reservoirs is done by CWC. The Ministry of Agriculture through field stations monitors the impact of drought on major crops.

Early warning of drought is an important requirement for tackling its adverse impacts. Currently the prediction or early warning of drought is carried out mainly based on following rainfall predictions: (i) Long range rainfall prediction of seasonal total rainfall by IMD using parametric and power regression models and dynamic stochastic transfer models; (ii) Medium range rainfall prediction by National Centre for Medium Range Weather Forecasting for 76 agro meteorological advisory service units and (iii) Short range rainfall predictions by IMD based on Indian National Satellite (INSAT) data supported with weather and agro-met observations. IMD issues Farmers' Weather Bulletins twice a day valid for 24 to 72 hrs from its forecasting stations located in the State capitals. IMD also issues agro meteorological advisories from its 17 Agro meteorological Advisory Service units established at State capitals. However, these forecasts are presently of qualitative nature. Crop yield predictions are carried out by the Department of Agriculture based on the rainfall and crop condition information received by the state agriculture department and is compiled at national level (High Powered Committee on Disaster Management Report).

Programmes

The Government of India has launched various development programmes to serve the long-term needs of the different sections of the drought-affected communities. Let us discuss some of the important programmes for combating drought.

The Desert Development Programme (DDP) initiated in 1977–78, covers both the hot desert regions of Gujarat, Rajasthan, and Haryana and the cold desert areas in Jammu, Kashmir, and Himachal Pradesh. It is functional in 131 blocks of 21 districts in 5 States covering an area of about 0.362 million sq. km. and a population of 15 million. The objectives of the programme include controlling the process of desertification, mitigating the effects of drought, restoring the ecological balance, and raising the productivity of land, water, livestock, and human resources. At least 75% of the allocation is earmarked for activities, which contribute towards combating the process of desertification. The programme is implemented with 100% central assistance. The Programme Evaluation Organisation of the Planning Commission has the task of evaluating this programme in order to assess its impact on the control of desertification, and on the improvements effected in productivity and income for the people living in these areas.

The Drought Prone Area Programme (DPAP) was launched in 1973 in arid and semi-arid areas with poor natural resource endowments. The objective is to promote more productive dry land agriculture by better soil and moisture conservation, more scientific use of water resources, afforestation, and livestock development through development of fodder and pasture resource, and in the long run to restore the ecological balance. The DPAP covers 615 blocks of 91 districts in 13 States. This is a centrally sponsored scheme where the allocations are shared between the Centre and States on a 50:50 basis. Preparation of development plans on a watershed basis, participation of people in planning and implementation of the programme, and developing effective liaison between research agencies and implementing agencies are some of the priority areas of the programme.

In low fertility status soils covered by Drought Prone Area Programme and Desert Development Programme, cultivation of surface spreading crops (e.g. groundnut, soyabean, mung and urad) helps in checking soil erosion. In non-arable lands 3-tier system of plantation (grasses, bushes and trees) would provides adequate biomass to meet fodder, fuel and timber needs of the local population.

Contour vegetative hedges should be planted on sectional earthen bunds of small cross-section. Sunken dugouts at suitable places need to be constructed for rain water harvesting. Small stone and live checks across rills and gullies would stabilise soils cape and reduce erosion. Under DPAP, particularly, earthen and concrete water impounding structures should be constructed to collect maximum run off. Diversion drains in upper reaches of a watershed need to be provided for safe disposal of rain water and prevention of soil erosion.

Some of the other related programmes are:

- Food for Work Programme (FWP)
- National Rural Employment Programme (NREP)
- Rural Landless Employment Guarantee Programme (RLEGP)
- Integrated Rural Development Programme (IRDP)
- Accelerated Rural Water Supply Programme (ARWSP)
- Indira Awaas Yojana (IAY)
- Jawahar Rozgar Yojana (JRY)
- Employment Assurance Scheme (EAS)—to assure 100 days of employment during lean agricultural season in drought prone, tribal and hilly areas.

Mitigation

The commonly adopted drought mitigation strategies in India include:

- Construction of check dams
- Watershed management
- Water rationing
- Cattle management
- Proper selection of crop for drought affected areas
- Levelling, soil conservation techniques
- Reducing deforestation and firewood cutting in the affected area
- Checking of migration and providing alternate employment for people in government sponsored relief schemes or village cooperatives and non governmental programmes.
- Education and training to the people.
- Participation in community programmes, e.g., Pani-Panchayat in Maharashtra, Sukhomajri experiment in Punjab, and Anna Hazare's work in Ralegaon Sidhi village in district Ahmednagar of Maharashtra.

The focus in mitigation is on:

(i) *Improvement in agriculture* through modifying cropping patterns and introducing drought-resistant varieties of crops.

(ii) *Management of rangeland* with improvement of grazing lands improved grazing patterns, introduction of feed and protection of shrubs and trees.

(iii) *Development of water resource system* with improved irrigation, development of improved storage facilities, protection of surface water from evaporation, and introduction of drip irrigation system.

(iv) *Animal husbandry* activities helping in mitigation with the use of improved and scientific methods, increasing outputs without destroying the echo-system.

The *response and recovery* components for drought are undertaken by various government departments, NGOs, universities and training institutions operating at central, state, district and below district levels. The High Powered Committee on Disaster Management in its report has highlighted the following gaps in the existing system and suggestions for improvement in drought management (High Powered Committee on Disaster Management Report).

Phase	*Agencies*	*Current status*	*Gaps/Constrains*
Drought-prone areas	IMD, Ministry of Agriculture	Climatic data/ water balance models to classify arid/semi-arid areas.	Categorisation only at broad level, not at micro-level (Village/Taluka).
Warning/ monitoring	IMD, DOS, Ministry of Agriculture	Monitoring based on reports from State Governments. Rainfall deficiency main criteria.	Assessment qualitative, at Met. Subdivision level, VI based monitoring by DOS.
Damage assessment	Ministry of Agriculture/ State Governments	Assessment based on field reports on crop loss.	Information on crop loss is inadequate and patchy.
Mitigation/ relief	Ministry of Agriculture/State Governments, NGOs	Based on drought memorandum by State Government, relief work coordinated by State Government.	Lack of information on severity/impact, inadequate planning of relief.
Drought proofing	Ministry of Agriculture/ State Governments, NGOs	Long-term plans made for only selected water sheds.	Lack of holistic plan for sustainable development of drought prone areas.

The recommendations include:

- With the frequent change in the land use, irrigation development, cropping pattern and agriculture practices it is necessary for frequent updating of drought prone area.
- Suitable integration of ground based information and satellite derived parameters on a geographical information system framework would lead to a capability to monitor drought conditioned and severity in a realistic manner and provide information required by the administrators for drought combating.
- Development activities proposed should use maximum possible information from remote sensing. Frequent monitoring and

evaluation should also be carried out. Areas requiring attention are:

(i) Response requirements to be extensive and prolonged, thus involving major commitment and expenditure of resources.
(ii) Prolonged drought to undermine self-reliance of affected communities, thus making it difficult to withdraw disaster management assistance.
(iii) Logistic requirements to exceed in country capability, particularly if large inputs of outside (international) commodities are involved.
(iv) Rainfall and other agro-met data being monitored are inadequate in certain agro climatic divisions to meet demands of drought monitoring. The rain gauge network is too coarse to provide the variations of rainfall within districts. Space borne measurements have to be integrated with computed aridity anomaly based on field measurements of rainfall and crop calendars to bring out real time drought conditions of a region.
(v) While new technological options are emerging as use of satellite data, modelling, etc., there is no operational procedure currently to forecast the impending drought conditions with respect to area of impact, extent and duration.

MAJOR DROUGHTS

All India Drought

Drought is a perennial feature in some States of India. 19% of India's total land area, with 12% of the population, is drought prone. About 68% of the total cultivated area of the country is prone to drought. Most of the drought-hit areas are to be found in arid, semi-arid and sub-humid regions of the country, which experiences less than average annual rainfall. The impact of drought varies from year to year. India faced severe drought in 1966–67, 1972–73, 1979–80, and 1980–87. In each case, food production average was below than average. There was large scale loss through starvation, depletion of areas and livestock as well as high mortality.

India witnessed a severe all India drought in 1982. It was quite bad with a monsoon rainfall departure of minus 13.7% and 46.4% of the area suffering from deficient rains. The seasonal rainfall deficiency was much more randomly distributed in 1982. The food grain production was adversely affected. A loss of 9.5 million tonnes was reported in the Kharif crop only. As the efforts to increase food grain production in the following Rabi crop succeeded, the year finally ended with a shortfall of only 3.8 million tonnes over the previous years' food grain production. The drought of 1987 has been also phenomenal. The monsoon seasons' rainfall departure was minus 19.3% and the area under deficient rainfall was 64.3%. The worst sufferers were the three meteorological sub-divisions of Saurashtra, Kutch and Diu (–74%), West Rajasthan (–67%) and Haryana and Delhi (–67%). In term of overall annual food grain production in the country for both Kharif and Rabi, the 1987 drought resulted in a loss of production of 3.0 million tonnes as against that of 3.8 million tonnes in 1982. The drought of 1987 halted the momentum of agricultural growth established during the early 1980s. It affected 15 States and 6 Union Territories, damaging crops on an area of about 59 million hectares subtract over 267 districts. Gujarat and Rajasthan were the worst affected states. Nearly 285 million people have been the direct sufferers of the adverse socio-economic impact of the drought and of these, around 92 million people belonged to economically weaker and socially deprived sections of the society. The drought affected about 168 million cattle. During the drought of 1987, imports had to be resorted to the tune of 200,000 tonnes of pulses, 30,000 tonnes of butter oils, and 22,000 tonnes of skimmed milk powder.

The damage due to drought in 1987 became more acute because there had actually been continuous drought from 1984 to 1987. The following table shows the number of districts affected, population affected (millions), cropped area affected (million hectares), cattle population affected (millions), relief cost (Rs. in billions):

Damage Due to Droughts, 1984–87

Damage	*1984*	*1985*	*1986*	*1987*
Number of districts affected*	151	109	280	263
Population affected (millions)	70.5	78.6	191.90	285.4
Cropped area affected (million hectares)	15.4	28.2	40.13	58.6
Cattle population affected (millions)	47.5	65.4	111.90	168.1
Relief cost (Rs. in billions)	1.7	4.6	6.0	17.3

*Total districts 418 as on 1987.

The response of the Governments towards mitigating its impact is evident from the fact that more than 6 million people were given employment every day during 1987–88 under 100,000 ad-hoc relief projects. In addition to the supplementary nutrition programme being provided to 7.72 million persons, including pregnant women and children, around 13 million tonnes of food grains were distributed in the drought affected areas and two million tonnes of fodder was transported from surplus areas to scarcity areas to save cattle. 1.7 million cattle were maintained in 2,000 cattle camps.

Major drought prone states of India are shown in the following map:

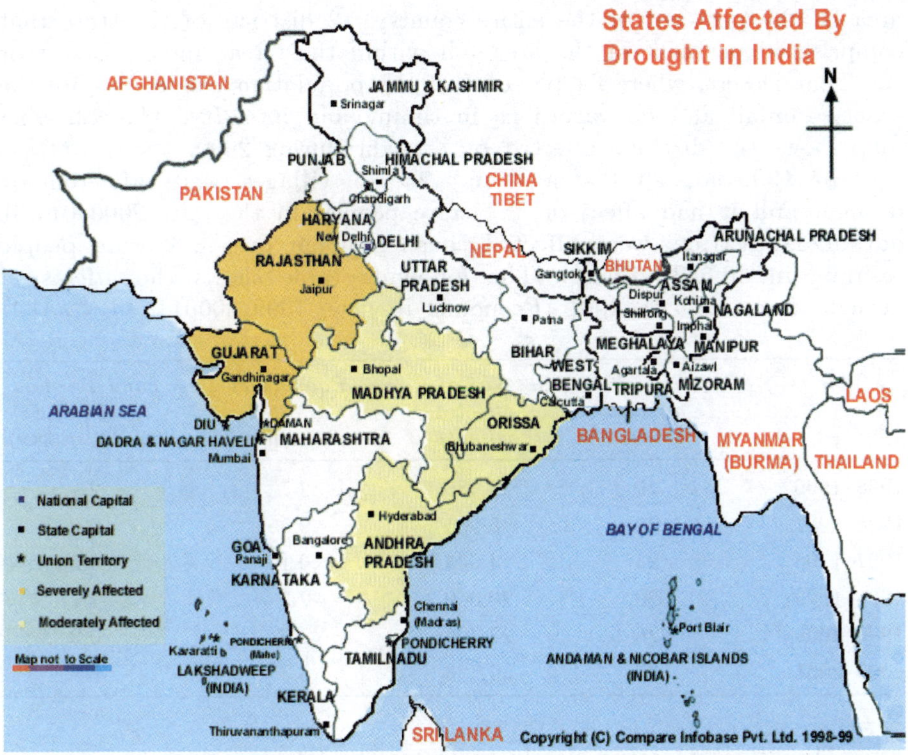

Drought in Rajasthan

Rajasthan is India's largest state with an area of 3.42 lakh sq. km. spread over 32 districts having 41,538 villages and around 56 million population. It has only one per cent water resources whereas the total

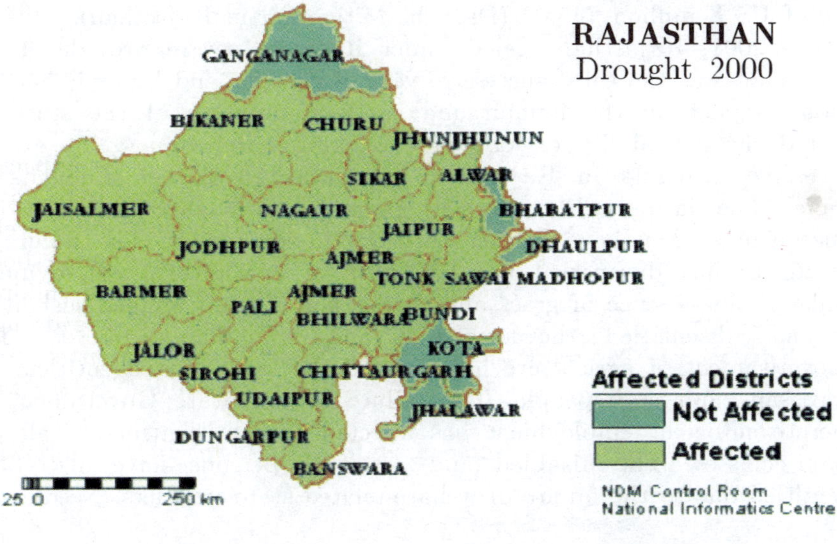

area is 10.7 per cent of the entire country. 12 districts of the state that comprise 60 per cent of the area fall within the Great Indian Desert or the Thar Desert, where 64 per cent of the population resides despite the scanty rainfall and its variations in timing and intensity. The following map shows the districts affected by drought during 2000.

In 1995–96, 29 districts and 25,486 villages were affected by drought and it had affect on 2.7 crore people. In the year 2000–01, 31 out of 32 districts were affected causing gloom over 3.3 crore people residing in 30,583 villages. The following table shows the affects of drought from 1995 to 2001 (Economic Review, 1999–2000).

Year	*No. of affected districts*	*No. of affected villages*	*Population affected* (in crores)	*Land revenue suspended* (in Rs. crores)
1995–1996	29	25,486	2.7	2.1
1996–1997	21	5,905	0.55	0.3
1997–1998	24	4,633	0.15	0.03
1998–1999	20	20,069	2.2	1.7
1999–2000	26	23,406	2.6	0.3
2000–2001	31	30,583	3.3	NA

The aforesaid table clearly indicates the fragile nature of the rural economy of Rajasthan. During a period of 30 years from 1970–71 to 2000–01, drought conditions prevailed in 29 districts, in 1995–96, they were in 31 district. During 2000–01 the damage to kharif crop was more than 75 per cent in 198,178 villages and in the remaining 10,766 villages, the damage ranged from 15 per cent to 75 per cent. The expenditure under relief operations in Rajasthan from 1950–51 to 2000–01 was to the tune of US $ million 745.02 (Drought Mitigation in Rajasthan).

In 2001, Rajasthan reeled under its most severe drought in 40 years. That was the third successive year of drought and hence it had far serious impact on the human and cattle population of the state. It affected the availability of food and drinking water to a great extent. Successive droughts in Rajasthan have affected the entitlements of people. The farmers have lost their crops over successive seasons. Consequently, they have neither their food, nor fodder. The failure of agriculture has disentitled the labourers of agricultural employment. Similarly, the absence of grass and fodder has affected animal husbandry and has disentitled the cowherds and nomadic shepherds. The impoverishment of agriculture has affected the trade-related entitlements of artisans and craftspeople. The failure of the State Government to generate sufficient employment has affected the entitlements of all the labour classes. The disabled and destitute persons have also been disentitled due to the failure of welfare-related state transfers. Needless to emphasise, women and children across all the sections are the worst affected lot.

The following measures can be undertaken for combating drought in Rajasthan:

1. Judicious use of limited irrigation by resorting to drip irrigation and sprinkler method.
2. Rain water harvesting which is a traditional way of life in arid regions from times immemorial. This practice is in the form of harnessing meagre runoff in different forms like rooftop rain water harvesting, Nadis (village pond), Tankas (underground storage).
3. Management of underground water by recharge of underground water by way of infiltration tanks, check-dams, injection wells, etc. has to be taken at all technically feasible sites in rural as well as urban areas. The need of the hour is to provide legal mechanism for regulating such extraction at least in critical areas, where the exploitation has reached a saturation point.
4. Improved agronomic practices considering that the major problem is that the farmers in such dry areas are poor and risk averse and their agronomic practices are said to be 30 years behind than those for the developed areas. There is need for prevent improvements in agronomic practices.
5. Contingency crop-planning for drought proofing whereby, in case of inadequate, rainfall, mid-season corrections like reducing plant population, spraying anti-transpirants, weeding and creating soil mulch are the strategies are to be adopted. Drought tolerant varieties of mungbeans and clusterbeans, etc., which performed well even in moderate rainfall conditions, need to be grown more.
6. Integrated Watershed Development.

Drought in Orissa

The state of Orissa having an area of 155,707 sq. km. spread over 30 districts borders West Bengal in the northeast, Andhra Pradesh in the southeast, Bihar in the north, Madhya Pradesh in the west, and the Bay of Bengal in the east. The state, which has a population of 46,706,920, has majority of its population, that is, 321,602, residing in the rural area with a density of 236 persons per sq. km., approximately 17% of the population is below the poverty line (1992–93 records). The following map shows the drought-affected districts of Orissa (UN Resident Coordinator Report).

Natural disasters like cyclones, floods, droughts, etc. strike Orissa year after year. There have been three major droughts in the state during the past five years. There was rainfall deficit of 34% of normal in 1996–97 consequent to which all the 30 districts of the state were

ORISSA MAP INDICATING CYCLONE-1999 ; DROUGHT- 2000 & FLOOD - 2001 AFFECTED DISTRICTS

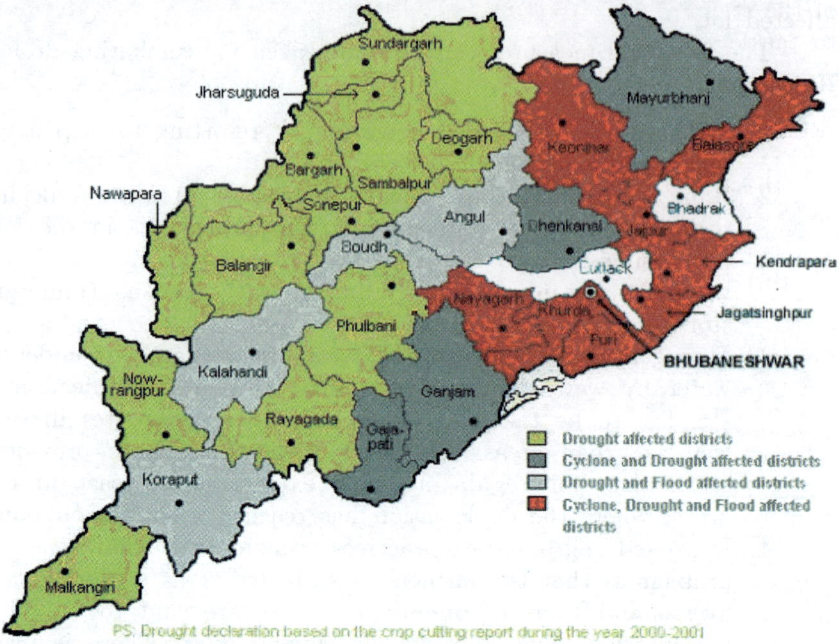

PS: Drought declaration based on the crop cutting report during the year 2000-2001

Source: UN Resident Coordinator, July 27, 2001.

drought-hit. In 1998–99, the rainfall deficit was 24% approximately. It affected 23 districts of the state hitting more than 15 million people. The crop area that could not be sown because of the drought was 477,000 hectares, and the crop sown was affected in 1,066,000 hectares. The production loss was approximately 2,000,000 MT and the total loss in aggregate was Rs. 880 crore. The year 2000–01 was again bad for the state as due to deficit rainfall of 26% of normal, 6 million people were affected. 100,000 hectares of crop area could not be covered on one hand and on the other, 10,700,000 hectares of crop sown area was affected causing production loss of 1,453,000 MT. It resulted in a total loss of Rs. 770 crore. The drought of 2000–01 affected 7.6 lakh farm households. The loss on account of person days was 3.6 crore. Due to lack of or less employment opportunities, there were starvation deaths and migration of families to other States.

The FAO/UNDP Mission made the drought damage assessment of three worst affected districts namely Bargarh, Nuapada, and Bolangir and found the following as the major causes for crop loss:

(i) Paddy seedling death in seed bed
(ii) Transplanted areas could not be fully covered
(iii) Beushaning could not be done in many areas
(iv) Delayed and/or improper fertiliser application due to non-availability of mater in paddy fields.

Due to drought there is often immense water scarcity in many areas of the state. The Government of Orissa has envisaged following measures to tackle drinking water scarcity in rural areas:

(i) Operation Trishna, an Action Plan for mitigation of drinking water scarcity was launched in 10 most drought affected districts.
(ii) In areas where there was likelihood of depletion of water table beyond suction level, worn-out "Riser Pipes" have been replaced.
(iii) PVC storage tanks have been procured for transportation of drinking water.
(iv) Defunct tube wells have been replaced.

The Government has planned to provide one tube well for 150 populations and for tube wells in all primary and upper primary schools. The state has also planned to gradually transfer maintenance of drinking water installations to community through Panchayat (local elected body in rural area).

The major impacts of drought on agriculture in Orissa was, higher percentage of fallow areas, low productivity and production, uncertainty in post-monsoon Rabi crops, and reduced wage rate due to availability of surplus man power, etc. UNDP Inter Sectoral Team supported from WHO, UNICEF, and FAO Orissa, in a study, assessed the nutritional status of the drought-hit community. It was found through the study that 58% of the children were chronically malnourished, and 11% had developed acute malnutrition as a result of the drought. With regard to food security at the family level, the study found that about 60% did not have food stocks for more than three days. Of this, 50% had no food stock (http://www.UNDP.Org.in/oriss/drought).

The Government of Orissa-UNDP-Department For International Development have envisaged the following drought mitigation measures (UNDP Orissa Document):

1. Immediate—livelihood interventions for targeted population.
2. Medium-term—livelihood support for Integrated System Approach in agriculture.
3. Long-term—linking Food or Cash for work with community asset creation for drought mitigation.

It has also been targeted to build and sustain capacity of the local youths, volunteers, etc. by constituting Community-based Rain Water Management Brigades in planning, technical skills development, and maintaining the initiatives for drought proofing.

Drought in Gujarat

Gujarat has 25 districts spread over an area of 1.96 lakh sq. km. Its population is approximately 43 million. Gujarat has experience natural calamities time and again in recent years. Saurashtra, Kutch, and north Gujarat are recognised rainfall deficient and drought prone areas. In the past 40 years, Gujarat has experienced 12 years drought and 5 major scarcity situations, in 1972–73, 1986–87, 1992–93, 1995–96 and 1999–2000. During 1999–2000, due to failure in Monsoon, the state was again hit by severe scarcity. Average shortfall in rain was around 35%, but in some districts like Patan and Kutch; the shortfall was of the magnitude of 70% and 60% respectively. After observing the pattern of rainfall and agricultural crop losses, the State Government had declared 8,666 villages in 17 districts as drought-hit on 22 December 1999. Later on, 755 villages were declared as Semi-Scarcity affected villages and 29 villages were declared as Semi-Scarcity affected after the Rabi Annawari assessment. Thus, a total of 9,449 villages were declared to be Scarcity/ Semi-Scarcity affected. The State Government incurred Rs. 602.07 crores to tackle the scarcity situation. The total expenditure on drought relief during 1999–2000 was Rs. 684.00 crores. The drought has manifested mainly as acute scarcity of drinking water and fodder together with other socio-economic consequences attendant on significant fall in agricultural production in the affected areas. In 2000–01, the State Government declared scarcity in 23 districts spread over 199 talukas comprising 13,148 villages.

States initiative to combat drought in 2000–01 are mentioned below (http://mapsofindia.com/stateprofiles/gujarat/index.html). The details below reveal the actions initiated by the Government of Gujarat in meeting the challenge through drought:

(i) Weather watch group analysed the data provided by the I.M.D. every week and alerted the State Government of the impending scarcity situation.
(ii) The District authorities were asked to provide the water tankers wherever drinking water shortage existed. In Junagadh, Porbandar, Amreli, Bhavnagar, Kutch, Rajkot, Surendranagar and Jamnagar, tankers were pressed into service.
(iii) In the month of September 2000, the State Government directed district collectors to prepare a Master Plan for scarcity.
(iv) On 30-12-2000, a cabinet sub committee decided on the policy relating to relief measures in the affected areas.
(v) The State Government requested the Ministry of Railways for extending facility of rail movement of fodder and drinking water free of cost.
(vi) State Government requested the Central Government to delegate power to the State Government regarding ban of movement of grass under the Essential Commodities Act 1955. Accordingly notification on stock declaration was issued and interstate and inter-district movement of grass was been banned.
(vii) The Government issued a policy directive to reserve water available in the reservoirs for drinking and not to release for

irrigation unless approved by a High Level Committee consisting of three senior ministers.

(viii) The Government also decided to stop supply of water from the dams and reservoirs to the industries so as to reserve water for human consumption. The Industries were to make their own arrangements for continuing their water supply through alternative sources. This decision specifically applies to the Mahi basin.

(ix) The Government decided that the Gujarat Electricity Board will give 4,000 numbers of electric connections to the farmers/ villagers to energise their water source if they agree to give water to the village for drinking.

It is worth mentioning here that the individuals and the communities have made some excellent attempts in Gujarat and also in Rajasthan to meet the challenges posed by droughts. Several innovative, traditional methods of water harvesting saved the day for a number of villages in these worst affected states. Deepening village ponds, recharging dried wells and construction of simple watersheds successfully have enabled some of the villages to face the acute shortage. Communities have demonstrated that rain water can be collected in dried up ponds, old village wells can be recharged and ponds with plastic lining can effectively hold water. The study entitled *Pani Ghano Amol* (water is too priceless) commissioned by *Charka* and the National Foundation for India has recorded such success stories. In Saurashtra, Gujarat, a crusader has taken upon himself the task of reviving dried wells in all nearby villages. The Gram Sabhas are now being asked to train villages in the technique of reviving dried wells. There are also some innovators in the region spreading the message of *Plant a Tree and get Rain.* Such initiatives of the community will definitely help in improving the situation being faced by the people almost every year.

RISK IDENTIFICATION, REDUCTION AND TRANSFER

Risk Identification

As mentioned earlier, a sizable part of the country reels under drought almost every year affecting substantive segment of population and livestock. The drought conditions affect the livelihood of the persons considerably. The risk in the form of losses and damages to the population in drought prone areas are mainly on the following counts:

1. Costs and losses to agricultural producers

- Annual and perennial crop losses
- Damage to crop quality
- Income loss for farmers due to reduced crop yields

- Reduced productivity of cropland (wind erosion, long-term loss of organic matter, etc.)
- Insect infestation
- Plant disease
- Wildlife damage to crops
- Increased irrigation costs
- Cost of new or supplemental water resource development (wells, dams, pipelines).

2. Costs and losses to livestock producers

- Reduced productivity of range land
- Reduced milk production
- Forced reduction of foundation stock
- Closure/limitation of public lands to grazing
- High cost/unavailability of water for livestock
- Cost of new or supplemental water resource development (wells, dams, pipelines)
- High cost/unavailability of feed for livestock
- Increased feed transportation costs
- High livestock mortality rates
- Disruption of reproduction cycles (delayed breeding, more miscarriages)
- Decreased stock weights
- Increased perdition
- Range fires.

3. Loss from fishery production

- Damage to fish habitat
- Loss of fish and other aquatic organisms due to decreased flows.

4. General economic effects

- Decreased land prices
- Loss to industries directly dependent on agricultural production (e.g., machinery and fertiliser manufacturers, food processors, dairies, etc.).

5. Damage to animal species

- Reduction and degradation of fish and wildlife habitat
- Lack of feed and drinking water
- Greater mortality due to increased contact with agricultural producers, as animals seek food from farms and producers are less tolerant
- Disease
- Increased vulnerability to perdition (from species concentrated near water)
- Migration and concentration (loss of wildlife in some areas and too many in others)

- Increased stress to endangered species
- Loss of bio-diversity.

6. Hydrological effects

- Lower water levels in reservoirs, lakes and ponds
- Reduced flow from springs
- Reduced stream flow
- Loss of wetlands
- Estuarine impacts (e.g., changes in salinity levels)
- Increased ground water depletion, land subsidence, reduced recharge
- Water quality effects (e.g., salt concentration, increased water temperature, pH, dissolved oxygen, turbidity).

7. Damage to plant communities

- Loss of bio-diversity
- Loss of trees from urban landscapes, shelter-belts, wooded conservation areas.

Risk Reduction

The Government of India, in keeping with the fact that natural disasters, particularly droughts, create unemployment and under employment problems in the rural areas, has started a number of programmes and schemes for risk reduction ultimately leading to people's benefit of the people. Some of the important measures through programmes and schemes initiated by the Government of India include:

1. Swaranjayanti Gram Swarozgar Yojana. This scheme, launched with effect from 1st April 1999, aims at establishing a large number of micro-enterprises in the rural areas, building upon the potential of the rural poor. A significant aspect of the scheme is that every family assisted under this programme is to be brought above the poverty line in three years and therefore it aims at creating substantial additional income for the rural poor. The scheme targets the most vulnerable people; at least 50% of the beneficiaries are Scheduled Castes/Scheduled Tribes. This is a credit-cum-subsidy programme envisaging a greater involvement of the banks, in the planning and preparation of projects, identification of activity clusters, infrastructure planning as well as capacity building and choice of activity of the Self-Help Groups. Subsidy under the programme is uniform at 30% of the project cost subject to a ceiling of Rs. 7,500 (for scheduled castes/scheduled tribes, it is 50% and Rs. 10,000 respectively). For Self-Help Groups, the subsidy is 50% of the project cost subject to a ceiling of Rs. 1.25 lakh. However, there is no limit on the subsidy for irrigation projects. The Central and State Governments share funds under the scheme in the ratio 75:25. The central allocation earmarked for the states is distributed in relation to the incidence of the poverty in the States.

2. Employment Assurance Scheme. This scheme was launched on 2nd October 1993. The scheme is implemented in the blocks falling in the drought prone areas, desert areas, tribal areas and hilly areas for providing gainful employment opportunities to the people. The scheme restructured in April 1999 is opened to all adult rural poor who are in need of wage employment. Another objective of the scheme is the creation of community, social and economic assets for sustained employment and development. The resources under the scheme are shared between centre and state in the ratio of 75:25. The scheme had central allocation of Rs. 1,990 crore during 1998–99. As per records available, the employment generated under the scheme up to March 1999 was Rs. 4,165.31 lakhs man-days. Since the inception of the scheme, the total number of registered wage seekers is 4.29 crore.

Besides the above, the Government of India has launched other programmes and schemes like National Social Assistance Programme; National Family Benefit Scheme; Indira Awaas Yojana; Pradhanmantri Gramodaya Yojana; Samagra Awaas Yojana; Innovative Scheme for Rural Housing and Habitat Development, etc. All these schemes aim at improving the economic condition of the people residing in drought prone areas. Since majority of the people in the rural areas have their livelihood based on agriculture profession, they get severely affected in the event of drought. Some Universities and NGOs in the country have also initiated efforts to search for alternative livelihood vocations for the people residing in drought prone areas. Some of these vocations include leather work; craft; weaving; knitting; dairying; sheep rearing and goat rearing; pattal and dona making; fruit plantation; vegetable cultivation; soap manufacturing; carpentry; masonry; tailoring; grading and grinding spices and condiments; preservation of fruits and vegetables; stone cutting and polishing; wood and metal craft work; sale of local crafts, etc.

Risk Transfer through Insurance

India is a multi-disaster prone country. A large segment of its population is subject to risk, which implies *hazard* × *value* × *vulnerability*. Insurance is one risk management technique available to those *at risk* that may reduce the impact of a hazardous event by spreading it among policyholders (Van Oppen). Insurance is the means by which the risks are shared between many individuals or institutions who face them, so that the event of contingency befalling on individual is compensated for his loss out of the premium paid by all the insured against it. It is also the pooling of enough small predictable risks so that the annual losses for the group are predictable. In keeping with the impact of natural disasters, especially drought in India, disaster insurance could be a critical instrument of development in the field of crop production, providing financial support to the farmers in the event of crop failure. It can encourage farmers to adopt progressive farming practices, better technology in agriculture, besides providing significant benefits not merely to the insured farmers, but to the entire community directly and indirectly through spill over and multiplier effects in terms of maintaining production and employment generation of market fees, taxes, etc., and net assertion of economic growth. Crop insurance could also streamline loss assessment procedures and help in building up huge and accurate statistical base for crop production (India: Agriculture Knowledge).

For risk transfer, the Government of India has introduced a scheme namely, National Agricultural Insurance Scheme, available to all the farmers—loanee and non-loanee—irrespective of the size of their holding. It envisages coverage of all the food crops (cereals, millets and pulses), oil seeds and annual commercial/horticultural crops.

The basic objectives of the scheme are to:

(i) Provide insurance coverage and financial support to the farmers in the event of failure of any of the notified crop as a result of natural calamities, pests and diseases.
(ii) Encourage the farmers to adopt progressive farming practices, high value inputs and higher technology in agriculture.
(iii) Help stabilise farm incomes, particularly in disaster years. Comprehensive risk insurance is provided to cover yield losses due to non-preventable risks namely natural fire and lightning, storm, hailstorm, cyclone, flood, inundation, drought, dry spells, pests, diseases, etc. The premium rates for different crops are as follows:

Season	*Crops*	*Premium rates*
Kharif	Bajra and oil seeds	3.5% of sum insured or actuarial rate, whichever is less
	Other crops (cereals, other millets and pulses)	2.5% of sum insured or actuarial rate, whichever is less
Rabi	Wheat	1.5% of sum insured or actuarial rate, whichever is less
	Other crops (cereals, millets, and pulses)	2.0% of sum insured or actuarial rate, whichever is less
Kharif and Rabi	Annual commercial/ annual horticultural crops	Actuarial rates

It is worth noting that 50% subsidy in premium is allowed in respect of small and marginal farmers, to be shared equally by the Government of India and State/Union Territory Governments. A small farmer is the one who has 5 acres, i.e., 2 hectares or less of land holding and a marginal farmer have land holding of 2.5 acres, that is, 1 hectare or less. The 17 States and 2 Union Territories, which have implemented the scheme during the year 2000–01. These include Andhra Pradesh, Assam, Bihar, Goa, Gujarat, Himachal Pradesh, Karnataka, Kerala, Maharashtra, Madhya Pradesh, Meghalaya, Orissa, Sikkim, Tamilnadu, Uttar Pradesh, West Bengal; Chattisgarh, Andaman and Nicobar Islands, Pondicherry.

Seed Crop Insurance. To boost production of seed by seed-growers and farmers, a pilot scheme on Seed Crop Insurance has been launched to cover the risk factor involved in production of seeds. The Plant Varieties Protection and Farmer's Rights Bill, 1999 aims at facilitating the growth of the seed industry, protect indigenous varieties and ensure the availability of high quality seeds.

Kisan Credit Card. A record number of 5,939,318 Kisan Credit Cards have been issued till March 2000, providing flexibility and security in the flow of credit and to strengthen the agriculture credit system.

Earthquake, flood, drought, landslide, forest fire, etc. lead to severe economic losses for rural people. The prevalence of such natural disasters is not unique and new for the rural people and institutions. They have, over generations, developed ways of reducing and coping with risks (Hazzel). India's poor have limited access to credit facilities and little and no savings. After a disaster, the poor become largely dependent on grants from the Government and other agencies. Micro finance can become the viable nucleus of a community-rebuilding effort, providing an enabling environment for the inherent resilience of people. AWARENESS, an NGO working in Orissa started a savings programme called *Safe Savings and Simple Credit*. The programme has not only provided timely financial assistance for rebuilding houses as well as livelihood sources, but has also helped spread the realisation among people that they can help themselves with their own money through small savings (Menon, 2000).

SEWA Insurance. In 1956, the Life Insurance Corporation of India became only authorised life insurer in India. Later in 1970, about 200 private non-life companies were amalgamated and four general companies were formed under General Insurance Corporation of India. The Schemes of these companies did not do much to help real poor because the schemes were designed without proper market survey; procedures were too complicated for illiterate poor, and the insurance companies did not know clearly how to reach out to poor. A Cooperative Group, *SEWA*, started its own Insurance Unit called *VimoSEWA*, which insures women for life, health, assets, widowhood, and accidents. The following lessons learnt by *SEWA* are rather encouraging and bring the fact to light that disaster risk transfer through insurance has vast scope:

(i) Insurance gives a solid economic support to poor people.
(ii) Poor people face multiple risks and multiple crises. Therefore, all the insurance services provided to them should be integrated and preferably available as a package.

(iii) Services need to be tailor-made to people's needs.
(iv) Insurance encourages forward planning among women.
(v) Insurance services need to be decentralised.
(vi) Insurance for the poor is not only possible but also financially viable. It needs proper advocacy and campaigning for spreading out the message among large numbers and across different income groups.

There is much felt need for well-conceived, area based insurance instruments to be offered for transferring risk of drought victims. The insurers can take better advantage by utilising the services of post offices, NGOs, Panchayati Raj Institutions, community-based organisations, credit and thrift societies, etc.

The Latur earthquake (1993), the Orissa super cyclone (1999), the Bhuj earthquake (2001), the drought (1986–87, 1999–2000, 2000–01), etc., are just a few examples to prove how natural disasters have been posing distinct challenges to the developing India. These have stalled the development process, disrupted social networks, and put people to a substantive amount of miseries and hardships. It has become almost impossible for the Government of India and State Governments to compensate loss of lives, buildings, crops, infrastructure, etc., after each disaster almost every year. The viable solution seems to be to put in practice the concept of risk transfer through insurance. More effective mechanisms are required for enabling disaster-hit population to manage its catastrophic risks, as it will definitely reduce the burden on Government if not completely ruling it out. In keeping with the frequency and intensity of disasters, economic conditions of the people living in disaster prone areas, etc., the insurance sector, which now is opened to private insurance companies, need to work out an insurance system which:

(i) People, especially the poor and ruralites, could afford.
(ii) Compensates for catastrophic income losses for repaying debt and protecting consumption.
(iii) Could be properly executed, given the problems of data collection and management in India, like any other developing country.
(iv) Could work for doing away with moral hazards belittling many of the hitherto agricultural insurance schemes.

It is a fact that each person is desirous of having better or at least safe future for self and family irrespective of the means and assets. There are a fairly large number of populations who could easily be persuaded to go in for insurance provided there are instruments available to their interest. There is a much-felt need for undertaking extensive market-based, risk-sharing and risk-transferring studies to come out with required insurance alternatives instruments and operating mechanisms.

REFERENCES

Annual Reports, Ministry of Agriculture, Government of India, New Delhi, 1994–95, 1996–97, 1999–2000, 2000–01.

Community-based Rain Water Harvesting in Drought Affected Areas through Livelihood Interventions, UNDP Orissa Document, 2001.

Drought Mitigation in Rajasthan, The Times of India, special supplement on drought.

Economic Review, 1999–2000, Government of Rajasthan.

Hazzel, Peter, *Potential Risk for Insurance in Managing Catastrophic Risks in Developing Countries* (http://www. IADB-ORG/SDC/DOC).

High Powered Committee on Disaster Management Report, October 2001.

http://mapsofindia.com/stateprofiles/gujarat/index.html

http://www.UNDP.Org.in/oriss/drought

India: Agriculture Knowledge, by Indiaagronet (http://161.58.78.178/Indiaagronet).

Menon, Meena, *Grants* vs. *Loans in Disaster Management*: A study on the Experience of Micro finance in Orissa after the Cyclone, a Paper presented in International Conference on Disaster Mitigation and Preparedness, Bhubaneswar, October 19–21, 2000.

UN Resident Coordinator, July 27, 2001.

Van Oppen, Charles, *The Role of Insurance in Disaster Eduction* (http://www.bghrc.com).

26

Disaster Risk Reduction through Development Orientation in Disaster Management: The Significance of People's Participation in India

— **Janki Andharia**

INTRODUCTION

India is one of the most disaster prone countries in the world with 85% of its area being vulnerable to a variety of disasters. About 28% of the total cultivable area is drought-prone, 57% of its geographical area lies in high seismic risk zone and on an average, 76 lakh hectares suffer the onslaught of floods every year. The state of development of a region directly determines the human consequences of a disaster. For example, if a settlement is well connected by road or rail, possesses telecommunication linkages and the administration has resources for early warnings and to evacuate people to safer areas, the impact of the disaster is likely to be lesser.

In India, the vulnerability to disasters is high, especially as people are lacking in basic facilities such as clean drinking water, adequate housing or shelter, schools, hospitals or roads. In addition, the way electricity is tapped illegally, the way radio-active and other hazardous chemicals are dumped into rivers or *nallahs* (drains), the flagrant violation of pollution control norms and the fact that safety and security rules are observed only in breach, all imply that our populations are highly vulnerable to disasters. In developing countries therefore, disasters are not to be viewed solely as emergencies. Recognising that the magnitude of a disaster and its impact on human life are determined by the extent of poverty and under-development is an important step in exploring the link between disasters and development. The marginalisation of remote villages, the absence of readily available maps showing locations of settlements or absence of motorable roads, all impinge upon quick deployment of person power for rescue, relief and rehabilitation.

Understanding the backward and forward linkages between disasters, the role of development and the significance of participatory approaches in disaster preparedness and management practices are key factors to fostering a culture of preparedness and prevention.

This chapter highlights the significance of people's participation in disaster management and the challenge of institutionalising participatory approaches.

PEOPLE'S PARTICIPATION

Individually and collectively, citizens have always acted voluntarily to improve their communities and societies. National and local governments, international donor agencies, academic institutions, political organisations, and charity and welfare groups—all refer to people's participation as an important component for the success of specific programmes or projects or for development in general. It is cited as a panacea to the noted failures of centralised, top-down development planning as it emphasises forms of bottom-up, grassroots and people-centred development. Complementary notions such as local organisation, capacity building and community-based initiatives accompany it. According to Cernea (1992), the recognition of a partnership between public programmes and people must be ensured through 'popular' participation. The following three key factors discourage popular participation:

- The expanding role of the public sector in launching such programmes, without a commensurate improvement of the mechanisms for the public's involvement.
- The growth of international aid, which amplifies government programmes, while increasing the distance between the programmes' 'centre' and 'periphery'.
- The recurrent failure in public programmes due to alienation of intended beneficiaries.

The term, 'Community Participation' has diverse definitions and represents a range of interpretations and approaches. In fact some argue that the subject is so broad that it can become ambiguous, diffuse and lacking in meaning (White, 1996; Gardner and Lewis, 1996). In the context of disaster management in particular, there is a need to recognise the differences and complementarities in the myriad approaches to community participation.

Historically, the terms Community Development (CD) and Community Participation (CP) have been inter-related in significant ways. Community participation, in the context of disaster management, has the focus on community development. In India, the term community development was used initially as part of our nation-building efforts to develop basic education, social welfare and rural development. The first Five-Year Plan regarded people's participation as a principal force and sanction behind the plan. It addressed the need for people's participation as follows: 'Planning in a democratic state is a social process in which every citizen should have the opportunity to participate and set the patterns of future development. It should embody the impact of public opinion and the needs of the community'. The relationship between participation, planning and democracy, viewed in this way is obvious, for the value that people should take action to improve the quality of their lives, that they have a right to guide such action, is embodied in the way these concepts are knit together in practice.

The box below provides a brief illustration of the various usages of the term 'participation' within a few UN agencies.

BOX 1

In 1955, according to the United Nations, CP was synonymous with CD as it stated, 'CD is a process designed to create conditions of economic and social progress for the whole community with its active participation'.

In 1979, UNRISD identified the real objective of participation as being, 'to increase control over resources and regulative institutions in given social situations, on the part of groups and movements of those hitherto excluded from such control'.

For the ILO (1986), people's participation implies collective effort by the people concerned, stimulated by their own deliberations within the following framework:

- Free and independent organisation
- Voluntary pooling of efforts
- Sharing of risks and responsibilities as well as resources and benefits to attain objectives set by the people themselves
- Having a say in decisions which may effect them.

The term, community development also has strong philosophic underpinnings and reflects a value base that stresses:

- Democracy
- Worth and dignity of every individual; his/her capacity to contribute to society
- Organised and cooperative efforts to improve living conditions
- Self-direction
- A multi-disciplinary approach
- People's capacity for participation.

Over the years, it has been argued that for development to be socially just, economically viable and environmentally benign, people's participation is very important. The Approach Paper to the Tenth Five

Year Plan (2002–2007, pp. 28–30) while discussing poverty-alleviation programmes acknowledges reasons for failure and provides several suggestions that embody the idea of people's participation in their own development, instilling in them, a sense of ownership for programmes meant to provide support for their development. Further, in Chapter 4 (p. 47) the Approach Paper argues that successful implementation of development programmes does require adequate funds, but the determining factor is the capability of the funding Ministries/State Governments to formulate viable schemes and a delivery system to implement these schemes on the ground. The serious deficiencies observed reflect poor governance and therefore, reform of governance is regarded as one of the centrepieces of the Tenth Plan. The Approach Paper also recognises that improvement in governance can take place only when countervailing forces in society develop the confidence to oppose inefficiency and corruption in government (p. 53). This calls for active involvement of people in development and in schemes and programmes that affect their lives. The notion of institutionalising people's participation is thus contained in the Tenth Plan.

Range of Meanings of Participation

One of the most obvious features of the literature explaining 'participation' is a description of the wide range of meanings the term implies. On the one hand, participation was endorsed and facilitated for specific development projects to support the emergence of more effective, efficient and responsive interventions. On the other hand, the radical implications of the oppressed or excluded masses organising themselves and having a voice in the process of development are inherent in the notion of participatory development. Social movements with emancipator expressions of people asserting their rights, requiring real 'transfer of power' from dominant, decision-making structures and institutions to people subordinated in the process, reflect a radical dimension of CP. In this context there is a need to recognise the inherent contradictions in institutionalised community participation (Chambers, 1994). The Government as well as some sponsors of participatory initiatives often refrain from considering community participation in terms of social class and power and seem reluctant to seek a precise definition of the term to clarify their own standpoint.

A community is the smallest social grouping with an effective social structure and potential administrative capacity. However, the term 'community' itself is not a homogeneous category. It comprises diverse groups representing various castes, classes, gender groups, age groups, religious groups and a host of other interest groups. Consequently, community participation means different things to different people/stakeholders.

Various studies, project documents and manuals have interpreted participation in different ways. These are summarised briefly below:

- Participation is the voluntary contribution by people in projects, but without their taking part in decision-making.
- Participation is the sensitisation of people to increase their receptivity and ability to respond to development projects.
- Participation is an active process, meaning that the person or group in question takes initiatives and asserts his/her or its autonomy to do so.
- Participation is the fostering of a dialogue between the local people and the project preparation, implementation, monitoring and evaluation staff in order to obtain information on the local context and on social impacts.
- Participation is the voluntary involvement of people in self-determined change.
- Participation is involvement in people's development of themselves, their lives and their environment (*see* Box).

BOX 2

Four main kinds of participation have been identified (Cohen and Up off, 1977) which are distinct from one another though they are interactive and have effects on one another:

a. Participation in *decision-making*, in identifying problems, formulating alternatives, planning activities, allocating resources, etc.
b. Participation in *implementation*, in carrying out activities, managing and operating programmes, partaking of services.
c. Participation in economic, social, political, cultural or other *benefits*, individually or collectively, and
d. Participation in *evaluation* of the activity and its outcomes, which should feed back into (a), (b) and (c) above.

Generally administrators tend to regard participation in implementation as the most important form and economists may equate participation with obtaining benefits, where the process itself is not given any significance.

Another way of conceptualising participation is distinguishing between instrumental participation and transformational participation. The former is when participation is viewed as a way of achieving certain specific targets—the local people participate in the outsider's project. Transformational participation is when it is viewed as an objective in and of itself and as a means of achieving some higher objective as self-help or concretisation or an assertion of one's rights. Instrumental participation predominates project situations, while transformational participation predominates people's movements.

In disaster management, experience in India indicates that relief and rescue operations are viewed as the responsibility of the Revenue

Department and the space for public support is not envisaged at all. Local people are regarded as 'spectators' thwarting rescue measures and at best expected to be passive recipients of relief. The notion of people as partners in disaster preparedness and management is relatively new and involvement of local population is not yet clearly delineated within the policy framework.

Experience of various civil society organisations suggests that providing space for participatory alternatives and outcomes improves post disaster recovery, the efficacy and overall quality of a reconstruction or rehabilitation programme. Therefore, community participation in emergency relief, in needs assessment, in policy negotiations and in planning, execution and evaluation/impact assessment of disaster management must receive considerable prominence to ensure that interventions for relief and rehabilitation translate into long-term development of the affected people.

Following the Latur earthquake the Government of Maharashtra (GOM) launched the Maharashtra Emergency Earthquake Rehabilitation Programme (MEERP) in which one of the most progressive features was the importance given to community participation in the programme (*see* Annexure I).

Objectives of Community Participation

In most disasters it is the community that provides initial rescue and first aid. Actions to protect their lives and that of their relatives and neighbours are first taken by the victims of disaster, whether it is digging out people from the rubble after an earthquake or removing bodies from inundated areas in case of floods. Local communities are thus, at the centre of immediate response and recovery activities. In fact, when transport and communication are disrupted, an external emergency response may not arrive for days. The ideal of community participation should reflect one or more of the objectives discussed below (Bhatnagar and Williams, 1992).

Empowerment. Empowerment involves becoming critically aware of cultural, political or even psychological conditions that reproduce marginalising conditions for groups and communities and then growing 'Transformative capacity' to alter power imbalances (Guijt, 1998). Participation could translate into a more equitable sharing of power and a higher level of political awareness and strength for disadvantaged people. The most important result of this process is the development of people's capacity to initiate actions and influence decisions of more powerful actors. This is the least emphasised objective in government programmes because sharing power is antithetical to bureaucratic ways of functioning.

Enhancing beneficiary capacity. This may be achieved not only in relation to an operation or a project but in ensuring that beneficiaries may share in management tasks by assuming responsibility for it themselves. People thus develop a sense of ownership and this contributes to the sustainability of a project due to enhanced beneficiary interest and competence in project management.

Increased effectiveness. Refers to the degree to which a given objective is achieved. Effectiveness increases when people's involvement contributes to better public design and implementation and when a better match of project services/benefits with people's needs and constraints is achieved.

Cost sharing. This may include sharing of financial costs or contributions of labour and capital. People may undertake 'self-help' activities to maintain a programme/an activity of the project. Participation facilitates a collective understanding and agreement on cost sharing and its enforcement.

Improved efficiency. Efficiency measures the relationship between a given output and its costs (input). Project planning and implementation becomes more efficient because of timely beneficiary inputs from local population. Participation can be used to promote agreement, cooperation and interaction among people. As a result, delays are reduced and there is a smoother flow of project services while overall costs are minimised.

It is evident that community participation with the above objectives must be an integral part of a relief and rehabilitation policy in the context of post-disaster intervention. A participatory approach to planning and implementation ensures that rescue, relief and rehabilitation plans reflect sensitivity to culture and tradition and the needs and aspirations of the local people. The box that follows (Box 3) summarises the World Bank's perception of these terms.

Participatory Approaches: Constraints and Contradictions

The objectives discussed in the previous section, illustrate the value of participation, which is undeniable. There is a sense of dynamic tension as the official thrust on people's participation is sought to be enhanced in policy documents. On the one hand institutionalisation of participation is an attempt to standardise or mainstream participation for reliability and uniformity. On the other hand, this conflicts with one of the original aims of participation, which is to move away from rigid, blue print methods for development planning and recognise the 'process' aspect of community participation, rather than view it as an end in itself. An appreciation of these contradictions and constraints of participatory approaches is useful.

Some participatory appraisal and planning exercises as promoted through a multitude of methodological manuals, reduce the process to a convenient one to two week planning exercise. In such exercises, the

BOX 3

The range of Participatory Mechanisms could include:

1. Information sharing
   - Translation into local languages and dissemination of written material using various media
   - Informational presentations and public meetings.
2. Consultations
   - Meetings
   - Field visits and interviews.
3. Joint Assessments
   - Participatory assessments and evaluations
   - Beneficiary assessments.
4. Shared decision-making
   - Participatory planning
   - Meetings, workshops and seminars to determine positions, priorities, roles
   - Meetings to resolve conflicts, seek agreements, engender ownership
   - Public reviews of draft documents.
5. Collaboration
   - Formation of joint-agency/stakeholder committees/task forces
   - Joint work with user groups, NGOs or other stakeholder groups
   - Stakeholder groups given principal responsibility for implementation.
6. Empowerment
   - Capacity building of stakeholder organisations
   - Hand-over and self-management by stakeholders
   - Support for new, spontaneous initiatives by stakeholders.

Adapted from *The World Bank and Participation, Operations Policy Department, September 1994*.

expectation of any significant micro or macro structural changes can never be met. From the perspective of an administrator, recognising some of the 'risks' or so-called 'constraints' will enhance the readiness of the Government to work towards participatory approaches.

Implies higher costs. While cost sharing is envisaged as an objective of participation, the entire process of making institutional arrangements for fostering people's participation, training and capacity building of existing officials and/or hiring trained and sensitive personnel may imply higher degree of financial commitment towards personnel.

Time consuming and can cause delays. 'Participatory Processes is Delaying Rehabilitation and Reconstruction' was a standard exclamation heard in the post earthquake reconstruction phase in Latur. Most government rehabilitation programmes are driven by targets of beneficiaries, of finances to be disbursed or spent within a stipulated time frame. This often limits and circumscribes the space for a full play of participatory processes.

Participation in decision-making can be a time consuming process. Consciousness raising, empowerment and enhancing the capacity to confront exploitation and oppression and a capacity to change social structures are process-oriented and not restricted to instrumental aspects of enhancing efficiency or cost sharing. It may, in fact be contrary to project priorities which may be to achieve physical targets (of reconstruction for example) within a stipulated period.

Conflicts, interpersonal dynamics including shifts in gender relations and community attitudes may be observed as an outcome of community participation. These dynamics are intrinsic to the rehabilitation process and often cause delays. For example, participating groups may have conflicting opinions about objectives and about how to achieve them. These differences must be made explicit and debated. Tension can develop between the state trying to promote participation and the hitherto excluded groups trying to increase their control over resources. Sometimes the rehabilitation programmes or the policy pronouncements themselves lead to intensification of conflicts making consensus building a long drawn out process causing delays.

Organisational structures/critical decision-making capacities may be absent in communities. This is generally due to lack of information or the absence of technical expertise and justification of procedures laid down. These capacities would have to be enhanced by presenting information through a media that local people can understand. Use of appropriate technology and simplification of procedures can help overcome this bottleneck. Besides, people and communities have an enormous capacity to learn and to contribute. This potential remains untapped in most policy frameworks.

Can raise expectations prematurely. 'People are expecting too much from the Government', such statements are often made with utter exasperation especially after a disaster. In fact an increase in expectations, self-confidence and a proactive role in decision-making can expand the participatory potential. This may find expression in a demand for more benefits and/or a greater role in local or government decision-making, which may displease the bureaucracy or the local politicians whose power is also likely to be challenged. It may result in governments crushing participation before it 'gets out of hand' making people's participation merely instrumental or rhetoric to make a programme appear to be politically correct on paper.

Can give a feeling of loss of control to project staff and those in-charge. When people begin to participate and give directions to a programme, local officials may feel threatened as they are used to maintaining total control over social processes through rules and regulations. Accepting people's participation is antithetical to their established ways of functioning.

Capturing of resources by local elites. This generally happens because local elites or leaders have advantageous ties and links with national elites, elected representatives, corporators and can often use force and threats against the disadvantaged. Communities are heterogeneous and therefore who participates when, in what direction and to what extent are important questions. Arriving at the right balance between macro and micro-level planning may be difficult.

However, this is very necessary when trying to institutionalise participation. It is often argued that the government is trying to legitimise 'top-heavy' policies through the rhetoric of participation. The gap between prevailing hierarchical social structures and the proposed progressive intent of the scheme must be acknowledged and addressed.

Prerequisites of Participation

Recognising centrality of people or individuals in a community, having faith in their capabilities and acknowledging their need to work towards their own well-being is of paramount significance as it is the local communities that are at the centre stage of immediate response and recovery activities. Institutionalising participatory approaches calls for certain pre-requisites.

Government Commitment

This refers to the readiness to examine closely how the government's own internal operational procedures may need to be modified in order to support participation especially of the marginalised and the will to implement the necessary changes in procedures. This actually requires a long process of negotiations, establishing a base with people, which include organising and attending meetings and extensive information dissemination. Typically government departments do not allocate adequate resources for effective information dissemination and for strengthening institutional mechanisms. The orientation, the previous positions and practices of governments in providing people opportunities and spaces for participation are critical. Further, problems of organised collective action in a heterogeneous community may require being resolved before advantages of CP can be visible and sustained. This requires faith and patience.

Clear Policy and Procedures

Development experience all over the world demonstrates that policies and institutions matter more than public investment.

An effective participatory approach requires a sound institutional framework. It is often argued that people do have scope to participate but the question is that does the machinery operate in a fashion that enables or facilitates people's participation? Enforcement of rules may be lacking and functionaries do not follow procedures as diligently. Therefore policies and procedures must be clear. Implementation must be monitored. Grievance redressal and feedback mechanisms must be laid down by policy itself. Clear channels of communication must be established especially, between agencies and departments. Finally policy is useless without the institutional machinery is capable of implementing it.

Information Dissemination

Partnership and responsiveness to micro-level needs and initiatives is possible only if there is adequate information sharing. In the absence of formal channels of communication, information about policies and projects filter down to people from a variety of sources and such information is distorted, misinterpreted and used by vested interests. Therefore systematic information dissemination requires a well-articulated communication strategy as well as training of personnel in communication and skills in using a variety of media. Various methods are available for information dissemination. It is important to choose the media appropriate to the audience.

Building Trust and Accountability

People express feelings of fear and resentment towards government bureaucracies when decision-making is not transparent and officials do not enjoy people's confidence. A process of dialogue and consultation builds trust between people and government, which is crucial for good governance. By ensuring that development is participatory the project enjoys greater confidence and gains greater credibility.

An atmosphere of transparency is important if the accountability to the public is enhanced. Accountability requires mediation amongst three distinct stakeholders:

- Citizens and local population concerned
- Political leaders and those officials responsible for project planning and implementation
- The investors in the project.

The goals and interests of these groups differ. Transparency and accountability allow for multiple perspectives to enter and influence public debate in the policy arena. Institutional reform in the direction of greater accountability is gathering momentum in the country.

Support to Local Groups

It is well acknowledged that people participate best through their own organisations or community-based organisations. These should therefore be included in policy dialogue or in the process of planning itself. Mechanisms for discussing issues and resolving differences must be strengthened (these may be weekly committee meetings or the gram sabha meetings). Experience in social mobilisation, managing group dynamics and use of participatory strategies may require the presence of adequately trained staff. Enabling different voices to emerge, recognising power differentials and ensuring a platform for those with less power to speak and consistently arbitrate conflicts and differences require capable facilitators.

Gender Sensitivity

Policy and officials must pay special attention to gender concerns of each area or region. Women take a number of decisions in the course of a typical day and these are ascribed to traditional roles and rarely accorded significance. Consulting formal and traditional leaders is not sufficient as most are men. Women's views must be legitimised in formal decision-making, as they are most concerned and knowledgeable about micro economy and local practices for handling change and continuity. Many efforts at participation overemphasise community cohesion and consensus at the cost of ignoring gender differences and fail to create space for dissenting voices to be heard. For example, pre-conceived notions about desirable gender relations may inadvertently result in ignoring views expressed by people during participatory initiatives and processes.

Flexibility

Flexibility in planning and implementation is essential if popular participation is to be built in. It permits operational modifications at various stages of the project. In India, the participatory spirit in most government programmes is virtually absent and the scope to include views of local people especially the disadvantaged is very limited. However one also observes that several rules and procedures are easily skirted to appease certain vested interests and/or to meet political exigencies. Participatory processes, however take time to develop.

Evaluation Procedures

These are critical for mid-course corrections. Monitoring and evaluation procedures must be institutionalised and incorporated in project plan itself. Built-in mechanisms for ongoing evaluation help strengthen accountability, both internal and external. It ensures that a project is moving in the right direction and structures are utilised to operationalise objectives.

Finally, there is no doubt that without government commitment there is little that any single stakeholder can do to broaden and sustain people's participation. Participation should not be simply viewed as an end in itself, nor should it be reduced merely to a means. Judgements of what constitutes an optimum level of participation should involve the participating people themselves. It is important to recognise that community participation in disaster management is not a technical fix or a mere add on, but has implications for power distribution and can imply a conflict of interest between different groups and individuals.

Community participation in the policy context with regard to the Tenth Plan is given in the box below.

BOX 4

Tenth Plan Approach to PRIs:

The Approach Paper on tenth Plan (Government of India, 2001) recognises the strengths and bottlenecks experienced in the constitution of Pańchayat Raj Institutions (PRIs). Its states, 'There is a mixed experience with the work done by PRIs at the block and district levels, and there is a feeling that there may be too many tiers leading either to ineffectiveness or excessive control. The financial conditions of local bodies too are precarious since in many cases neither is there effective devolution of financial resources by the states nor adequate revenue raising powers. Indeed, the near absence of revenue raising powers with the PRIs is leading to a sense of dependence rather than empowerment, which was the original intent of the Constitutional amendment. Control which is exercised by the sarpanch and Block Level officials over the village panchayats and gram sabhas (which rarely meet) has not only buttressed corruption, but it has also led to pessimism that villagers at their own level cannot change and improve performance because of heavy dependence on elected functionaries and Block officials. Though providing a framework for decentralised rural development, trends so far suggest that the panchayat raj system has not been able to enhance participation and empowerment as effectively as would be desired' (pp. 53–54).

DISASTER MANAGEMENT STRUCTURE IN INDIA

In India, disaster management is the responsibility of State Governments. Research, surveys, guidelines and provision of financial assistance to the States are the responsibility of the Central Government. There is a Crises Management Group headed by the Cabinet Secretary and consists of nodal ministries in charge of various disasters. For natural disasters, the Ministry of Agriculture is the Nodal Ministry and the other Ministries are supportive. In the event of a disaster, a multi-disciplinary Central Government team, at the invitation of the affected state, conducts a disaster assessment and also makes recommendations for assistance. However, disaster management is primarily the responsibility of the State Governments. More recently, a National Committee on Disaster Management under the Chairpersonship of the Prime Minister has been set up and a decision to make the Ministry of Home Affairs the Nodal ministry for disaster management in India is under consideration.

State Level Organisation

Disaster preparedness and response in the State is usually delegated to the Relief and Rehabilitation Department or the Department of Revenue. The Chief Secretary with participation of all the related agencies heads the Crises Management Group at the State level.

District Level

In the event of a disaster, a District Level Coordination and Review Committee is constituted, which is headed by the Collector as Chairman with participation of all other related agencies and departments.

It is clear that the Panchayati Raj institutions do not have any role in the present system of disaster management and the bureaucracy takes most decisions. The 73rd and 74th constitutional amendments have sought to revitalise Panchayati Raj Institutions (PRIs). The structural pattern of Panchayati Raj, composition of panchayats, organic linkages of PRIs, direct elections to PRIs, devolution of powers and functions of PRIs and reducing bureaucratic control over PRIs—are important aspects of decentralised governance.

The state legislature is expected by law to provide the panchayats with powers and authority as may be necessary to enable them to function as institutions of self-government and devolve powers and also facilitate functional strengthening of different tiers. This however is progressing at a very slow pace in most states. Detailed executive instructions have not been issued in most states and bureaucratic control over panchayats continues to be strong. Even where such control is not visible, PRIs have been placed in such a position that the chairpersons have to make repeated trips to Government offices located far away, for getting approvals and sanctions (Meenakshisundaram, 1997).

The decision-making structure in MEERP, Latur is summarily presented in Box 5.

In disaster management, the institutional structures for relief and rehabilitation should take into account not only the development thrust of the village (or the region) but also build on the capabilities at the local levels and provide scope for participation in decision-making.

The National Policy Framework on Disaster Management, December 2001, is an effort in that direction. A planned response through administrative, financial, legislative and techno-legal reforms in tandem with public awareness and networking with local groups and PRIs is expected to go a long way towards prevention and mitigation.

CP in Disaster Management

For the reasons discussed earlier, community participation must be an integral part of a recovery or rehabilitation policy and a truly democratic government should enable such participation. However, it has been observed that in disaster management, relief and rehabilitation programmes are formulated outside the community by the State Government with a degree of inflexibility, which is observed in most government run development programmes. Therefore, CP is minimal at

BOX 5

The general institutional framework for the implementation of the rehabilitation programme comprised of a five-tier structure:

1. A cabinet sub-committee chaired by the Chief Minister, to formulate broad policy.
2. A Central Implementation Group (CIG) headed by the Chief Secretary, to monitor, coordinate and approve specific projects.
3. A Project Management Unit (PMU) with executive and financial powers, headed by Secretary and Special Commissioner, Earthquake Rehabilitation to expedite implementation.
4. PMUs with similar powers, at the District and Block levels and comprising elected representatives, district officials and representatives of beneficiaries, NGOs and donor agencies.
5. Village Level Committees (VLC), which were to serve as fora for participatory planning and grievance redressal for beneficiaries.

Limitations of VLCs in MEERP: The MEERP incorporated community participation as a critical area of focus within the rehabilitation process. The Village Level Committees (VLCs) were envisaged as the facilitating institutional mechanism to ensure people's participation in the rehabilitation programme. The VLC was chaired by the Tahsildar or Nayab Tahsildar and included the Gram Panchayat members, the Gram Sevak and the Talati of the concerned village, a technical representative nominated by the Deputy Engineer (Works) of the area and an NGO representative, co-opted by the Collector. In order to protect the interests of women and the backward classes two persons from each of these groups were to be nominated by the beneficiaries of the village. The VLCs were expected to decide the method of plot allotment in the new villages, to select the plan and layout for relocation/construction and repair from among those approved by the government, give wide publicity to the rehabilitation programme and ensure peoples participation. It was also expected to monitor the progress and quality of work in the village and redress public grievances.

In principle, the idea of constituting a local or village body to monitor the rehabilitation was excellent. The VLC had the potential to become a vehicle for the community's involvement in the rehabilitation programme. However it failed to function as stipulated due to a number of factors. According to the rehabilitation policy the VLC was the apex of the participation structure (Earthquake Rehabilitation Policy of GOM, 1994). At the same time the organisation of the programme was such that the VLC itself was at the lowest level in the programme implementation structure. The VLCs were only mandated to decide on a limited number of options that the government put before them regarding relocation sites, design and layout of village and the houses, technology and material to be used in house construction and repair, etc. The most critical decisions regarding these and other options were taken at the State level and did not involve the beneficiaries (TISS, Terminal Report, 1997).

the stage of formulation of rehabilitation packages. The expectation is that people should voluntarily accept and appreciate these measures and implement these with the spirit in which they are formulated. However, implementation is organised in a top-down manner with a large number of government employees and advisors or consultants expected to supervise and advise the community, as was the case in MEERP (*see* Annexure II).

CP and NGOs

There has been substantial international experience of people's participation in NGO initiatives in disaster management. However, recognising the difference between CP elicited by NGOs and donor agencies (such as Save the Children, OXFAM) and multilateral agencies (such as UNICEF), and the challenge of institutionalising participation through government agencies is critical. First, most of these agencies give grants rather than loans. Second, NGOs specialising in relief and rehabilitation employ highly trained specialists, both nationally and internationally. Third, they operate with a sectoral focus such as health and sanitation, or a target group focus such as the aged, poor women and children requiring support. Fourth, they tend to work on small-scale innovations, which can be replicated, rather than large-scale coverage of entire districts or states. As a result their efforts at CP are more fine-tuned and community specific. On the other hand, governments are expected to reach out to all the disaster affected people and also garner financial resources. The World Bank or ADB generally provide financial support to governments in the form of loans. The onus of providing a broad-based, comprehensive policy framework and the execution or implementation of the programme is on the Government.

Experience shows that despite a good administrative set-up and a well formulated disaster management plan, disaster mitigation measures may not find the required direction and yield desired results in the field. Generating data, accurate assessments and a circular flow of information from the state administration to the local, affected population are critical factors. The structure and the process of decision-making, devolution of powers and responsibilities, the integrity of key actors, the formal and informal communication channels—all impinge upon the kind of recovery and rehabilitation process witnessed.

The growing awareness of the need for participation in governance combined with disenchantment with performance of the Government and recognition of its limited capabilities has contributed to the growth of civil society organisations. They are the response to a bureaucratic and unresponsive state machinery.

In the absence of a clear policy and an ethos of participatory approach in governance, most administrators project the involvement of NGOs as representing people's participation. However NGOs are of various shades and hues and include residents' associations, religious trusts, donor/funding agencies, development-oriented organisations and local groups. Therefore there is no parity in the level and extent of people's participation they seek to elicit in their operations. The broad participatory framework must be laid down and upheld in spirit in relief and rehabilitation programmes of the Government. This is essential to avoid reducing people to mere recipients of relief and rehabilitation packages, and to ensure that the assistance people need during a crisis does not dehumanise them, does not breed a dependency syndrome but in fact treats them with the dignity and respect that all people deserve. The affected people must be viewed as partners/collaborators in the process of disaster management. The orientation may not be easy to institutionalise in the current practices but the country has sufficient experience to assist such institutional shift.

The Challenge of Institutionalising Participation

It is necessary to consider the orientation of the Government toward participation. To the extent participation contributes to greater resource mobilisation and efficiency in meeting technical objectives, it is likely to be accepted by administrative as well as political leaders. In principle, governments of practically all orientations aim to benefit the rural majority. Yet they may fear that new capacities for participation will lead to broader claim making and to voicing criticisms. Similarly some political leaders may fear that participation by rural people will lead to a loss of their political 'power'. However this view, at least in its general form, misconstrues the essence of power, which is the ability to achieve what one desires. If a government wants for rural people what they want for themselves, and if participation enhances their capability to determine the course of development and to accelerate progress along that course, then 'power' devolved to rural communities will add to the power of the government in a positive-sum manner. Only when governments want to initiate activities that are contrary to the interests and needs of its populations will power become zero-sum, and a gain for the public will then be a loss for their rulers (Upoff and Esman, 1974). Project planners and implementers have to consider various factors for achieving their objectives (*see* Box 6).

The plan for disaster preparedness as also for rehabilitation and the implementation of the plan must uphold the policy in spirit and substance. The translation of participatory and developmental goals into institutional mechanisms (the structures established and the operational directives issued) is the real arena of challenge. For example, regular orientation, training and feedback mechanisms across participating bodies and organisations and at various levels of district functionaries are vital. Institutionalising forums for participation and providing scope for creative problem-solving are aspects that pose a challenge in a context where most rehabilitation and developmental planning is guided by the Revenue

BOX 6

Some Suggestions for Project Planners and Implementers:

1. Policy instruments must reflect the realities faced by the local communities that are going to be affected, most especially any planning process from above.
2. Every policy/plan/project should have institutional mechanisms and a start-up period where informed views of local people are sought. The participation of local population, of MLAs, people's representatives, of CBOs, NGOs of both men and women as agents and equal partners in directing their own development, must be ensured (an inclusive village level committee is an example).
3. Mechanisms of ensuring such participation must be identified and put in place. Nature of institutional arrangements, the degree of transparency and the extent of public participation elicited are critical. Sharing of responsibilities by all actors, particularly the beneficiaries must also be envisaged.
4. Mechanisms for grievance redressal are very critical in translating any objective into an operational plan.
5. There is a need to be alert to and access and quantify likely adverse effects or fallout, if any.
6. Greater focus on long-term development of local populations, particularly the disadvantaged and the vulnerable must be attempted.
7. Greater thrust on gender and environment sensitive development, and strengthening of linkages between local action and policy processes must be attempted.
8. Government programmes must provide or create an active forum for people working at the grassroots to consolidate their ideas into a feasible vision, into a partnership, a collaboration rather than view them with suspicion as hurdles/enemies or as adversaries having nuisance value to be sidelined or 'bought of'. The gram sabha as an institution must be viewed as a strong forum for participatory democracy.
9. Clear and adequate financial allocation for participatory rehabilitation plan.

Department which typically functions in a 'top-down' manner. Further, participatory approach must translate into a clear and adequate financial committee in instituting forums, in demonstrating creative 'bottom-up' approach initiatives and in capacity building of community-based organisations.

Review of international experience especially with rural development suggests that governments, which are not very participatory 'at the top', may nevertheless find it valuable to promote participation at lower levels, through voluntary sectors, NGOs and other civil society organisations.

Finally, organisations working on rehabilitation must acknowledge the existing vulnerabilities of certain sections of society and reflect sensitivity to the fundamental inequities in social and economic structures in order to be effective and inclusive in their interventions.

Recognising the multiplier effects resulting from institutionalised participation in decision-making, implementation, benefits and evaluation is vital.

CONCLUSION

There is a need to systematically and continuously document how community participation in practice was actually implemented. Trying to generalise about communities or the nature or quality of participation is inadequate because a wide diversity of factors such as the context, the nature of communities, the gender issues, and the agencies—all impinge upon the extent of community participation in disaster management.

It is well known that in post-disaster rehabilitation work, external influences bring about complex changes in a society, all of which cannot be predicted or fully grasped without examining and monitoring the broader social, economic and historical context within which the rehabilitation programme is carried out. It has been argued that the sudden flow of material resources, the nature of assistance and the manner in which it is provided, itself generates mistrust, tensions and values that are alien to a traditional community. Are there mechanisms that are built in within the structure of programme management to deal with these consequences? Who assumes responsibility for these social implications? Regular monitoring of the social impact of both, the disaster, and the rehabilitation measures, must be an important component that should feed into policy on an ongoing basis. Scope for mid-course corrections, reflecting people's wishes and aspirations and a will to incorporate participatory processes even in decision-making at the highest levels, should ideally be provided in a truly participatory structure. Providing spaces to take people's suggestions on board is necessarily a challenging task.

To conclude, while acknowledging that a series of economic exigencies compel the Government to access loans from multilateral aid agencies, social actors and operational issues of public administration deserve equal attention. Who assumes responsibility for the social processes generated at the community level as a result of imperfect, myopic policies of aid and rehabilitation? The approach of multilateral agencies such as the World Bank generally reflects an overall dissociation with the social fabric of a country and it follows its own idea of a structure that it considers most appropriate to monitor the progress of rehabilitation and appropriate utilisation of finances. The focus on micro level social processes can be sustained through PRIs and village level communities and vibrant grievance redressal mechanisms. Decisions are generally contingent upon utilisation of finances and figures indicating how many beneficiaries are covered. Quantifying parameters of progress in this manner is inadequate. It reflects very poor will and commitment to the social processes that are vital for development, capacity building or empowering of people, especially the weaker sections of society affected by the disaster (Andharia, 2001).

Although it is acknowledged that administration takes place within broad political contexts, the gap between policy statements and actual practices must be minimised. There is a need to examine the observed interplay between professional norms, values and expectations embedded within policy frameworks and the emotional and political issues of class, gender and ethnicity that impinge daily on the public life of bureaucrats and on the decision-making process. Relief and rehabilitation in Latur and in Gujarat earthquakes have highlighted the fact that administrative mechanisms need to ensure flexibility of responses to the field realities and problems. A development orientation and a people-orientation must be mandatory for the government and it must not be limited to merely exhausting budgets within a time frame, for a predetermined purpose, for records, or to appease multilateral aid agencies. Albeit, the rehabilitation process itself, like development, is one of dynamic change, throwing up new problems and challenges as well as demonstrating the possibilities of new solutions.

REFERENCES

Andharia, J., *Reflections on MEERP Some Social and Institutional Aspects of Disaster Management* (Unpublished Paper) TISS, 2001.

Bhatnagar, B. and A.C. Williams (Eds.), *Participatory Development and the World Bank: Potential Directions for Change*, World Bank Discussion Papers 183, Washington, The World Bank, 1992.

Cernea, M., *The Building Blocks of Participation: Testing Bottom-up Planning,* World Bank Discussion Papers 166, Washington, The World Bank, 1992.

Chambers, R., *The Origins and Practice of Participatory Rural Appraisal*, World Development. Vol. 22 (7), pp. 953–969, 1994

Chambers, R., *Who's Reality Counts? Putting the First Last*, London: Intermediate Technology Publications, 1997.

Cohen, J.M. and N. Uphoff, *Rural Development Participation: Concepts and Measures for Project Design, Implementation and Evaluation*, Ithaca: Cornell University, 1977.

Esman, Milton J. and N. Uphoff, *Local Organisations: Intermediaries in Rural Development*, Ithaca: Cornell University Press, 1984.

Gardner, K. and David Lewis, *Anthropology, Development and the Post-Modern Challenge*, London: Pluto Press, 1996.

Government of India, Planning Commission, *Approach Paper to the Tenth Five-Year Plan (2002–2007)*, Sept. 2001.

Guijt, Irene and Kaul Shah, *Waking up to Power, Conflict and Process*, in Guijt and Kaul Shah (Eds.). *The Myth of Community: Gender Issues in Participatory Development*, London: Intermediate Technology Publication, 1998.

Korten, David, C., *Getting to the 21st Century: Voluntary Action and the Global Agenda*, West Hartford: Kumarian Press, 1990.

Meenakshisundaram, S.S., *The 73rd Constitution Amendment—A Case for Further Amendment,* Journal of Rural Development, 1997, Vol. 16 (4), pp. 549–556. India, Hyderabad: NIRD.

Nederveen, Pieterse, *My Paradigm or Yours? Alternative Development, Post Development, Reflexive Development*, Development and Change, Jan. 1998, Vol. 29, pp. 343–373 Oxford: Blackwell Publishers.

White, Sarah, C., *Depoliticising Development: The Uses and Abuses of Participation*, Development in Practice, 1996, Vol. 6, No. 1, pp. 6-15.

World Bank, *The World Bank Participation Sourcebook*, Washington, DC: The World Bank, 1996.

TISS (URCD) Terminal Report (1997) submitted to MEERP, Earthquake Rehabilitation Cell, Government of Maharashtra.

ANNEXURE I

HIGHLIGHTS OF SIX COMPONENTS OF MEERP

Housing

- Relocation of 52 severely affected villages
- Reconstruction of 27,415 houses in 52 relocated villages
- Acquisition of land for relocation sites
- In-situ reconstruction and repair and strengthening of 190,000 houses in 2,500 villages
- Pilot strengthening of 5,000 undamaged houses
- Construction of 475 model building

Infrastructure

- Construction of access to relocated villages
- Repairs/reconstruction of roads and bridges
  - — Repairs to culverts and minor bridges
  - — Construction of missing culverts and bridges
  - — Strengthening of bridges
  - — Repairs to access road to relocated villages
- Repairs/reconstruction of public buildings/school buildings
- Water supply
  - — Repairs to water supply infrastructure and provision of temporary water supply to relocated villages
  - — Interim water supply to relocated villages
  - — Permanent water supply to relocated villages
- Irrigation
  - — Repairs to lower Tema dam and Talni aquaduct
  - — Strengthening of embankments of percolation tanks
  - — Strengthening of Kolhapur Type weirs
  - — Repairs/reconstruction of Offices and other buildings
  - — Strengthening of embankment
- Repairs and strengthening of historic monuments

Economic Rehabilitation

- Replacement of farm implements
- Replacement of implements and minor equipment
- Replacement of bullocks
- Repair/reconstruction of dug wells
- Replacement of milch cattle
- Replacement of sheep and goats
- Rehabilitation of artisans

Social Rehabilitation

- Physical infrastructure
- Repairs/reconstruction of roads and bridges
  - — District trauma centres, District resource centres,
  - — Homes for the handicapped, Community centres for women, Anganwadis
- *Other Components:* Supporting 500 Self Help groups for women, 192 seats created in it is specially for girls from earthquake affected areas, creation of village development fund
- *Social Forestry:* Block plantation, homestead plantation, Roadside plantation

Community Rehabilitation

- Construction of transit shelters
- Provision of services for the transit shelters
- Replacement of medicine stocks for human and veterinary services

Technical Assistance, Training and Equipment

- Programme management consultancy
- Supervision consultancy
- Seismic consultancy
- Community participation consultancy
- Water supply consultancy
- Disaster management plan
- Earthquake monitoring and research
- Training of engineers and artisans
- Training of other staff

Source: MEERP, 1998, Information Brochure.

ANNEXURE II

TISS INVOLVEMENT IN LATUR QUAKE

This can broadly be divided into two components or phases. First was a voluntary response to a national calamity. The second was appointment of TISS as Community Participation Consultant in MEERP. Activities during both phases are described here.

Phase I

Tata Institute's Rural Campus at Tuljapur, located about 40 km from Sastur, one of the most seriously affected villages, swung into relief work from the morning of September 30, 1993. The Rural Campus staff was among the first to reach the seriously affected villages in Osmanabad. During the relief phase, for a period of about 4 months the Institute was engaged in the following activities:

1. Deputed three faculty members to coordinate the activities of the NGOs from the central coordination centre at Latur as required by the collector.
2. Deputed a few faculty members to help the Latur district administration to conduct the panchanama work in some affected villages, as requested by the collector of Latur.
3. Conducted a comprehensive survey covering 34,446 households in 69 seriously affected villages to assess the impact of earthquake on human lives, cattle and property in order to develop the necessary parameters for the rehabilitation programme. Individual reports for 69 villages and a comprehensive report covering all villages were completed and given to all agencies and individuals engaged in reconstruction and rehabilitation activities.
4. Deputed a few faculty members to analyse data related to organisation and administration of relief and write a report for Osmanabad district at the request of the collector, Osmanabad.
5. Deputed a faculty member for six months (November 1993 to May 1994) to facilitate and coordinate the activities of the NGOs in the villages of Osmanabad and Latur. This was done in response to the request from Osmanabad and Latur Collectors.
6. Directly involved in social, economic and environmental rehabilitation work in Rajegaon village of Osmanabad district, where the Tata Relief Committee was engaged in reconstruction of houses and infrastructure.

Phase II

In the reconstruction and rehabilitation phase, the Institute was engaged by the Government of Maharashtra as Community Development Consultant to facilitate the participation of the community in reconstruction and economic and social rehabilitation in 52 villages selected for relocation and reconstruction under the MEERP. This involvement lasted for a 3-year period.

The TISS as CP consultants were involved in a wide range of activities, from building a strong enabling presence in the field, demonstrating the participatory process wherever possible building the capacities of government and village level committees as well as monitoring and assessing the rehabilitation programme on an on going basis. The CP consultants were active in a cyclical process of gathering people's views on the various rehabilitation packages, clarifying issues of concern to them, making periodic recommendations to the government and actively assisting the state authority in conflict resolution in the villages. The responses of the PMU to these recommendations, the negotiations, are documented less extensively and less known.

All this meant understanding village specific problems, issues and social dynamics. Eliciting community participation and institutionalising it was by no means an easy process and many of the efforts aimed at securing a village level consensus on housing or socio-economic resettlement were painstakingly slow.

Focus Areas of CP Consultants in Relocation Villages

Key Areas of CPC Intervention:

1. Rapid appraisals for policy inputs
2. Suggestions on participatory implementation of MEERP (through field demonstrations, information dissemination).
3. Gender sensitivity in policy and implementation
4. Land acquisition
5. Bifurcation of villages
6. Disputes related to beneficiary lists
7. Liaising between CCs and VLCs (and with NGOs)
8. Redressing Grievances (when formal channels and attempts failed)
9. Conflict resolution
10. Ongoing monitoring and feedback to PMU.

Source: TISS, Terminal Report, 1997.

27

Stakeholders' Response in Disaster Risk Reduction

— E. Vayunandan

INTRODUCTION

Disasters have always been mankind's constant companion. Generations of people have had to withstand disasters. They suffered from the consequences and recovered from them, and life continued. But somehow, over the ages, the scenario has changed quite a bit, yet there has not been much reduction in the traditional disaster threat. Natural disasters like earthquakes, cyclones, volcanic eruptions, tsunamis, wildfires, floods, landslides and droughts continue to strike. While we have learned to cope with these problems to a certain extent, we have neither eliminated nor contained them. So, whilst their effects may have been modified, they continue to inflict unacceptable pressure on the world population. The largest sufferers are the least developed nations and economically weaker sections of the society (Natural Disaster Reduction Report, 1999).

New disaster threats have also developed in the modern world. Increased social violence has drastically affected many nations and communities. Instances of hijacking, terrorism, civil unrest and conflict with conventional arms have become commonplace. Instances of cross-border terrorism in parts of India continuing for many years and the organised simultaneous multi-target terrorist attacks in USA on September 11th, 2001 are the most despicable instances of wilful man-made disasters. These inflict heavy burdens on governments and societies, more so in developing countries whose existence is already precarious because of poor economic and social conditions.

The South Asian region faces various kinds of natural disasters. The South Asian countries have diverse agro-climate regions, each subject to particular natural disaster. Long coastal regions are prone to cyclone, arid and semi-arid regions are prone to droughts, the Himalayan mountain terrain and parts of the continental crust are prone to earthquakes and landslides; and the near perennial rivers of the region are prone to floods. The countries in this region are densely populated and are low-income economies. Recurrent natural disaster inflict set backs to their efforts in development and aggravate poverty conditions. Hence there is a need for coordinated efforts towards disaster mitigation among these countries (Carter, 1992). It should be noted that India faces the largest number of disasters among the countries of South Asia.

This chapter will briefly discuss the efforts towards disaster risk reduction worldwide, focusing particularly on the role of stakeholders in disaster reduction in India.

DISASTER RISK REDUCTION: INTERNATIONAL EFFORTS

In the era of advanced satellite and other remote sensing techniques, the magnitude of damages wreaked by natural calamities can be reduced considerably by building a *Culture of Prevention* through awareness, knowledge and appropriate use of such technologies. We may not be able to eliminate the occurrence of such disasters but certainly; action can be taken to reduce their impact. The most significant global effort made in recent times to mitigate the effects of disasters was declaration of the International Decade for Natural Disaster Reduction programme by the United Nations.

International Decade for Natural Disaster Reduction (IDNDR)

The objective of IDNDR was to reduce through concerted international action, especially in the developing countries, the loss of life, property damage and social and economic disruption caused by natural disasters such as earthquakes, windstorms, tsunamis, floods, landslides, volcanic eruptions, wildfires, drought and desertification and other calamities of natural origin (Mohan, et al., 1995).

The world conference on Natural Disaster Reduction held in Yokohama (Japan) in May 1994 evolved a plan of action for disaster reduction called the Yokohama Strategy. It gave guidelines for natural disaster prevention, preparedness and mitigation and plan of actions for a safer world.

The plan of action was to be based on:

- Development of a global culture of prevention as an essential component of an integrated approach to disaster reduction.
- Adoption of a policy of self reliance in each vulnerable country and community comprising capacity building as well as allocation and efficient use of resources.

- Community participation in the disaster reduction process, and improved risk assessment.
- Broader monitoring and timely communication of forecasts and warnings.

Furthermore, the strategy called upon all countries to express political commitment to reduce their vulnerability through appropriate means. It also recommended that donor countries should upgrade the priority on disaster prevention, mitigation and preparedness in their assistance programmes and budgets.

The International Strategy for Disaster Reduction (ISDR) has been adopted as the successor to IDNDR. The ISDR has been programmed to take advantage of the network and experience gained under IDNDR. While the main achievement of IDNDR was to create awareness among the people and policymakers worldwide, ISDR is aimed at upgrading this awareness into realistic action plans (IDNDR, 1999).

The main focus of ISDR will be on:

- Continuing the efforts to increase awareness
- Obtaining commitment from public authorities
- Creating disaster resistant communities
- Reducing socioeconomic losses.

EFFORTS OF DISASTER RISK REDUCTION IN INDIA

India is highly prone to natural and man-made disasters. Because of adverse weather conditions, population growth, urbanisation and industrialisation there is an increase in the number of disasters every year. In this context we have to see how the Indian government organises for managing natural and man-made disasters, its capacity for effective response, and its ability to achieve unity of efforts among government and non-government organisations.

Under India's federal system, the states have responsibility for disaster response and the Central Government supplements the efforts of the state governments by providing financial and material assistance.

In 1988, National Commission on Disaster Management was constituted to prepare a National Disaster Management Plan. In 1995, the Government of India established a National Centre for Disaster Management to encourage effective training, coordination and planning for disaster response (Sharma, 1997).

In India, various stakeholders are involved in disaster risk reduction. The stakeholders are the government including central, state, district and local administration, the non-governmental organisations, international agencies, community and various other agencies. Efforts pertaining to disaster risk reduction by various stakeholders are discussed in the ensuing paragraphs.

National Level

At the national level, depending on the type of disaster, a nodal ministry is assigned the task of coordinating all activities of the state and district administration and the other support departments/ministry. The nodal ministries form part of the National Crisis Management Committee. Part of their tasks is to prepare detailed Contingency Plans for each type of disasters falling in areas of their responsibility.

Basically the responsibility for undertaking rescue, relief and rehabilitation measures in the event of natural disasters is that of the concerned State Government. The role of the Central Government is supportive in terms of providing financial and other resources. The Department of Agriculture and Cooperation (DAC) of the Ministry of Agriculture is the nodal department in the government at Central level that deals with the subject of natural disaster management.

Type of Disaster and Nodal Ministry

Type of Disaster/Crisis	*Nodal Ministry*
Air Accident	Ministry of Civil Aviation
Civil Strife	Ministry of Home Affairs
Major breakdown of any of the essential services posing widespread protected problems	Concerned Ministries
Railway Accidents	Ministry of Railways
Chemical disasters	Ministry of Environment
Biological disaster	Ministry of Health
Nuclear accident inside or outside the country which poses health or other hazards to people in India	Department of Atomic Energy
Natural Disasters	Ministry of Agriculture

In the DAC, the Central Relief Commissioner functions as the nodal officer to coordinate relief operations for all natural disasters. (In recent decisions, the Government of India proposes to make the Ministry of Home Affairs the nodal agency for disaster management. The decision is still under consideration.)

The Central Relief Commissioner receives information relating to forecast/warning of natural calamity from the India Meteorological Department (IMD) or from the Central Water Commission (CWC) on a continuing basis and keeps the Secretary (Agriculture and Cooperation) and through him the Agriculture Minister and the Cabinet Secretary and the Secretary to Prime Minister and through them, the Prime Minister, the Cabinet and the National Crises Management Committee (NCMC) informed. He may, whenever required, also disseminate the information to different Central Government ministries/departments and the state governments for appropriate follow-up action (Sharma, 1997).

The Central Relief Commissioner also monitors the development of the situation on a continuing basis and provides the necessary information through the Agriculture Secretary, to the Agriculture Minister, Prime Minister and the Cabinet.

While the Ministry of Agriculture is the nodal ministry managing disaster situations, other ministries also support it. The Ministry of Health and Family Welfare through the Emergency Medical Relief Division of the Directorate General of Health Services makes an important contribution. In a typical Disaster situation, the Medical Relief Division gets in touch with the Central Control Room in DAC and obtains feedback on the extent of disaster situation on a particular day, the population affected and the health profile of the victims like the number of patients, type of injuries/diseases.

State Level

The State government in India is autonomous in organising relief operations in the event of natural disaster and also for purposes of long-term preparedness and rehabilitation measures. The Central government's role is limited to supplementing the efforts of the state governments.

Every State has a Relief Commissioner who is in charge of disaster relief. In states where there is no designated Relief Commissioner, the Chief Secretary or an officer nominated by him is given overall charge of relief operations. Every state has a number of secretaries who head various departments connected with administration and they all function under the overall supervision and control of the Chief Secretary, who ensures that their combined efforts are coordinated in disaster management. The states deal with natural disasters through their Revenue Departments or Relief Departments.

The state also have a State Crisis Management Group (SCMG) which function under the chairmanship of Chief Secretary/Relief Commissioner. The Group comprises senior officers from the departments of revenue/relief, home, civil supplies, power, irrigation, water supply, local self-government (panchayat), agriculture, forests, rural development, health, planning, public works and finance. The SCMG is expected to take into consideration, the guidance received from the Government of India from time to time and formulate action plans accordingly for dealing with different kinds of natural disasters.

The Relief Commissioner of the state establishes an Emergency Operations Center (Control Room) as soon as disaster situation develops. The Center collects and disseminates the latest information on forecasting and warning of disaster, and functions as the focal point for coordinating disaster relief efforts with the other concerned departments.

Several States have been active in training disaster management personnel and preparing contingency plans. A good example is the Centre for Disaster Management, at the Yashwant Rao Chavan Academy of Development Administration, Government Training Institute in Pune, Maharashtra. It is a research and training institution, which has focused on assisting the districts with developing their disaster management, plans. Among the states, Maharashtra has taken lead in preparing a comprehensive multi-hazard Disaster Management Plan (DMP). The plan has three components namely risk analysis and vulnerability assessment, response planning and mitigation strategy.

The Risk Analysis and Vulnerability Assessment depict the present picture for each disaster-exposure, loss of life, property damage, etc. It also shows geographic distribution of each hazard. The various monitoring facilities, regulatory regimes, and the countermeasures available for each disaster have been depicted in the analysis.

- The Response Plan gives the organisational structure of all the state, central and non-governmental agencies to effectively deal with disaster in a coordinated manner to mitigate its impact. It identifies the functional areas such as relief, communications, information, transport, health services, etc. and the proposed assignments to the various departments. The plan also lays down preparedness checklists, operating procedures and reporting formats.
- The Mitigation Strategy focuses on the long-term planning for disaster reduction. It deals with issues of continued commitment to hazard identification and risk assessment, applied research and technological transfer, investment-incentives of mitigation, and leadership for mitigation. The strategy argues for better land use management, building codes, traffic standards, health standards, etc. These objectives are to be secured through disaster legislation, mitigation regulation, and incentives for the same.

District Level

As rightly quoted in the World Disaster Report, 1998, 'Effective and accountable local authorities are the single most important institutions for reducing the toll of natural and human induced disasters'.

The country's day to day administration centers around the District Collector (in some states is also known as District Magistrate or Dy. Commissioner) who heads the administrative organisation in a District. District administration is a focal point for field level organisations and implementation of all government contingency plans. Every state is divided into a number of districts. The District Collector is in charge of all the relief measures at the district level. The State Government routes its instructions through the District Collector who ensures that the total efforts of the district are geared in a coordinated manner to providing disaster relief to the people.

A district is further divided into sub-divisions and tehsils. The head of the sub-division is called a Sub-Divisional Officer or Deputy Collector, while an officer called the Tehsildar heads the tehsil. The lowest unit of administration is the village whose revenue matters are attended to by a functionary known as Patel or Patwari. All these tiers of administration—the Patwari, Tehsildar, Deputy Collector and the District Collector—function as a team to provide saccour to the people in the event of a disaster (Singh, et al., 2000).

The various measures undertaken at the district level are:

- **Contingency Plans.** The District Magistrate develops a district level plan, which is submitted to the state for approval. The plan assigns measures to be taken by the different District departments and their functionaries, and it identifies the areas of coordination.
- **District Relief Committee.** The district level relief committee consisting of official and non-official members and including the local representatives of the people, members of the legislative assembly and the parliament reviews the relief measures.
- **District Control Room.** A control room is set up in the district as soon as a natural disaster occurs. The control room monitors the rescue and relief efforts on a continuing basis.
- **Coordination.** The District Magistrate also coordinates with the central government authorities and defence forces that may be posted in the district. He synchronises the voluntary efforts of NGOs with the efforts of the district administration. The communication channels are maintained through the police wireless network in the event of breakdown of the normal means of communication, the telephone and the telegraph.

ROLE OF NGOs IN DISASTER RISK REDUCTION

Many of the non-government organisations' (NGOs) in India have come to play a very useful role in disaster risk reduction. They operate at the grassroots along with the community. The various functions performed by NGOs are briefly enumerated below:

- Pre Disaster
  - — Awareness and information campaigns
  - — Training of local volunteers
  - — Advocacy and Planning
- During Disaster
  - — Immediate rescue and first aid and psychological help and counseling
  - — Supply of food, water, medicine, other materials

— Ensuring Sanitation and hygiene
— Damage assessment

- Post Disaster
  — Reconstruction aid
  — Financial aid
  — Monitoring

There are different types of voluntary organisations functioning in India, classified generally as International organisations, private volunteers, grassroots (local) organisations, and so on. They offer immediately available communications within the disaster affected community, technical services, manpower, and financial support. There are NGOs that deal with commercial research and development. Besides there are interest groups such as, Rotary Club or the Lions' Club which make resource contributions during disaster events. Various associations such as Medical, Trade, Army Wives Association provide specialised services and generate resources. There are NGOs who enjoy international support and can respond quickly with large amount of supplies and services. Apart from this associations there are religious, charitable and education institutions, which come to aid immediately for providing shelters, mass feeding services and providing public information.

NGOs have played very supportive role in various disasters in India. Some examples are noteworthy (Singh, et al., 2000).

Gujarat Cyclone, 1998

- CARE-India provided food in the Khambhalia and Kalyanpur blocks of Jamnagar districts for a period of two months and also gave 350 tents worth Rs. 700,000 in Jamanagar district.
- The OXFAM (India) Trust gave immediate relief in the form of food grains and plastic sheets and also provided clothing to over 1,000 families.
- The World Vision concentrated in Kutch District where the salt workers were hit badly. They identified 250 families and provided them with roof tiles, wooden material, etc. to reconstruct their houses.
- The Discipleship Center reached the Ropar and Bachau blocks of Kutch district within 72 hours of the earthquake and covered around 870 families with emergency relief material. They also provided utensils and distributed plastic sheets, bed sheets, saris, dhotis and children's clothing.
- CARITAS stepped into the areas within the first few hours of the calamity and concentrated on reconstruction of houses.
- The CRS, which works through the church organisations, provided US $ 10,000 for providing clean drinking water and shelter. 161 tones of food were also provided in Jamnagar, Porbandar, Dwarka and Kandla district.

Apart from the above-mentioned immediate response by the NGOs they also have long-term mitigation plans to prepare the community for preparedness and prevention.

- CARE (Cooperative for Assistance and Relief Everywhere) started working in India in 1950 when the Indo-CARE agreement was signed between the Government of India and CARE. The organisation has since then supported many Government of India programmes including the mid-day meals, the special nutrition programme and food for work. It has also been extending humanitarian assistance to save lives and reduce suffering caused by disasters.
- CASA (Church Auxiliary for Social Action) was formed in the wake of the partition of India and constituted a Christian response to the human suffering engendered by that event. CASA's Peoples' Action for Transformation (PAT) programme involves empowering people at the village level through Grass Root Level organisations such as Village Development Associatons, Youth Groups and Economic Groups. To achieve sustained advancement of disadvantaged groups, village-level groups attach such issues as environmental degradation, deforestation, drought, migration, child labor, casteism, land encroachment, illiteracy, unemployment, indebtedness, pollution and so on.
- VHAI (The Voluntary Health Association of India) contributes to disaster mitigation and management in the field of public health. It has branches in each state and does its own disaster management planning at central and state levels. During the Maharashtra earthquake of October 1993, it provided to the government a health status report, which helped to guide public health recovery efforts. It publishes disaster management materials.
- IRC (The Indian Red Cross) provides disaster relief as one of its services, rendering aid through a network of societies at the state level. The IRC maintains its own disaster assistance plan. It includes a survey of past disasters and training for staff and volunteers in disaster preparedness and relief operations. The IRC has established regional disaster relief operational headquarters at Guwahati in Assam and Ahmedabad in Gujarat with additional centers forthcoming at Vijayawada, Andhra- Pradesh and in northern Punjab. Other IRC disaster relief operational headquarters are located at New Delhi, Mumbai, Kolkatta and Chennai. Each headquarters will have a central warehouse of about 50 MT capacity, transportation fleet, communications center, and operations center.
- OXFAM (India) (Oxford Committee for Famine Relief) has served in India for over 25 years. Its goals include increasing food

security, improving access to health and education, improving the lives of women, and improving India's disaster preparedness and mitigation.

- Sambhavana Trust (Bhopal People Health and Documentation Clinic) is an example of a locally run grassroots organisation. It is tending to the long-term physical and psychological injuries of Bhopal disaster victims. It is also collecting data about the Bhopal victims that may be useful to disaster management professionals. Especially if these data can be properly recorded and analysed in a computer-supported data base program.

Besides, these international agencies provide assistance at various stages of disasters viz., pre-disaster assistance, assistance in responsive operations, assistance in recovery prgorammes and assistance in future development (Mohan, et al., 1995). A brief description of role played by U.N. bodies is given below:

- Department of Humanitarian Affairs (UN-DHA), Office of Disaster Relief Coordinator. An agency that assists in disaster assessment and relief management. It also advises on hazard risk assessment, mitigation planning and implementation.
- UNDP (United Nations Development Programme). It is responsible for coordinating disaster response operations that United Nations organisations would support or conduct in India.
- UNICEF (United Nations Children's Fund). With a child population of over 300 million, India is UNICEF's largest country programme. UNCEF funds programmes in mother and child health, water and sanitation, nutrition, primary education, and elimination or child labour.
- UNESCO (United Nations Educational, Scientific and Cultural Oranisation). It funds research in mitigation, strengthening in heritage structures against flood and earthquake damage. It is supportive of flood management programmes and runs a publication programme pertaining to disasters.
- FAO (Food and Agriculture Organisation). It offers technical advice on the reduction of vulnerability and monitors and advises in food production.
- UNIENET (United Nations International Emergency Network). Through a network of computers, it places members of the world-wide disaster management community in direct communication with each other and provides them instantaneously with both background and operational disaster related information.
- WFP (World Food Programme). It provides targeted food aid, sometimes linked to 'food for work' programme for construction of flood protection structures and coordinates pre and post disaster emergency food aid. It also runs its own publications programmes on disasters.

- WHO (World Health Organisation). It provides rapid response in post-disaster situations. It promotes 'health cities' programmes and is supportive of disaster mitigation measures.

Besides these, there are other international agencies outside U.N. System which are also providing the assistance in disaster management viz., Asian Development Bank, Asian Disaster Preparedness Centre, European Community Humanitarian Office, The International Institute for Environment and Development, Organisation for Economics Cooperation and Development, World Bank and Department for International Development, U.K.

ROLE OF VARIOUS SERVICE AGENCIES IN DISASTER RISK REDUCTION

Various service agencies like medical, fire, civil defence, home guards, police, paramilitary and defence forces are involved in disaster risk reduction (Singh, et al., 2000).

Medical and Public Health Services

Though health is a state subject under the Constitution, the Central Government's intervention is needed in the areas of control and eradication of major communicable and non-communicable diseases, national health policy formulation, medical and para-medical education and international health. It also oversees drug control, prevention of food adulteration, and activities concerning the population including safe motherhood, child survival and immunization programmes.

The major health schemes for control and eradication of communicable diseases include the National Programmes for Eradication of Malaria, Leprosy, and Tuberculosis and AIDS including Blood Safety Measures and STD Control. The National Programmes for control of non-communicable diseases like cancer; cataract blindness, iodine deficiency disorders and mental health are also being implemented. The Directorate General of Health Services, Ministry of Health and Family Welfare, Government of India provides technical assistance to the states to improve the health sector responses. The Emergency Medical Relief Division of the Directorate General of Health Services discharges the responsibility. The Medical Stores Organisation maintains the depots in Mumbai, Chennai, Kolkatta, Karnal, Hyderabad and Guwahati. These depots maintain civil defence stocks and mobile hospitals units for emergency use.

As in other countries, in India also steps are being taken to educate medical professionals about their role in disaster relief. The training activities were conducted under the WHO funded programme by the disaster management collaborating centers with a view to creating awareness among the officials of the health and other departments on various aspects of health sector disaster management. This has contributed to building up a cadre of officials in the health and other departments who are sensitive to natural calamities and are conversant with the drill to be adopted and the procedures to be followed in the event of a disaster so as to provide necessary relief to the affected population.

In the states, the district and sub-divisional and the primary health care centers provide medical relief in case of disasters. Some charitable, voluntary and private institutions also provide medical relief.

Fire Fighting Services

The states and Union Territories administer the fire services in the country, as fire is a state subject. The Ministry of Home Affairs renders technical advice to states and Union Territories and the Central ministries on fire prevention, fire protection and fire legislation. The National Fire Service College, Nagpur, conducts different types of courses for the training of fire officers in India. It is the only college of its kind in South-East Asia and it trains fire officers of several countries.

Civil Defence

Civil Defence aims at saving life, minimising damage to property and maintaining continuity of industrial production in the event of a hostile attack. The Director General of Civil Defence (usually a police officer) works under the Ministry of Home Affairs. Each state and territory has a Director of Civil Defence who works under the Home Department. Central financial assistance to the states for Civil Defence measures is confined to categorised towns only. Civil Defence is primarily organised on a voluntary basis except for a small nucleus of paid staff and establishment, which is augmented during emergencies.

Police

The police are entrusted with the responsibility of preventing and detecting, crime and maintaining law and order. Public order being a state subject under the Constitution, police is maintained and controlled by the states. The police functions cover a very wide spectrum. In times of disaster, the police is first to respond and it maintains security and law and order at disaster locations where there might be chaos and miscreants may take advantage of the situation, and prevent occurrence of cognisable offences including all offences against property, human body and public tranquillity. The other important thing is the police communication system, which is made available for transmission and receipt of messages in connecting with the disasters site. The police also regulate the movements of victims, rescue and relief, medical assistance and supplies. The enormous manpower of the police and its resources enable it to provide the first healing touch in any emergency.

Paramilitary Forces

The Government of India has a large paramilitary force structure, which is employed to assist the states in their responsibility to maintain public order. The paramilitary forces include Border Security Force (BSF), Central Reserve Police Force (CRPF), Indo-Tibetan Border Police (ITBP), Central Industrial Security Force (CISF), National Security Guard (NSG), Railway Protection Force (RPF), Homeguards and Assam Rifles (AR). Each of these forces has a well-defined role in disaster situation.

In the event of an emergency, the paramilitary forces often get drawn into the situation. The Central Reserve Police Force and the Border Security Force have quite frequently been utilised in disaster situations. Recently, during the landslides in he Uttar Pradesh Hills, the Indo-Tibetan Border Police came forward in a big way to rescue the trapped victims (pilgrims) and provide relief to the affected population.

Defence Forces

Within the State, the District Magistrate can requisition services of the Army in the role of 'aid to civil power'. India defence forces have made a significant contribution to the country in their principal roles of national defence, stability and integration. Today they continue operations in border areas (viz. Pakistan and China) and counter insurgency operations in Kashmir and the Northeast states. Under the role of ensuring national stability, they have been used to provide disaster assistance during floods and earthquakes. The defence forces are most useful in providing timely assistance such as dropping food, supplying water and medicines to affected areas, erecting military bridge equipment and other services that save lives immediately. They have played a vital role in times of disaster, providing prompt relief to the affected people even in the most inaccessible and remote areas of the country.

In the recent past, Government response to natural disasters has progressively improved in terms of its effectiveness. This is chiefly due to the emergence of a well-organised administrative machinery, presence of Relief Manuals at district level, predetermined allocation of duties and recognised public private partnerships. However, the absence of an integrated policy at national level has led to overlooking of some of the vital aspects of disaster management. Presence of such a policy helps clearly define the Government's approaches on a continuing basis. It also provides for an appropriate legislation and associated regulations in this regard besides an overall national competence and self-reliance vis-à-vis international initiatives.

Due to the increasing frequency of natural disasters and their severe impact on the individual, society, economy and environment, The

Government of India constituted, in August 1999, a High Powered Committee (HPC) on disaster management to suggest means to bring about institutional reforms in the field and planning of disaster management. The committee was also required to prepare comprehensive plans for national, state and district levels. Soon after its formation, the scope of the committee was enlarged to include man-made disaster like chemical, industrial, and nuclear and others. The committee submitted its report in October 2001 (Report of High Powered Committee, 2001).

The major recommendations of HPC as follows:

- First the item of disaster management should be specifically mentioned in any of the three lists.
- A suitable legislation to provide appropriate legal frame at national and state level.
- After legislation it will be necessary to evolve detailed regulations to help in the enforcement of law.
- Setting up of Cabinet Committee on Disaster Management and the all party national committee under the chairmanship of Prime Minister and the working group set up under the guidance of vice-chairman need to be institutionalised as permanent standing bodies as the former would help generate the necessary political will, consensus and support while the latter, that is the working group, being body of experts, will evolve appropriate strategies for implementation of broad policy guidelines of the cabinet committee.
- National Center for Calamity Management, as suggested by the Eleventh Finance Commission need to be set up at the earliest and a National Institute for Disaster Management also to be established as a Center of Excellence in the area of creation of knowledge and its dissemination including training and capacity building.
- A networking of training institutions led by a national level disaster management institutions with symbiotic linkages with other national and state level institutions need to be forged and developed.
- There should be full cooperation between various governments, voluntary agencies, and relief workers with a commitment to perform to the best of their abilities and they should refrain from the public criticism of each other (code of conduct).

The Committee pleads strong case for ushering in a new culture of disaster management placed on the four aspects namely:

- Culture of preparedness
- Culture of quick response
- Culture of strategic thinking
- Culture of prevention.

Culture of Preparedness

It is not possible to do away with the devastation due to natural hazards completely. However, experience has shown that destruction from natural hazards can be minimised by the presence of well functioning of warning system, combined with preparedness on the part of vulnerable community, which reduce and modify the scale of disasters. A community that is prepared to face disasters, receives and understanding warning of impending hazards and resorts to precautionary and mitigatory measures, and is able to cope better and resume normal life sooner.

Culture of Quick Response

In accordance with the need for the Central Government to respond promptly and in the most appropriate manner, an appropriate organisational set up at the state level to cope with incoming relief and rescue measures is an urgent necessity, so that in disaster situations of colossal magnitudes no time is lost in directing incoming relief and rescue measures to the exact locations where they are required.

Culture of Strategic Thinking

The crucial importance of strategic thinking to combat disasters, and the need for networking of institutions engaged in the pursuit of knowledge has been emphasised which will be useful in forecasting future disasters.

Culture of Prevention

The communities actively involved in working on prevention of natural disasters belong to all groups of society; international and regional organisations, national governments or private firms, local administration and specialised associations. It is important to instill a culture of prevention in disaster managers and all communities. Action must be taken at all levels so as to save lives before the disaster strikes. Early warnings and conscious developmental planning are key elements to preventive planning.

CONCLUSION

The Central and State and Local governments and the NGOs have stressed more upon relief. Now it is high time that all the agencies, which are involved in the disaster, should direct themselves towards the disaster risk reduction management. They should also evolve an integrated and holistic approach in dealing with disasters. The initiative should come from the government and the community itself as the community is the first responder in case of disasters. The other important instruments of good governance i.e., local bodies viz.

Panchayati Raj and Urban local bodies have to be integrated in the disaster management and their plans also should have the linkage in mitigating disasters, as these bodies are nearer to the communities. Apart from all these, the recommendations of the HPC should be implemented in true spirit that will create a disaster free India through the confluence of culture of preparedness, quick response, strategic thinking and prevention.

REFERENCES

Carter, Nick W., *Disaster Management: A Disaster Manager's Handbook*, Asian Development Bank, Manila, 1992.

International Federation of Red Cross and Red Crescent Societies, *World Disasters Report*, Geneva, 1999, 2000.

Keith, Smith, *Environmental Hazards: Assessing Risk and Reducing Disaster*, Routledge, London, 1996.

Mohan, Munasinghe and Caroloine Clarke (Eds.), *Disaster Prevention for Sustainable Development—Economic and Policy Issues*, IDNDR and World Bank, Washington, 1995.

Natural Disaster Reduction—South Asian Regional Report, Ministry of Agriculture, Government of India, New Delhi, 1994.

Report of the Geneva Mandate on Disaster Reduction, July 5–9, 1999, International Decade for Natural Disaster Reduction Programme Forum.

Report of High Powered Committee on Disaster Management, Ministry of Agriculture, Government of India, New Delhi, 2001.

Sharma, Vinod K. (Ed.), *Disaster Management, National Centre for Disaster Management*, Indian Institute of Public Administration, New Delhi, 1997.

Singh, Prakash Mendel, W. William, and Graham H. Turbiville, Jr., *Disaster Response in India*, A joint study initiative by the Center of Excellence in Disaster Management and Humanitarian Assistance and the Foreign Military Studies Office, August 2000.

Sinha, P.C. (Ed.), *Encyclopaedia of Disaster Management*, Vol. 1, Introduction to Disaster Management, New Delhi, Anmol Publicatons Pvt. Ltd., 1998.

Index